HOW TO DESIGN, SPECIFY, INSTALL, AND REPAIR A SLATE ROOF

THE SLATE BOOK

BY BRIAN STEARNS, ALAN STEARNS, AND JOHN MEYER

Copyright © 1998 Brian Stearns, Alan Stearns, and John Meyer. All Rights Reserved. No part of this publication may be reproduced, stored in a retrieval system, or transmitted in any form or by any means, electronic, mechanical, photocopying, or otherwise without prior written permission of the publisher and the authors.
ISBN-0-9661363-0-6

TABLE OF CONTENTS

FOREWORD

PART 1: PLANNING AND ORGANIZATION

    Chapter 1   Understanding Slate 9
    Chapter 2   Design Considerations 17
    Chapter 3   Project Planning and Estimating 35
    Chapter 4   Ordering and Handling Slate 61
    Chapter 5   Roof Construction and Preparation 67

PART 2: INSTALLATION

    Chapter 6   Roof Layout 83
    Chapter 7   Slate Tools. 95
    Chapter 8   Installing Slate 101
    Chapter 9   Flashings 139
    Chapter 10   Reroofing 151
    Chapter 11   Repairs, Maintenance, and Troubleshooting. 153
    Chapter 12   Advanced Situations 169

AFTERWORD

    180

PART 3: APPENDICES

    Appendix A   Chemical Composition of Slate 182
    Appendix B   Checklists 183
    Appendix C   Slate Specifications 187
    Appendix D   Snow Guard Specifications 199
    Appendix E   List of Tables and Figures. 201
    Appendix F   ASTM Specifications 204
    Appendix G   Roof Parts Nomenclature 205

BIBLIOGRAPHY

    206

FOREWORD

Slate is a natural stone that comes in a wide range of colors and textures. As a building material, its appeal derives mainly from the beauty, quality, and old-world elegance that buildings with slate seem to possess. The natural beauty of stone on a roof enhances the structure itself as it blends with the natural earth tones of the surrounding landscape. Many people may still remember stately buildings from their childhoods, the most beautiful of which had slate roofs that had been on the buildings forever. Others may recall growing up in a house with a slate roof—while the neighbors were replacing their roofs, the slate roofs stood the test of time.

The use of slate in building construction has a long history. The first recognized slate quarry in the United States was established in 1734 in the Peach Bottom district on the Pennsylvania-Maryland border. From this beginning, the slate industry grew slowly until the mid-1800's. By the 1850's, slate roofing had emerged as a growing

■ While slate, as a natural rock, is not subject to the precise control that governs fabricated materials, representative men in the industry have long been of the opinion that the ambiguity surrounding the terms used in specifying slate for roofing purposes could be clarified. Furthermore, the great variety in sizes and dimensions of this commodity have convinced those concerned that a substantial economy of time and money would attend the elimination of the superfluous and seldom-called-for sizes of roofing slate.

As a result of a survey of practices current in 1923, the slate industry prepared tentative recommendation covering sizes and nomenclature of roofing slate and submitted them to the annual meeting of the National Slate Association on January 21, 22, and 23, 1924, in New York, N.Y. At the last day of the session, which was held under the auspices of the division of simplified practice, United States Department of Commerce, it was pointed out that only those block sizes of slate which satisfy the greatest demand should survive this proposed elimination. Not only would the program eliminate waste, but it would also operate to conserve a natural resource. A great variety of sizes of blocks is obtained in quarrying roofing slate, and any recommendation should be so drafted as to include sizes which would make it possible to utilize, with a minimum of waste, as many of these quarried blocks as possible. These proposals were accordingly discussed, and it was resolved that they should be adopted as a simplified practice recommendation.

It was the opinion of those who participated in the New York meeting that the elimination of excess varieties of roofing slate, and the clarifying of descriptive terms used in connection therewith, would prove of marked benefit to the manufacturers, distributors, and users of this commodity. Slate, being a natural product, can be cut to any desired size without the employment of special dies, patterns, molds, casting, etc., to meet any unusual conditions, but the recommended sizes are suggested for the purpose of encouraging a demand that will justify some production for stock. Such predetermined stocking will provide employment for quarry and mill workers during slack periods and will facilitate prompt shipments on orders for these standard sizes.

The action of the industry effected a reduction from 60 sizes of roofing Slate to 30, from 21 thicknesses to 10, and from 17 descriptive terms to 8, constituting elimination of 50, 52, and 52.9 percent respectively.

After selecting July 1, 1924, as the effective date, the meeting adjourned.

—*A Department of Commerce Simplified Practice Recommendation document (R14-28) written in 1924 and revised in 1928, describes the creation of slate industry standards.*

FOREWORD

industry, and the production of slate reached its peak in 1902. According to a U.S. government survey, 1.4 million squares of slate shingles were produced in that year.

In the 1920's, the federal government initiated a broad-based program, led by then Secretary of Commerce Herbert Hoover, aimed at assisting all industry through cooperation rather than regulation. The goal of the program was to create simplified standards for various industries, including the slate roofing industry. The National Slate Association was formed in 1922 in response to this call for standardization. This group of slate producers, distributors, and users worked with the government to create slate industry standards. A Department of Commerce Simplified Practice Recommendation document (R14-28, on previous page) written in 1924 and revised in 1928 describes the process.

Members of the National Slate Association involved in this 1924 meeting were also responsible for publishing the book that is still a primary source of information about the slate roofing industry. *Slate Roofs*, or the "Slate Book," as so many people refer to it, was first published in 1924 and later revised in 1927. The "Slate Book" has served the industry for over 70 years and is still an excellent source of information about certain aspects of the slate roofing industry. Much of the information in the "Slate Book" originally came from U.S. government bulletins, and we have incorporated some of that information here.

By 1924 when the "Slate Book" was published, the production of slate had already begun to wane. Shortly after the book's revision in 1927, several factors began to affect the slate roofing industry adversely. Foremost among these was the growing popularity of asphalt shingles. (Ironically, the first manufacturers of asphalt shingles used crushed slate granules to coat the shingles and give them a more slate-like appearance.) Then in 1929 during the Great Depression, building construction practically came to a standstill. As a result, very little slate was installed during the 1930's. Later in 1941, when the United States entered into World War II, the nation's industrial focus shifted to the production of military material.

Following the end of World War II, another building boom took place. However, the demand for huge volumes of inexpensive, easily installed roofing materials spelled the further demise of slate. A decreasing demand for slate roofs led to a dwindling labor force of experienced slate roofers. By the 1960's, very few of the old slate craftsmen were still active in the trade. Those who had learned the trade from previous generations of slate roofers had probably been between 18 and 20 years old during the Depression; by the early 1960's, they were 48 to 50 years old. There really was not enough demand to justify teaching a younger generation of roofers the dying trade of slate roofing. Those few remaining craftsmen who understood slate could keep up with the demand, such as it was. As a result, a considerable store of knowledge and expertise faded into history.

The revival of slate began in the 1970's, once again as a result of several different factors. Asphalt roofs installed in the 1950's had begun to deteriorate. Also, new discoveries in science and medicine called into question the use of asbestos as a building material. The oil crisis of the mid-70's raised the cost of producing petroleum-based products such as asphalt, and steel had not yet reached an acceptable level of aesthetic appeal and cost-effective field production.

While searching for new alternatives, building professionals began to realize that slate roofing had lasted two to three times as long as most other roofing materials, and the life-cycle cost of slate roofs became much more appealing.

Ironically, at nearly the same

FOREWORD

time, the flashings on many of the slate-roofed buildings from the turn of the century had reached the end of their service life. The roofing industry was faced with the dilemma of either reroofing with composition shingles, which had proven to have a short life cycle, or installing a new hundred-year roof. Preservationists insisted that the slate roofs be repaired or replaced with traditional slate, but the biggest problem was finding roofers qualified to do the work. Composition shingles had been the industry standard for a long time, and the principles and installation techniques for composition shingles were completely different from those of slate.

The slate industry has staged a strong revival since the mid-1970's. Unfortunately, the five decades between slate roofing's heyday and its revival have created an information gap: the number of experienced craftsmen who understand the entire slate industry from production to finished roof has greatly diminished in the last 50 years. Nevertheless, because of the growing demand during the revival period, slate has been quarried, specified, ordered, and installed by many individuals with only limited knowledge of the various facets of the industry.

The dilemma faced by today's slate roofing industry parallels that of the slate industry in the early 1920's. In the 1920's, the slate industry adopted common standards for slate roofing widely practiced during that era. Today, the quarries still follow the standards set in the 1920's, but slate roofing practices are no longer standard. Modern slate roofers are largely self-taught and, although many of their practices are sound, there is no clearly defined benchmark by which to distinguish a proper installation from an improper one.

The main purpose of this book is to provide updated information about slate roofs and their installation. In many ways the basic principles of slate production and installation have not changed since the writing of the 1927 "Slate Book," but some of the tools, handling equipment, and related roofing materials have changed significantly. By updating the nomenclature and reviewing the technological changes that have occurred over the last 75 years, we hope to help establish a common language for communicating about slate and a set of standard practices for working with it.

In writing this book, we have made every attempt possible to be fair, objective, and impartial. This text is not intended as a censure of those who use different installation approaches or less common nomenclature. Rather, we hope that it will help restore a basic level of understanding, cooperation, and communication within the industry.

FOREWORD

How to Use This Book

This book is organized in a logical sequence to parallel the planning and installation of a new slate roof. (The steps for reroofing an existing building, although similar, may require a different approach.) The book's twelve chapters are presented in two parts.

Part 1 discusses the steps involved in planning and organizing a slate roof project. It begins with a chapter on "Understanding Slate," which explains where slate comes from, how it is quarried, and the terminology used when discussing slate. There are many other texts which provide a wealth of information about the geology of slate and the quarrying process, and they cover these facets of the industry more completely than space allows in this text. (If you are interested in further researching these areas, you will find a list of sources in the bibliography.) The rest of Part 1 includes information on design considerations, estimating, ordering and handling slate, and roof construction and preparation.

Part 2 explains installation techniques, from roof layout to installing slate shingles to repair and maintenance. Drawings and photographs found throughout the book help clarify important points.

At the end of the book, you will find appendices, a bibliography of sources, and an index. Appendix B provides a series of checklists which correspond with specific portions of the book. Each checklist is intended to simplify the tasks of organizing different facets of a slate roofing project. These checklists also provide standard forms for communication among all of the parties involved with the project, which may be useful in clarifying decisions during the planning stages and resolving any potential disputes that may arise later on.

PART I
PLANNING AND ORGANIZATION

The first part of this book has five chapters covering background information and the initial steps in planning and organizing a slate roofing project.

CHAPTER 1   UNDERSTANDING SLATE: PAGE 9

A description of slate and where it comes from, how slate is quarried, and the terminology used in the slate roofing industry

CHAPTER 2   DESIGN CONSIDERATIONS: PAGE 17

An explanation of aesthetic and technical aspects of slate that should be considered in designing a slate roof

CHAPTER 3   PROJECT PLANNING AND ESTIMATING: PAGE 35

Procedures and factors to consider when estimating costs and materials for slate roofing.

CHAPTER 4   ORDERING AND HANDLING SLATE: PAGE 61

Procedures for ordering and handling slate in preparation for construction

CHAPTER 5   ROOF CONSTRUCTION AND PREPARATION: PAGE 67

Technical considerations for building construction to support slate roofing

CHAPTER 1

Understanding Slate

Slate is a rock that has more or less perfect cleavage or, in more technical terms, a parallel molecular structure. This cleavage allows slate to be split into thin shingles that can then be used for a wide variety of products, including shingles for roofs. Slate is a dense, durable material that is virtually non-absorbent. The texture of the slate varies with its chemical composition. Slate that is too textured is called **schist**. Due to its coarser particle make-up or wavy structure, schist is unacceptable as a roofing material.

This chapter explains how slate is formed, where it comes from, and the terms used to describe it. It also provides a brief explanation of the quarrying process. All of this information will prove very helpful later during discussions of various slate roofing projects.

Where Slate Comes From

Geologically speaking, the slate we see today has been formed in one of two ways. The most common types of slate were formed from marine deposits of clay and sand called **aqueous sediments**. The second and much more rare types, called **igneous slates**, were formed from layers of volcanic ash. Both types of slate were formed millions of years ago by the heat and pressure within the earth bearing down on the sediments, and either type can contain slate suitable for roofing material.

Cleavage is the direction in which slate is most easily split. The **cleavage plane** is created at a 90-degree angle to the pressure placed on the sediment from which the slate was formed. It is this unique cleavage that makes slate different from most other rock.

Beds of slate, or layers of rock, formed by the settling of aqueous sediment, mostly clay, lie parallel to the surface of the water in which they were formed. Clay deposits that split along the bedding planes are shale. However, some of the beds will have become inclined and broken over time due to tectonic forces in the earth. If the deposit was tilted up at a new angle and compressed so that it spread out and produced cleavage planes at right angles to the direction of pressure, slate was formed. Slate from the same bed will have similar qualities. By following a particular bed, it is possible to continue quarrying slate of similar characteristics. However, because slate is the result of natural conditions, different layers—and to a much lesser extent, different areas within a bed—can produce slate with different qualities.

Chapter 1

There is also a **grain** within a bed of slate. Grain is the direction that the particles making up the slate lie. Slates should be split so that the grain runs parallel to the long edge of the roofing shingles.

Slate comes in several different colors, ranging from black to green to purple to red. The color of a particular slate is determined by its chemical composition (see Appendix A).

In addition to classification by color variations, slate is also classified into two categories of **weathering** based on the amount of color change that can be expected as the slate is exposed to the elements. Unfading, permanent, or non-weathering slate will not discolor significantly as a result of prolonged exposure. Fading, weathering, or semi-weathering slate, however, contains a sufficient amount of $FeCO_3$ (ferrous carbonate) to cause discoloration over time.

Ribbons are narrow bands of rock that run through slate and that differ from it in chemical composition and color. If the ribbons do not weaken the slate or detract from the color, then the slate may be used as roofing slate.

Beds of slate have been found and quarried in many parts of the world, including the United States. In the early 1900's, Arkansas, California, Colorado, Georgia, Maine, Maryland,

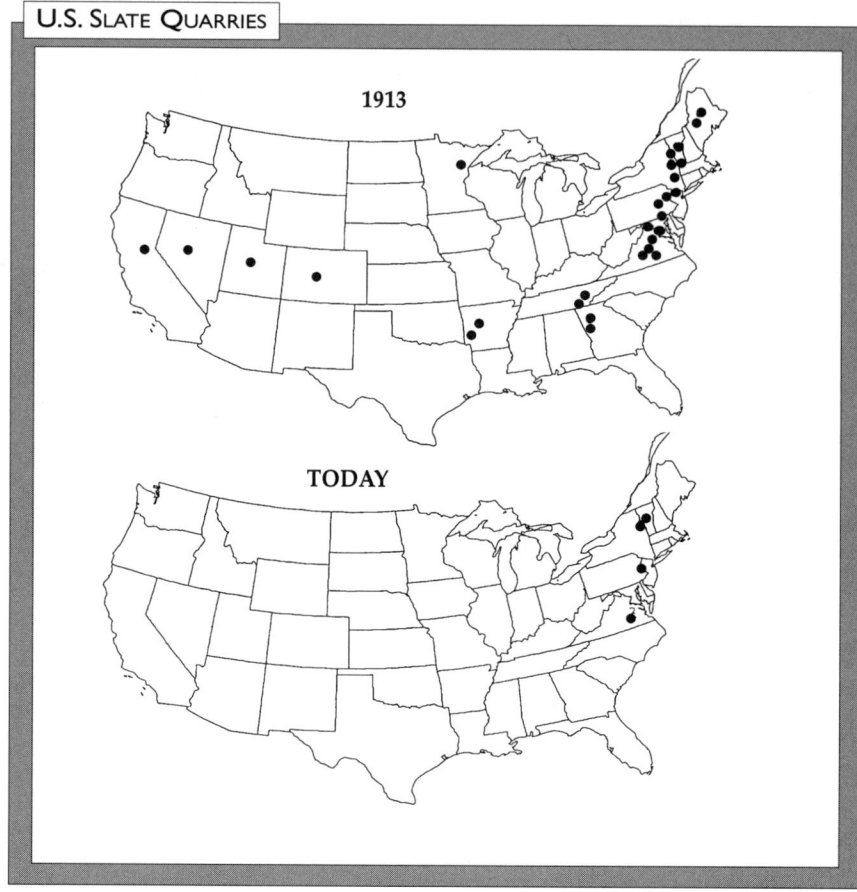

U.S. SLATE QUARRIES

Massachusetts, Minnesota, Nevada, New Jersey, New York, Pennsylvania, Tennessee, Utah, Vermont, and Virginia all had active slate quarries. Today, slate quarried in this country comes primarily from Vermont, New York, Pennsylvania, and Virginia.

Many other countries besides the United States produce slate. Large producers and exporters include Canada, Wales, Germany, Austria, France, India, New Zealand, Spain, Portugal, Italy, China, and South Africa. Some imported slates are better than others, and all should be tested and evaluated carefully for quality —just as any domestic slate should be evaluated. Simply because slate is imported to the United States does not necessarily make it as good as or better than the slate produced here.

The American Society for Testing and Materials (ASTM) has established several tests for determining the quality of slate that is intended to be used for roofing. Document C406-89 (re-approved in 1996) contains the primary specifications for roofing slate. In addition, it lists related ASTM documents which describe specific test procedures. (For a copy of Document C406-89, contact the ASTM.) (See Appendix F for more information.

Understanding Slate

The Quarrying Process

Unlike man-made materials which are produced in a factory, slate is a natural stone which must be mined or quarried from the earth. Understanding the basic process by which slate shingles are produced will aid in the understanding of their thickness, size, cost, and availability. The method used to remove rough slate blocks from the ground varies depending upon the nature of the vein being quarried. Today, all of the quarries operating within the United States are open pits. Depending upon the geological formation of the slate, it is either blasted loose from the ground or cut away with large sawing equipment. The sawing technique is by far the best approach because it minimizes the stress fractures caused by blasting. However, the geological formation of some veins will not permit sawing because the veins vary in slope from 15 to 60 degrees. Unlike those veins which run horizontally, angular veins cannot be worked with a sawing apparatus. Instead, the slate must be blasted loose. Regardless of the approach used, less than 50% of the slate removed from the quarry actually becomes roofing slate. Evidence of this loss or waste can be seen throughout the various slate districts in the veritable mountains of unusable material that have been created.

After the large rough blocks are removed from the ground by either sawing or blasting, they are loaded into trucks and hauled to the mill. From there they are either moved by hand or placed on pallets and moved by forklifts from one step to the next during various stages of production.

Splitting the Slate

Blocks dropped off at the mill often have to be split to a size and weight that a forklift can handle. This splitting is done by hand using a series of chisels or with jackhammers.

Blocks small enough for a forklift are then hauled into the mill and cut to a more manageable size with a large diamond-blade wet saw. This wet-sawing process leaves a milky residue on the blocks when they dry. Also, slate mills are very dusty, and dust tends to accumulate on the slate surfaces, especially when they are wet. Although this may not seem significant, it does sometimes cause some alarm when the slate is later received on a job site. The residue makes the slate look different from the samples that were washed and wiped clean before being sent out earlier by the supplier. However, both the milky residue and the dust will wash off naturally during the course of several rainstorms and leave a clean roof.

Cut blocks of slate are split with a chisel to the desired thickness. This is done by first splitting the block in half, then continuing to halve the subsequent pieces until the desired thickness is reached. It is very important to understand that this splitting process is carried out by hand with a chisel, so the thickness of the final piece is determined by the eye of the splitter. If the splitter is working on an order

FIGURE 1-1: SPLITTING SLATE BLOCKS TO A MANAGEABLE SIZE

Chapter 1

FIGURE 1-2: HAND-SPLITTING FOR DESIRED THICKNESS

that requires 1/4-inch thick slate, he does his best to achieve that thickness. However, due to the imperfections in the stone and human error, it's very common for thicknesses to vary. These variations will be noticeable from piece to piece and within individual pieces. A piece of slate 1/4 inch thick at the top could in fact vary to 3/8 inch thick on one of its lower corners and 3/16 inch thick on the other corner. This is hardly the norm, but it is not uncommon. You will see this more often in the dense, rough-textured slate, such as some of the Vermont Blacks, than you will in smoother varieties. You will also see that as the slate order calls for thicker material, the tolerances will vary more. *(See Appendix F for more information about ASTM Specifications.) The geological formation of the slate chosen will affect the texture, and it is important to remember that slate is a natural stone, complete with natural imperfections.

Some of the confusion related to roofing slate thickness may originate from the slate flooring or structural slate industry. Interior flooring slate is gauged and milled for thickness. An architect or designer who is familiar with flooring slate may not realize that roofing slate is gauged but not milled for thickness. Roofing slate is produced and sold in nominal thicknesses. Variations are expected within specified tolerances. This is a major point of confusion in the field.

Trimming the Slate to Size

The slate pieces split for thickness, called "chips," go next to the trimmer, who is responsible for cutting the chips to standard sizes. If the quarry is in the process of filling an order for 18" × R (random width) slate, the trimmer will do his best to trim the split chips to that size. However, the chips are odd in shape, so the trimmer may be able to trim only 75% of the chips to 18" × R. He may be able to trim 5% of the chips to 20" × R, and the remaining 20% will be 16" × R or less.

During the trimming process, some of the material may break. The trimmer then does his best to make the largest piece possible from the broken chip. In some cases, the pieces that break do so because of imperfections within them. Smaller pieces of slate often appear more textured and contain more variation within the piece since some of them began as larger pieces and continued to be whittled down as imperfections were trimmed out.

As the slate chip is trimmed, the trimming action creates a chamfered edge. This chamfered edge will be laid face up on the roof.

Sorting for Size

Within any given production run, the resulting product will be of many different sizes. The quarry will actually mine, mill, and process several sizes unrelated to one particular order. Remember that slate is produced by hand. When an order is placed for 100 squares

Understanding Slate

(one "square" is defined as the amount of material required to cover a 100-square foot area) of a given length, the quarry may actually produce 135 squares (35 squares of which do not fit the size requirements) of various lengths to fill that order. This is the primary reason most slate is sold today in "random" widths. If an order requires slate of one length and one width, the quarry may have to produce 175 squares to fill the same 100-square order. This will obviously affect the cost.

Creating Nail Holes

Once the chips have been trimmed to size and sorted, they are ready for nail holes. In most cases, the slate is punched for two nail holes, but any slate over one inch thick requires four nail holes. Some manufacturers will drill their slate, but for the most part it is punched.

By the time the trimmed slate gets to the puncher, the front and back sides have been established. With the exception of the starter slate pieces (see Chapter 8), the slate is always punched from the back. Just as the trimming process creates a chamfered edge, the punching process creates a similar chamfered edge around the perimeter of the hole. When the slate is installed face up, this chamfered hole allows the installer to set the nail head flush with the top surface of the piece of slate. If the manufacturer drills rather than punches the holes (which is sometimes necessary with thicker slate), any nail holes drilled in slate less than 3/4 inch thick should be countersunk.

During the punching process, the puncher evaluates each piece to determine which end the nail holes should go in and makes sure the holes are properly punched. Just like in the trimming process, some of the slate will break when punched. If this broken material is large enough to be trimmed down, it goes back to the trimmer; otherwise, it becomes scrap. The trimmer attempts to trim the

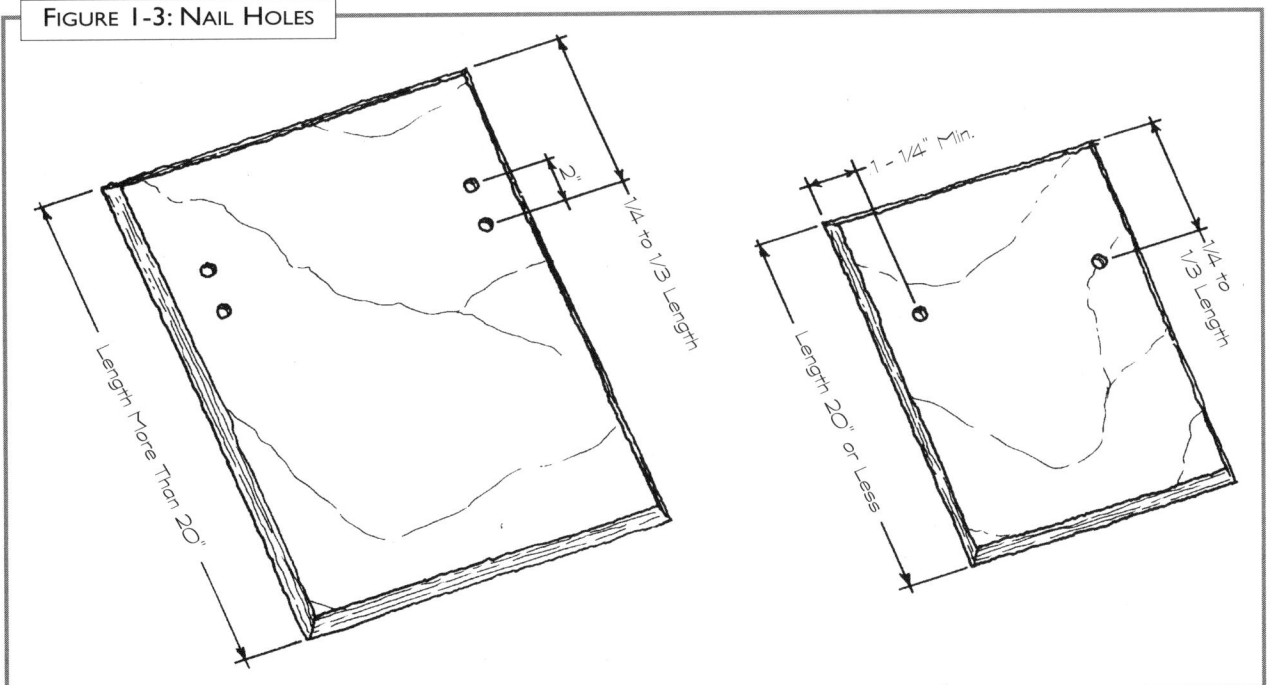

FIGURE 1-3: NAIL HOLES

By the time the trimmed slate gets to the puncher, the front and back sides have been established (chamfered edge up is top side). With the exception of the starter slate pieces (see Chapter 8), the slate is always punched from the back. Just as the trimming process creates a chamfered edge, the punching process creates a similar chamfered edge around the perimeter of the hole.

Chapter 1

> **FIGURE 1-4: CHAMFERING OR COUNTERSINKING NAIL HOLES** 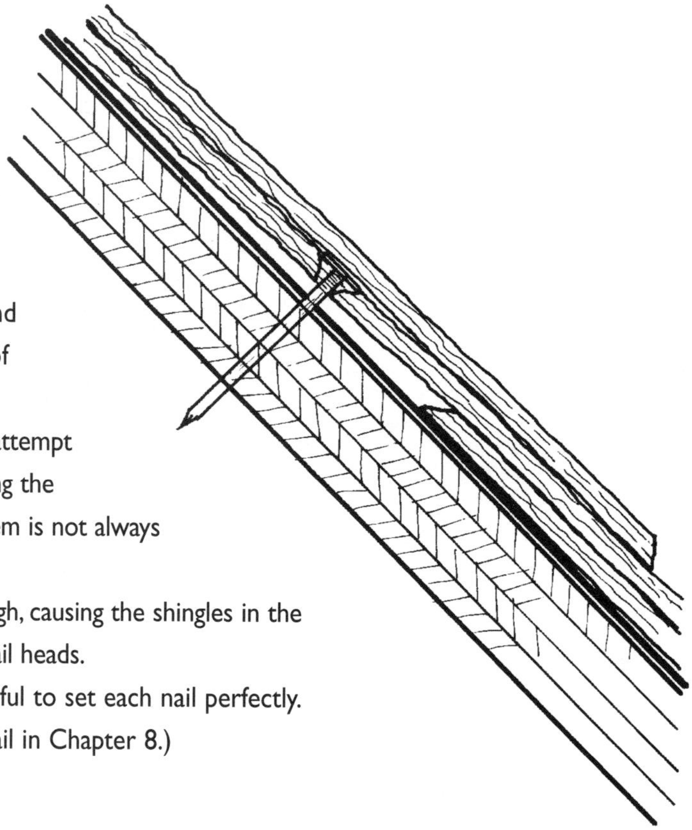
>
> Because the nail heads can be set flush with the top surface of the shingles, the shingle on the course above will not break over the nail heads from the previous course. If the slate is punched from the wrong side and does not have countersunk holes, one of three things could happen:
>
> 1. The nails will be driven too tight in an attempt to keep the heads flush, thereby breaking the shingle along the nail holes. (This problem is not always evident during the installation process.)
> 2. The nails are not driven down far enough, causing the shingles in the successive courses to break over the nail heads.
> 3. The installer slows his pace and is careful to set each nail perfectly. (Proper nailing is covered in more detail in Chapter 8.)

largest piece possible at his station, so some pieces may end up with small broken or missing corners. During the punching process, the puncher tries to keep these corners at the top of the piece where, once installed, they will be covered by subsequent courses. The puncher, who is the last person at the quarry to handle each piece individually, is responsible for quality control within the mill.

The figures on the following pages illustrate the terms used in this description of roofing slate and how they apply to slate roof installation.

FIGURE 1-5: SLATE PARTS

FIGURE 1-6: SLATE PARTS IN SECTION

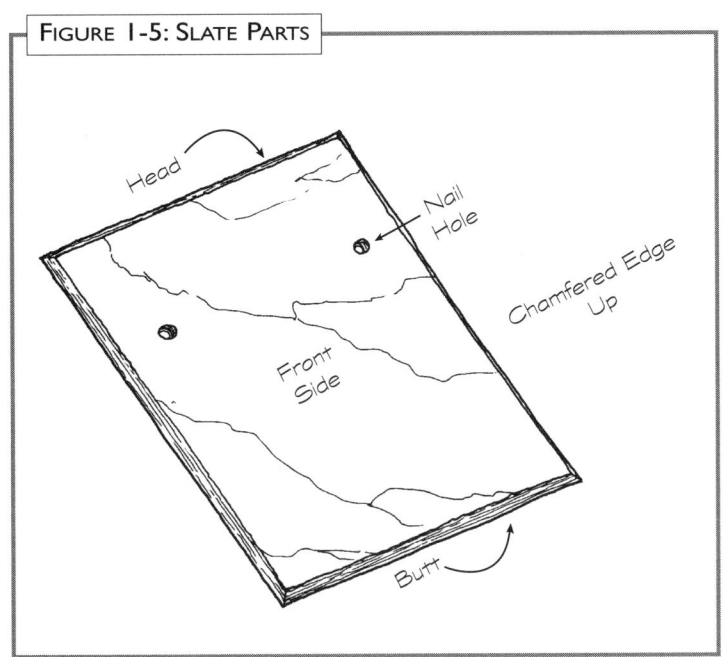

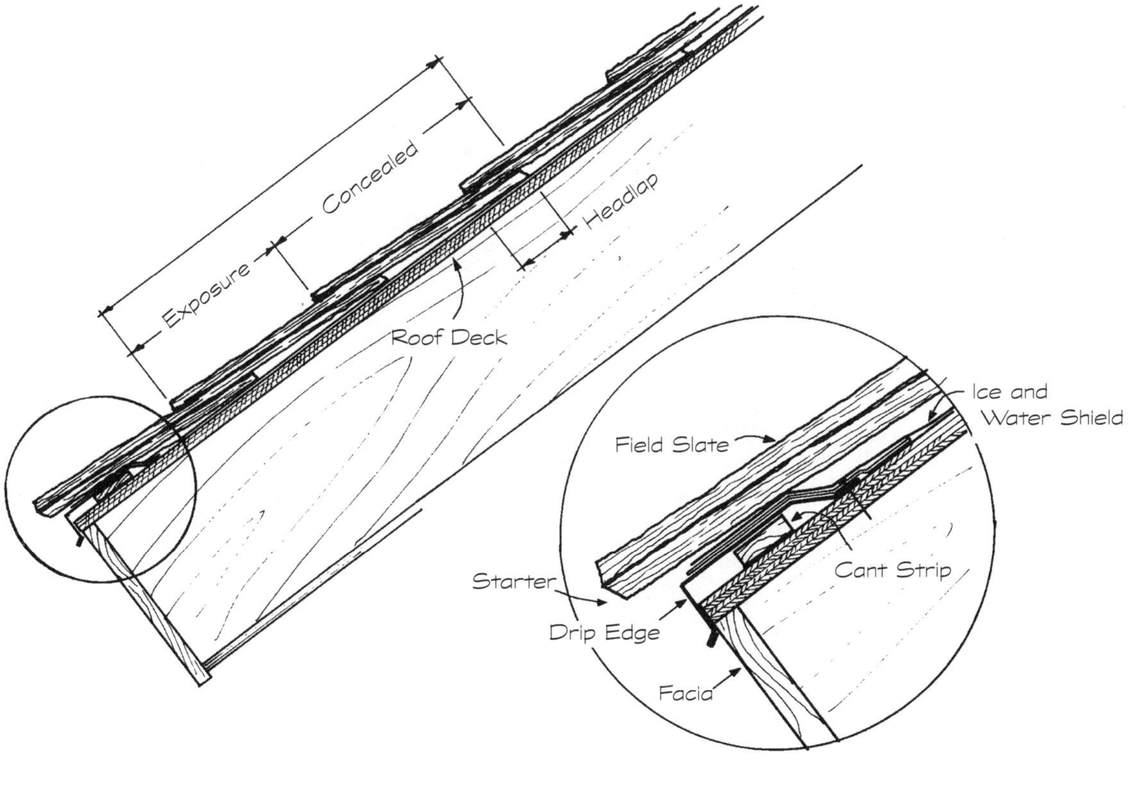

Chapter 1

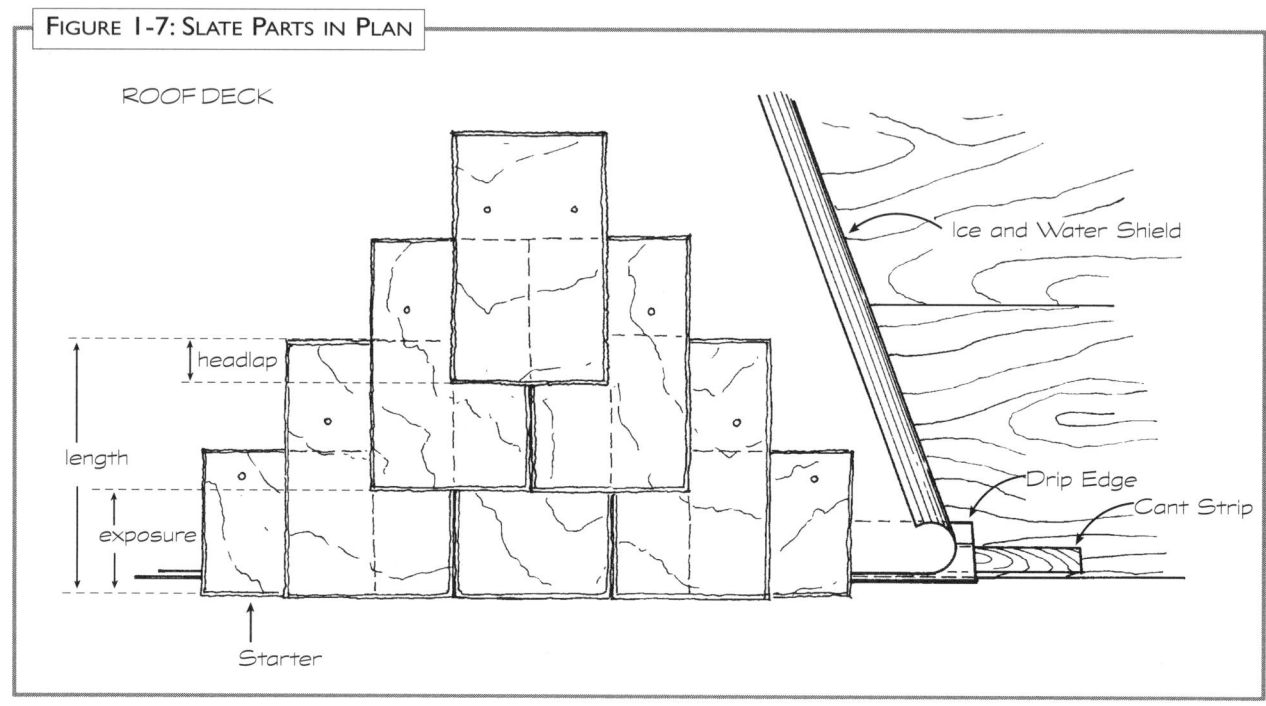

FIGURE 1-7: SLATE PARTS IN PLAN

FIGURE 1-8: SLATE PALLET READY FOR SHIPPING

Storing Finished Product at the Quarry

Finished shingles are stacked edgewise on a pallet, which is then wrapped with steel bands to keep it from shifting. The pallet is then ready for shipment.

CHAPTER 2

Design Considerations

Building owners may choose slate for a variety of reasons, but foremost among them will be its feeling of elegance, its stateliness, and its long life cycle. Beyond the initial appeal and quality of a slate roof, there are many other factors that may influence design choices. From a design perspective, consider that slate is a natural stone with color that permeates naturally all the way through the stone. As such, it allows for a wide range of architectural effects. It is also easy to work with. As a roofing material, it is non-combustible, fire-resistant, waterproof, practically non-absorbent, impervious to fungus and mold, wind-resistant (some manufacturers have obtained Dade County-approval), resistant to climatic change such as freezing and thawing, and resistant to rot. For practical purposes, a slate roof has a long life cycle, requires minimum maintenance and no treatments, and is durable as well as dimensionally stable. In addition, slate is resistant to chemicals and is environmentally friendly. It does not curl and is easy to repair.

Where to Start

In the initial planning stages, you may want to look at some slate roofs on other houses or buildings and start collecting resources. It is helpful but not essential to find an architect who has experience with slate. Many architects have libraries or prior projects that can provide useful examples and pictures to aid in defining the desired appearance. Engineers and consultants can also provide a wide range of information that you may wish to take advantage of, especially if you are considering reroofing an existing structure. Speaking to slate quarry personnel and qualified roofers can also provide a wealth of information.

One of the first steps in designing a slate roof is to determine the basic kind of roof needed for the structure. You may want to review the Roof Parts Nomenclature in Appendix G to familiarize yourself with the terminology used to describe different roof styles and features. The figures show some standard kinds of roofs and design options.

Slate Roof Designs

The next step is to make some preliminary decisions about the design of the slate roof itself. Historically, slate roofs have been categorized into three classifications—Standard, Textural, and Graduated—which refer to the type and thickness of the slate itself. Since these classifications were developed at a time when everyone involved in construction had a general knowledge of slate roofing, they were understood as generalizations. Over time, specific names and descriptions have been applied to these classifications in an attempt to define or describe them more clearly. As general slate roofing knowledge has faded, some of these added descriptors have caused some confusion.

Today, the classifications have remained essentially the same, but the definitions have evolved. Table 2-1 on the next page contains an expanded list of slate roof types and variables within those types. An explanation of each type and variable follows.

Slate Roof Types

The first type of slate roof listed is also the one most misunderstood. **Commercial Standard Thickness** is defined as 3/16-inch slate of nominal thickness. Today this roof is a special order that must be explicitly specified in order to receive the correct material. The shingles are hand-picked to be of uniform thickness with no more than 1/8-inch variation in surface face dimensions (length and width). The surface texture will be smooth with no knots and knurls. Commercial Standard Thickness slate could almost be described as dimensionally perfect pieces of slate. In terms of weight, this is the lightest type of slate roof under most circumstances. However, due to tight tolerances, it can be very expensive. The appearance of this type of roof is flat, even, and uniform. This roof type can be designed with the following variables: single width, color variation, patterns, and cut butts. It is rare to see this type of roof specified with more than one size.

Standard Thickness slate is defined as 1/4-inch slate of nominal thickness, varying by up to 1/16 inch. The shingles are hand-picked to be of uniform thickness with no more than 1/8-inch variation in surface face dimensions (length and width). The surface texture will be smooth with no knots and knurls. Although this is the most common type of slate roof, it is not always the least expensive. In appearance, this type of roof is flat, even, and uniform. It can be designed with the following variables: random widths, single width, staggered butts, color variation, patterns, cut butts, and graduated lengths.

Textural slate roof thickness is defined as 3/16-inch to 3/8-inch slate of nominal thickness. This is essentially the quarry-run material. The surface may have varying texture with thickness variations possible within one piece. Thickness variations should be anticipated but not counted on. The appearance of this type of roof may be textured with more noticeable thickness variation between individual pieces. Although a greater degree of thickness is acceptable, slates may still appear uniform. Compared to standard roofs, the textural roof is likely to have more character. This is due mostly to more pronounced shadow lines. (For a guaranteed textured appearance, see "Intermingled Thickness" roofs.) The textural slate roof type can be designed with the following variables: random widths, single width, staggered butts, color variation, patterns, and cut butts.

The **Intermingled Thickness** slate roof is defined as a roof with at least three distinct variations in slate thickness; for example, slate of 1/4-inch, 3/8-inch, and 1/2-inch thicknesses used in one roof. Any number of combinations of slate thicknesses can be used. However, with too great a variation in thickness, the slate may not lie properly. The textured appearance does not rely on an equal distribution of thicknesses. The most pronounced texture will be achieved by using a greater percentage of the thinnest material and a lesser percentage of the thickest material. This will allow the two thicker sizes to act as accents. The surface may have varying texture with thickness variations possible within one piece. With careful planning, the designer can create a rough appearance without adding excessive weight

Design Considerations

Slate Roof Types	Slate Roof Variables
Commercial Standard Thickness *(Special order only – must be explicitly specified)*	Color
Standard Thickness	Width
Textural	Length
Intermingled Thickness	Cut Butts
Heavy	Patterns
Graduated	
Special Order (thickness)	
Used or Salvaged	

TABLE 2-1: SLATE ROOF TYPES AND VARIABLES

Roof Types

FIGURE 2-1: COMMERCIAL STANDARD THICKNESS

FIGURE 2-2: TEXTURAL

FIGURE 2-3: INTERMINGLED THICKNESS

FIGURE 2-4: STANDARD THICKNESS

FIGURE 2-5: HEAVY

FIGURE 2-6: GRADUATED LENGTH AND THICKNESS

or cost as compared to that of a textural roof. The intermingled thickness roof has a textured or "cobbled" appearance, with noticeable thickness variation between individual pieces. This greater degree of texture is achieved by specifying thickness variation. The intermingled thickness roof can be designed with random widths, single width, staggered butts, and color variation. Patterns and cut butts are rarely seen in this type of slate roof.

A **Heavy** slate roof has a thickness greater than 3/8 inch. The thickness of a heavy slate roof needs to be clearly specified, and the specified thickness will be the thinnest slate provided with allowable tolerances on the larger side up to 1/8 inch. A heavy slate roof will typically be requested with the lowest and highest tolerances included; for example, an order might specify 3/8 inch to 1/2 inch with 3/8 inch being the lowest tolerable thickness. In appearance, this type of roof is thick, heavy, rough, hand-hewn, and textured. Heavy slate roofs up to one inch thick are not uncommon. However, the thicker the material specified, the more surface texture you can anticipate, as well as thickness variations within individual pieces. As the desired thickness increases, so too will the structural needs of the building. The thicker the slate, the more expensive the roof is. This includes not only the slate but also shipping, handling, and installation. This roof type can be designed with random widths, single width, staggered butts, and color variation. Patterns and cut butts are not usually seen in this type of slate roof.

A **Graduated** slate roof is defined as a roof in which the slate diminishes in length and/or thickness as it progresses from eave to peak. These roofs will often range in thickness from one inch thick at the eaves to 3/16 inch thick at the peak. Exposure may vary from 10 1/2 inches or greater at the eaves to 4 1/2 inches at the peak. Sometimes referred to as the "custom-made roof" of the industry, it allows the designer and installer the most creativity and greatest range of effect. The appearance of this type of slate roof is unique. The diminishing size of the slate progressing up the slope of the roof creates the illusion of a longer rafter and a more massive building. One does not necessarily need to use thick slate at the eaves to create this illusion. Nor does one necessarily need to diminish the thickness if the desire is to have a very thick appearance throughout the roof. Most of the aforementioned types of slate roofs can be turned into a graduated length slate roof while still retaining their defining characteristics. It is important to remember that to achieve the graduated effect, either the length or thickness, but typically both, must diminish in size from the eave to the peak. "Variable" describes this entire roof type, as variations are integral to creating the graduated effect. Special consideration must be given to the installation of this slate roof type (see Chapter 8: Installing Slate).

Special Order slate is any slate that is needed for a project but that isn't normally produced. This can include thicknesses over 3/8 inch, lengths over 24 inches, widths over 14 inches, specifications requiring one specific size, cut butts, special fastening considerations, and sometimes specific color requirements related to shades within a color. Each of these circumstances require extra handling by the quarries. The more specific the request, the greater the additional cost to the customer and the more difficult it is for the quarries to provide what is being requested.

A **Used or Salvaged** slate roof is one in which the designer has chosen to reuse slate that has already been installed on a roof. This practice has become more common in the late 1900's as major roof replacements related to deteriorated flashings have developed. One reason for considering this approach is the reduced material cost. Other considerations include the aesthetic appeal of an aged roof, the fact that its colors are not likely to change, and the need or desire to retain historical

Design Considerations

material. This practice is not uncommon, but slate quality varies. All slate ages differently, so it would be prudent to have salvaged or used slate professionally evaluated. Typically the salvaged slate required to do a project will come from several sources. The key to getting the desired appearance is to purchase all of the slate required for the project and mix it before the installation begins. The used or salvaged slate roof type can be designed with random widths, staggered butts, color variation, patterns, and cut butts.

Slate Roof Variables

When designing a slate roof, you may choose any of the eight roof types defined above, from Commercial Standard Thickness to Used or Salvaged. In combination with the roof type, you may introduce any of five variables: color, width, length, cut butts, and patterns. (However, as explained in the previous section, some of these variables are not feasible with certain roof types).

Color

One of the appealing characteristics of slate is the wide variety of natural colors. (See inside back cover for swatches of slate colors available.) These color variations are governed by the slate's particular chemical and mineral composition. Although there is a basic pallet of colors, the shades may vary within the same vein and from region to region. The designer also needs to understand that all slate will change in color to some degree when exposed to the elements. The eight basic slate colors cover the spectrum from black to red.

Each of these colors should be further described by the term **non-weathering** (or "unfading") or **weathering** (or "semi-weathering") to indicate the degree of color change that can be expected. With eight basic colors and two descriptive terms, there are actually sixteen basic colors of slate.

Weathering or semi-weathering slate will develop brown and buff tones as it ages. On average, 10% to 40% of weathering slate will develop these tones. The actual amount that a slate weathers will vary with the particular slate chosen. This weathering continues to some degree throughout the life of the slate but does not diminish the longevity of the material. The predictability of these changes will have to be estimated by the quarry producing the material. Rarely will an entire roof show weathering. (If this is a desired effect, consult with your supplier.)

When slate is labeled as non-weathering or unfading, this does not mean that the slate will not change at all. Unfading means that over time the slate will have essentially no weathering tones. The slate may lighten or deepen in color and may have varying tones during prolonged exposure to the elements. Consult your supplier for the characteristics of the material currently being quarried.

When specifying colors, the designer needs to understand that each producer may have a different trade name for slate that will have the desired appearance. Indeed, quarries in the same geographic region may be providing material with the same general characteristics but with subtle differences in shade. Colors can and do vary within any given vein. The color prefix indicated by the manufacturer is based upon that manufacturer's observation of what has happened over time with its material. In some instances, there are three or four trade names commonly used in the slate industry to describe material exhibiting similar characteristics.

Quarries use the designations "weathering" or "semi-weathering" because they have no way of knowing or controlling exactly how much weathering will occur. Each quarry will be able to estimate this weathering phenomenon based on its own prior records, but it is impossible to predict exactly how each and every stone will weather.

Keep in mind that slate is a sedimentary rock, as explained in

Chapter 1. As the miners work down through a quarry, they will encounter different colored layers of slate. Furthermore, the color that a particular quarry is mining today will probably be somewhat different 20 to 30 years from now. You will have to rely upon the knowledge and integrity of your source to best estimate what will happen to the slate you are looking to purchase. If you wish to have an unfading color, that must be specified. But again, the unfading designation is based upon the quarry's prior experience and is difficult, if not impossible, to guarantee.

During the design process, be sure to consider the degree to which a chosen material is likely to weather. This is especially important when working with multi-colored roofs. When a semi-weathering, multi-colored slate is chosen, the result a few years down the road will be a roof with more colors in it than when it was originally installed. Be sure that the colors are properly mixed by the installer to achieve the desired effect.

Once the desired color has been chosen, be sure that all of the parties involved in the roofing process agree upon and acknowledge in writing that they are bidding on the chosen material. Typically, the unfading colors are more expensive than the weathering colors. It is not uncommon for someone along the way to substitute a weathering slate for an unfading slate to reduce costs.

Because slate is a natural material, the colors are natural earth tones. The designer has the opportunity to design a roof that will blend with its natural surroundings. This is an important consideration when choosing between weathering and unfading slate. Be sure that the color fits the structure as well as the surrounding landscape.

When obtaining slate, it is important for a buyer to deal with a reputable source. Slate is available in the United States from quarries all over the world, but some of these sources supply material which is better than others. Some slate is just not suitable for roofing applications. The chemical composition of the stone and how it stands up to the ASTM testing procedures are good initial indicators of the stone's quality (see chemical composition chart and information about ASTM specifications in the Appendix). However, a supplier's previous track record, references of prior projects, and number of years in business can also prove helpful when choosing a reliable source for materials. Once the source has been qualified, it is important for the buyer to understand that the seller will do his best to provide the chosen color. However, even the best suppliers in the world cannot guarantee the exact shade of every piece of slate.

Width

A second variable associated with slate roofs is the width of the slate shingles. In a **Single-Width** or **One-Width** slate roof, all of the slate is the same width. At the turn of the twentieth century, slate was normally provided in one width. However, ordering a single-width slate roof today will increase the cost of the material for the roof. Slate is commonly produced in widths from one half of its length to 14 inches. Wider slates can be special ordered. In appearance, a single-width slate roof is one in which the alternating vertical joints line up in a symmetrical pattern over the entire roof surface. Single-width slate can be used with any of the roof types and with other variables. Most patterned and cut butt roofs are done in single-width slate.

In a **Random-Width** roof, the widths of the slate vary. A variety of widths will generally be provided by the quarry. The largest percentage of widths shipped will be approximately 2/3 the length of the slate being used; the smallest percentage will be of the largest widths. However, when specifying "random-width" this will vary and percentages cannot be specifically ordered. The use of random-width slate will generally reduce the material cost of the roof. In appearance, the alternating vertical joints between shingles on a random-width roof will not always line up with the vertical joints from the courses below. If you want to achieve a symmetrical appearance, you will want to avoid random widths. In some cases, the use of random widths will make an otherwise standard appearance look more textured or cobbled. Random-width slate can be used with most of the roof types and with other variables, but creating patterns in a random-width roof is not common.

Length

In addition to varying widths, slate comes in varying lengths, ranging from 10 inches to 24 inches in even-numbered increments (longer by special order). There is no one standard slate length; the length or lengths must be specified for each project. The length of the slate determines how much of each shingle will be exposed, and it will affect the cost of a slate roof. The shorter the slate, the more pieces of slate there are per square, thus the more labor required to install the roof. However, due to the way slate is made, longer slate is more difficult to make and therefore is more expensive.

Slate roofs commonly have even horizontal lines. This effect is achieved by using slate shingles of equal length. The designer needs to select slate lengths in relation to the size of the structure and the effect he hopes to achieve.

For some roofs, designers use **staggered butts**, which refers to the use of slate in varying lengths to achieve a random horizontal line appearance. This effect is achieved by mixing slate of two different lengths. This will add cost to the roof because of the additional material required to produce the desired effect as well as extra labor to mix the sizes. Percentages of each length will vary depending on the desired effect. On a staggered butt slate roof, as the name implies, the butt ends of the slate will form uneven horizontal lines, achieving a more or less random appearance.

Graduated slate roofs, as explained earlier, generally use thinner slate at the peaks and thicker slate at the eaves. However, the effect of a graduated roof can be achieved without graduating thicknesses. Instead, the shingles are set with a diminishing exposure as the courses progress up the slope of the roof. Creating a graduated length slate roof will increase the cost because of the expense of the longer material, extra labor needed to install the shorter material, and the added labor in layout and handling.

Cut Butts

A Cut Butt is a special quarry-manufactured piece of slate that is cut to a specific shape at the butt end. For example, the butt may be scalloped, diamond-shaped, or octagonal. The appearance of a cut butt roof is more sculpted and delicate than that of a regular slate roof. It is not uncommon for only a few courses of slate in a roof to have cut butts, thus creating a "band" effect in the roof. Headlap must be maintained no matter what the shape of the piece, and the cost for cut butts is higher due to the increased amount of cutting required to shape the piece.

Patterns

A patterned roof is one in which different colored slate is used to create one or more designs. These designs may include names, dates, stripes, zigzags, or quilted effects. The appearance of a patterned roof is more delicate and intricate than that of a regular slate roof, since patterns can be used to create many visual effects. Patterned roofs will increase the labor cost and may increase the material cost, depending on how elaborate the pattern is and which colors are chosen.

Requesting Slate Samples

When you have made preliminary decisions about the thickness, color, width, and length of the slate you want to use, it might be helpful to request slate samples from one or more suppliers. Most quarries

Chapter 2

will provide samples of their slate, and the samples will be fairly representative of the material from the beds they are currently working.

However, the samples you receive will be washed clean of the debris from the quarrying procedure. They will be handpicked by the salespeople for quality and color, and they will probably not be weathered.

If you order material long after you have received samples, be sure to confirm with quarry personnel that the material they are currently working will have characteristics similar to those of the samples that you have.

It is impossible for any quarry to guarantee that the slate it provides will exactly match the samples or color plates from brochures. Slate is a natural stone, and color will vary somewhat. When requesting samples, ask for several pieces of the same color so you will be able to get a real feel for the variations within the vein.

Alternative Installation Methods

In addition to the eight roof types and five variables discussed above, there are other design factors to consider in designing a slate roof, such as installation methods. There are some alternative methods of laying slate. The common denominator seems to be "stretching" the slate by using less overlap, which requires less material per square.

One alternative approach is the "French" or "45-degree angle" method, which provides an obvious aesthetic appeal. All of the basic principles of slate installation apply, but the layout will be somewhat different.

A second alternative is the "economy" installation, which uses an absolute minimum overlap in covering the roof surface. This approach, however, is only as good as the underlayment which is laced in between courses. The only benefits of this method of application are the reduced weight of the slate roof and lower cost of materials. The use of economy lap should not be considered standard practice.

Constructing Mock-ups

Sometimes, even after slate color and size have been chosen and samples received and approved, it may still be difficult to visualize what the roof will look like. In this case, it is very helpful to construct a mock-up with

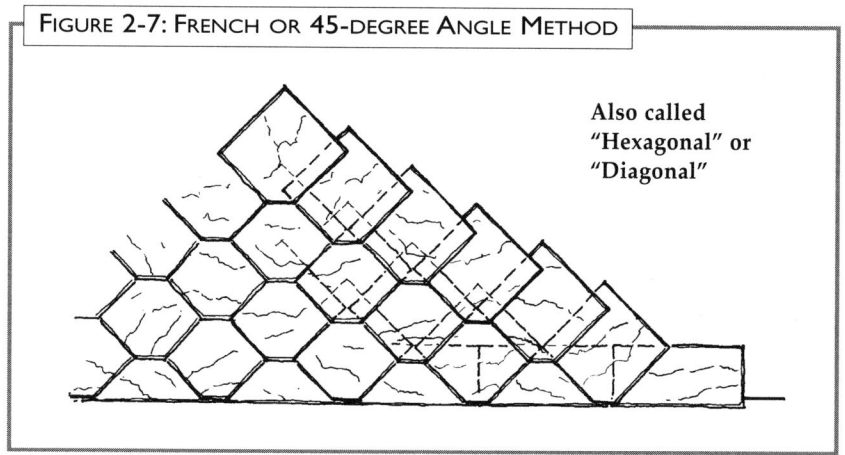

FIGURE 2-7: FRENCH OR 45-DEGREE ANGLE METHOD

Also called "Hexagonal" or "Diagonal"

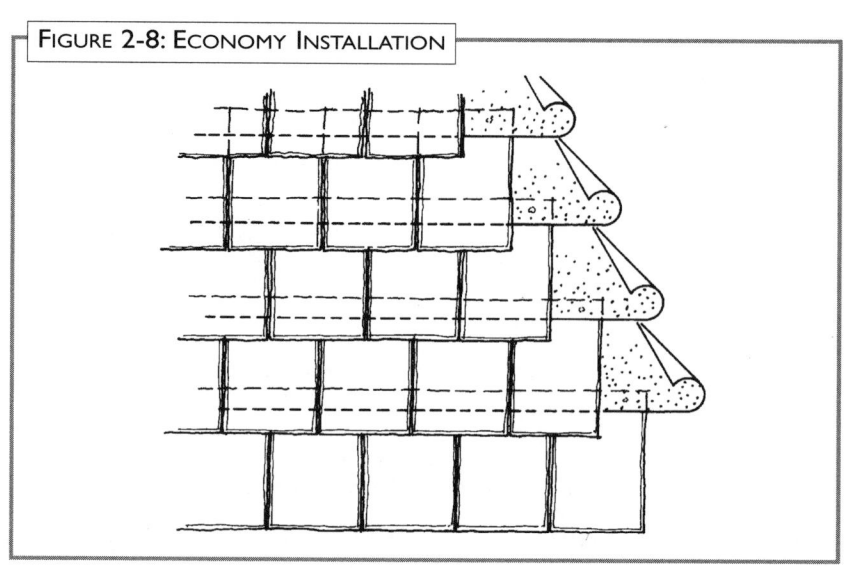

FIGURE 2-8: ECONOMY INSTALLATION

Design Considerations

several courses of slate, showing both horizontal and vertical lines and using the selected installation method. Whenever possible, this should be done right on the roof. If this is not possible, a half sheet of plywood (4' × 4') will work fine. The plywood sheet can then be placed on the roof or near the building and observed from a distance typical of the actual roof.

The use of a mock-up helps everyone involved, especially the owner, to visualize what the final roof will look like. In particular, the mock-up will show the design of the roof and its texture, and it will begin to give all parties a sense of scale related to the particular structure. Also, color mix for a multi-colored roof can be manipulated within the mock-up. It is not uncommon to have a situation in which a color mix is chosen based on percentages of each color per square, only to discover that when it is actually installed, it does not give the desired appearance. In this case, a mock-up may help resolve this issue before the roof is installed.

Hiring someone to construct a mock-up gives the owner and/or architect an opportunity to interview and evaluate potential roofers. It is far more practical, and may prove less expensive in the long run, to prequalify the roofing contractor on a 4' × 4' mock-up than it is to award an entire roofing project with limited knowledge of the contractor's capabilities. However this should not be the only criteria used to qualify your roofer.

If you are considering the construction of a mock-up, keep in mind that this slate will have to be purchased in addition to the material that you order for the roof. In most cases (unless it is a very large project), it is not reasonable to ask the slate supplier to provide a half-square of slate as a sample. In addition, this material will end up being shipped via common carrier, which can be costly in comparison to the relative cost of shipping an entire trailer load. However, you are installing a roof that should last 100 years. The additional cost in the beginning is minimal when considering the life of the roof.

One other thought when considering a mock-up: you might ask your roofer if he happens to have any of the desired material in inventory at his shop. This can be another way to evaluate the experience of your roofer. A qualified slate roofing contractor will most likely have material on hand or has a good enough relationship with a supplier that he can provide the material for the mock-up.

After a mock-up is constructed, it should be washed and wiped dry. It's not necessary to wash and dry the roof after installation, but the purpose of the mock-up is to give the owner a realistic look at the finished roof product after it has been naturally cleaned.

Once the mock-up is completed and approved by the owner, it is wise to either photograph it or, if possible, save the mock-up until the roof is nearly complete. This will give all parties a reference as to the agreed-upon look. Photographing the mock-up when it is completed will provide you with a good reference later with respect to weathering that may have occurred during the roofing process. For future reference, be sure to record the chosen color mix, slate size, texture, and exposure.

Roof Details

As the design of the slate roof progresses, you will need to consider roof details. Details are the parts of the roof, such as valleys, hips, and ridges, that require special treatment. The following are a number of different styles for finishing the detail areas of the roof.

FIGURE 2-9: OPEN VALLEY - FINISHED

Open Valley (Figure 2-9)

A valley in which the chosen flashing material is exposed. This is probably the most common valley treatment.

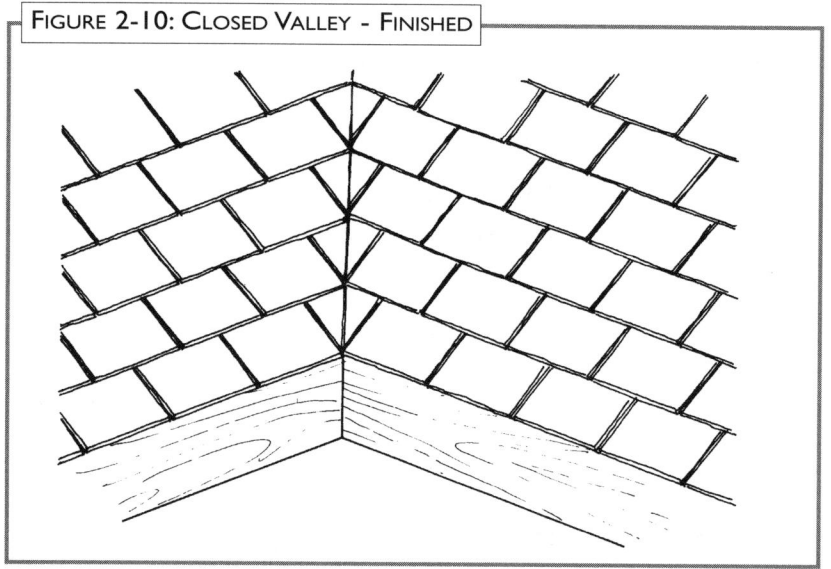

FIGURE 2-10: CLOSED VALLEY - FINISHED

Closed Valley (Figure 2-10)

A valley in which the chosen flashing material is concealed. With this application, the slates from the two sides of the valley touch.

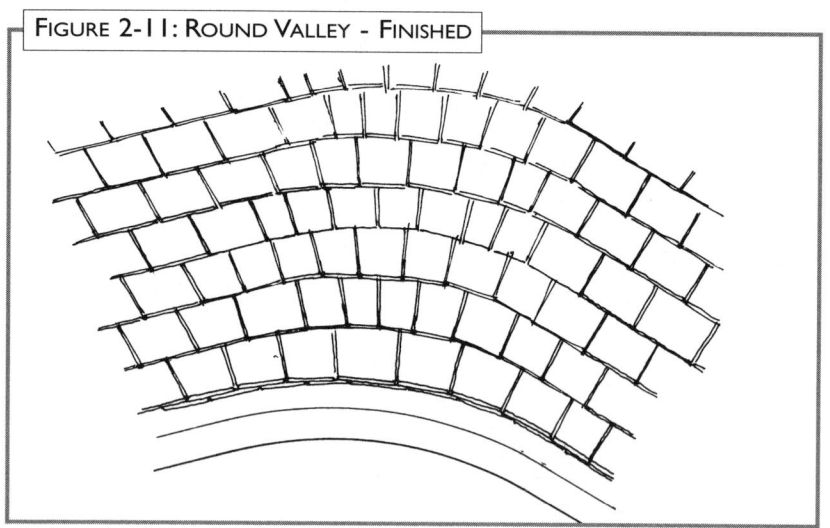

FIGURE 2-11: ROUND VALLEY - FINISHED

Round Valley (Figure 2-11)

A valley in which the slate is laid in a radial technique. There is no flashing showing with this application. These are extremely difficult to install and should not be done by unqualified laborers.

Design Considerations

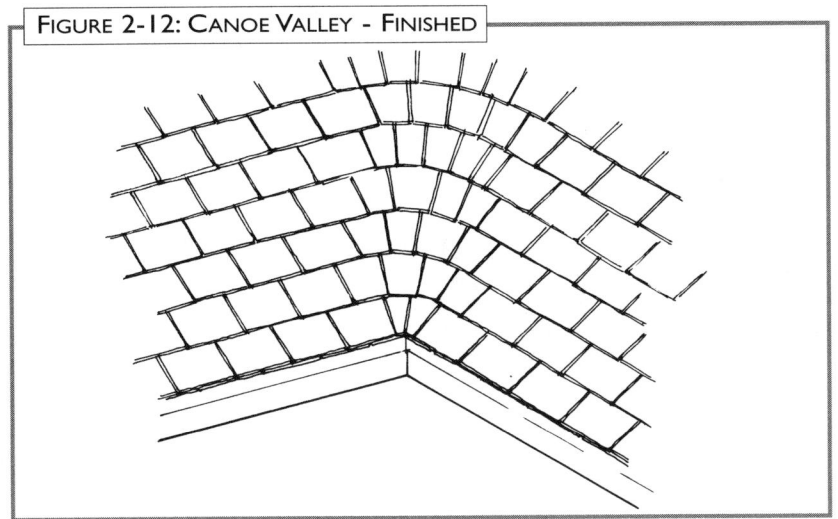

FIGURE 2-12: CANOE VALLEY - FINISHED

Canoe Valley (Figure 2-12)

A valley in which the slate is laid using a radial pattern progressing from a minimal radius at the eaves to its largest radius at the mid-point in the valley and back to a minimal point at the peak. No flashing is exposed. These are extremely difficult to install and should not be done by unqualified laborers.

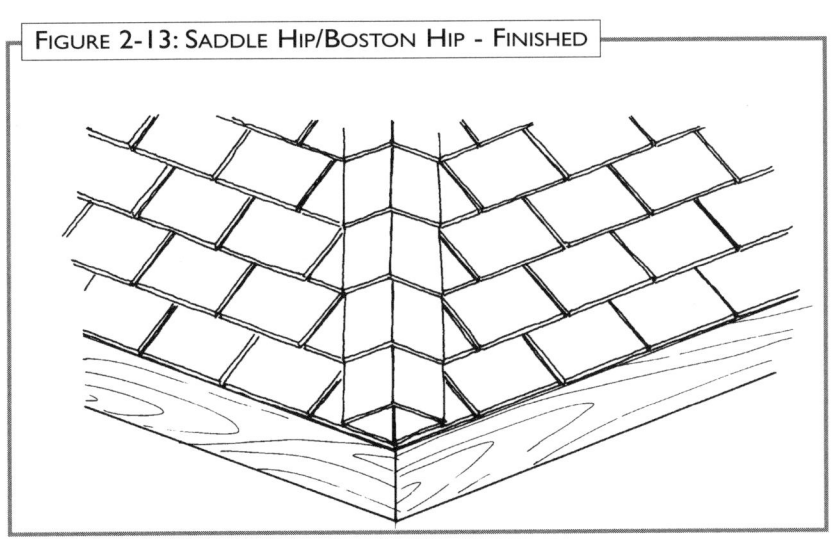

FIGURE 2-13: SADDLE HIP/BOSTON HIP - FINISHED

Saddle Hip (Figure 2-13)

A hip in which the cap slate is installed over the finished slate roof on either side creating a saddled effect. This is the most common hip detail used.

Boston Hip (Figure 2-13)

A hip in which the hip slate is woven into the field slate, creating a smooth appearance with a hint of a saddle hip. This detail is rarely used because of its additional labor costs.

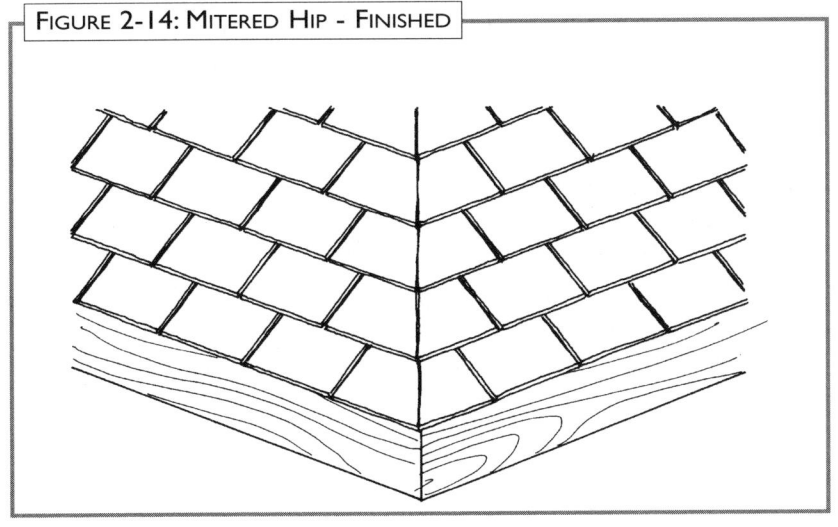

FIGURE 2-14: MITERED HIP - FINISHED

Mitered Hip (Figure 2-14)

A hip in which the last field slate on either side is miter-cut to form the hip. The miter cuts create a smooth transition with no pronounced hip cap. If this detail is used on adjoining surfaces of different pitches, the courses will not line up. For this reason, mitered hips are not recommended where two roof surfaces of different pitches meet at the hip.

Chapter 2

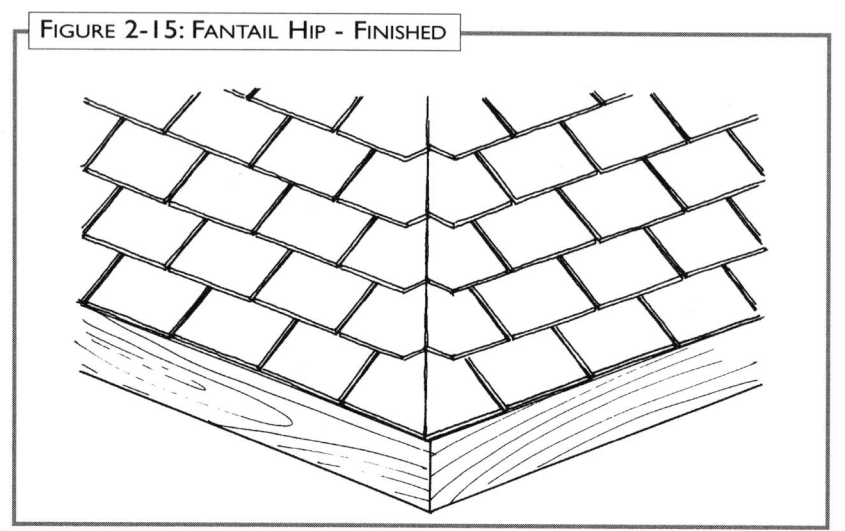

FIGURE 2-15: FANTAIL HIP - FINISHED

Fantail Hip (Figure 2-15)

A fantail hip is a variation of a mitered hip. In this application, the points of the mitered pieces are cut off in a consistent pattern to mimic a saddle hip.

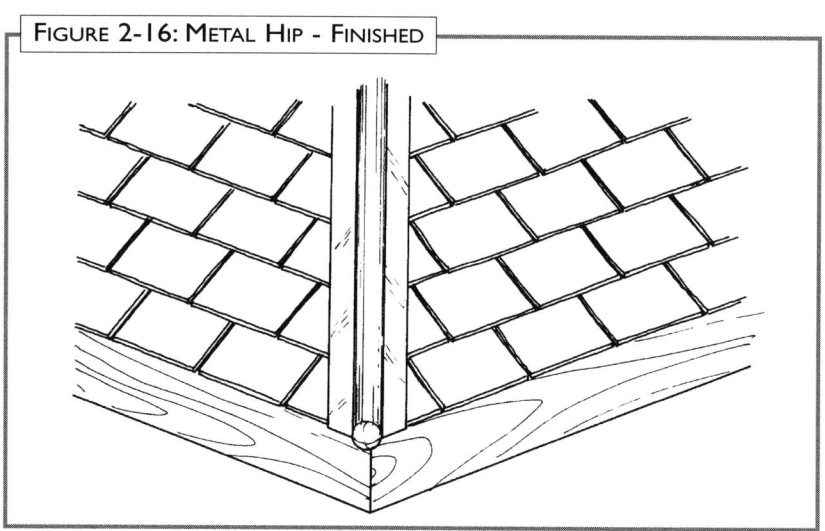

FIGURE 2-16: METAL HIP - FINISHED

Metal Hip (Figure 2-16)

A hip in which a metal cap is used to cover the hip area. This metal cap will vary from a plain flat design to an ornamental cresting detail. The metal usually matches that used on the rest of the project. The more ornamental the metal, the more expensive it is.

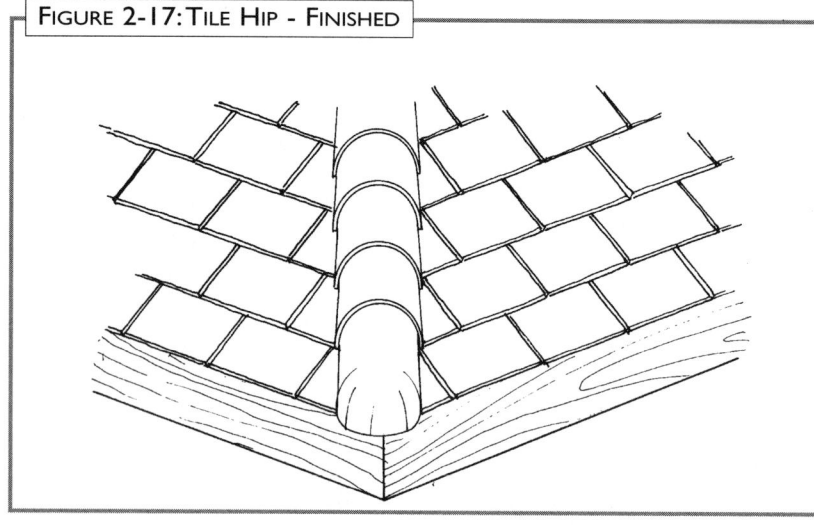

FIGURE 2-17: TILE HIP - FINISHED

Tile Hip (Figure 2-17)

A hip in which ceramic tiles are used to cap the hip. This detail is more common in Europe than the United States. Styles will vary from round, to 90 degrees, to ornamental. Cost will vary depending on the material and design of the tile pieces.

Design Considerations

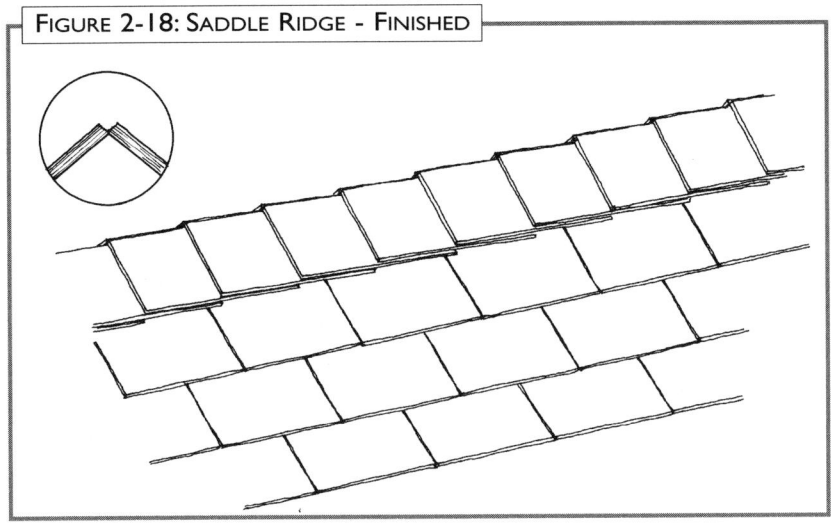

FIGURE 2-18: SADDLE RIDGE - FINISHED

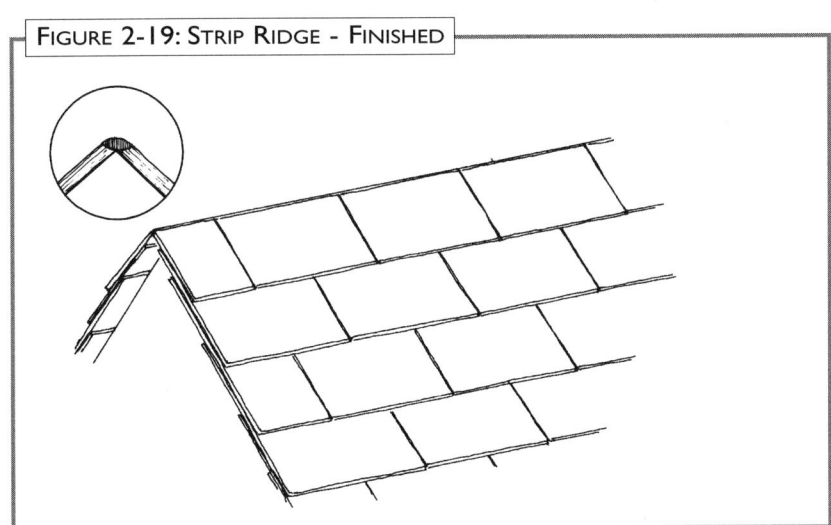

FIGURE 2-19: STRIP RIDGE - FINISHED

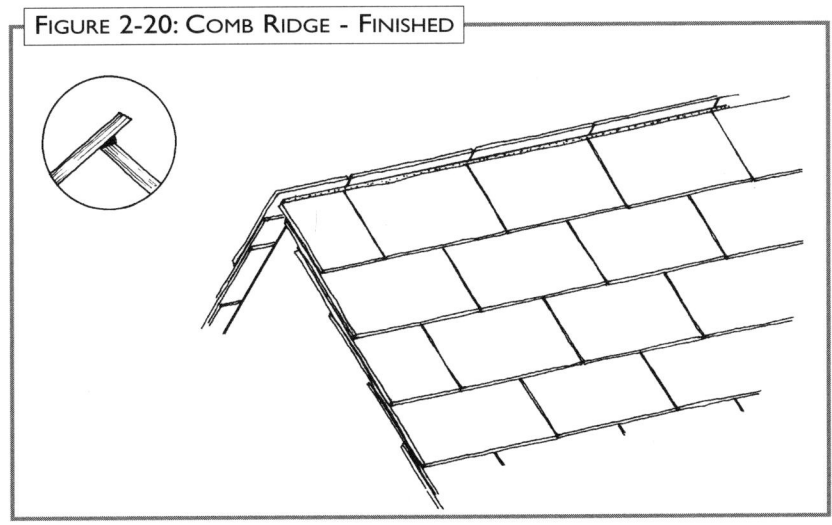

FIGURE 2-20: COMB RIDGE - FINISHED

Saddle Ridge (Figure 2-18)

A ridge in which cap slate is installed over the finished slate roof on either side, creating a saddled effect. The cap slate will be laid with overlap perpendicular to the field slate, creating a distinct elevated cap effect. This is the most common ridge detail used.

Strip Ridge (Figure 2-19)

A ridge in which cap slate is installed over the finished slate roof on either side in the same plane as the field slate, creating a smooth finished detail. Typically, the strip ridge slate will be field slate rotated 90 degrees and laid with no overlap side to side. However, this unusually wide appearance can be avoided by cutting the cap slate to match the field slate width. Properly done, this detail will resemble the smooth transition of a mitered hip. A strip ridge will affect the layout of the roof.

Comb Ridge (Figure 2-20)

In this variation of the strip ridge, the ridge slate on the windward side will be installed with 1/8" to 1" protruding beyond the peak of the roof. This method was developed to allow the installer to point the ridge with elastic cement and thus better waterproof the ridge of the structure. However, with modern waterproofing materials, this approach is not necessary and is rarely seen in new applications.

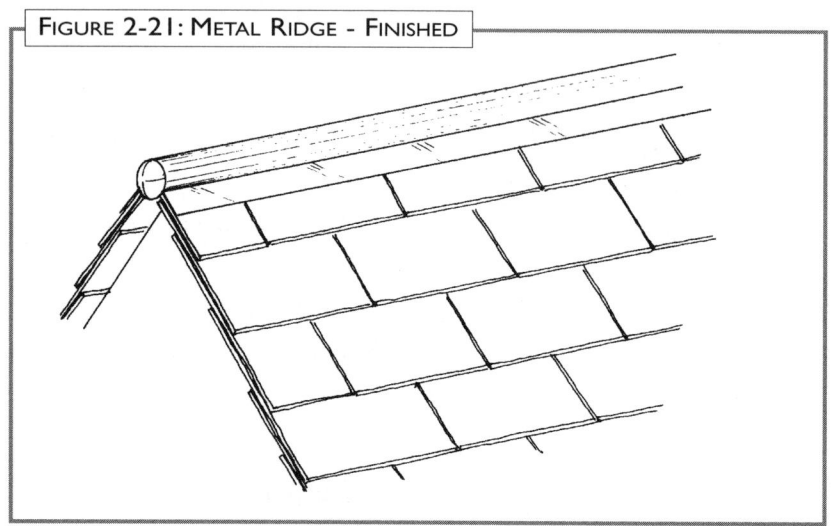

FIGURE 2-21: METAL RIDGE - FINISHED

Metal Ridge (Figure 2-21)

A ridge in which a metal cap is used to cover the ridge area. This metal cap will vary from a plain flat design to an ornamental cresting detail. The metal usually matches that used on the rest of the project. The more ornamental the cap, the more expensive it is.

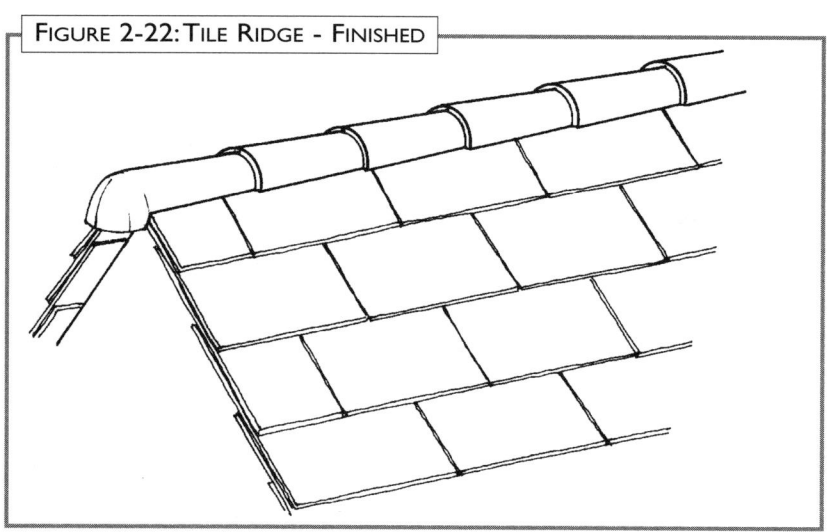

FIGURE 2-22: TILE RIDGE - FINISHED

Tile Ridge (Figure 2-22)

A ridge in which ceramic tiles are used to cap the ridge. This detail is more common in Europe than the United States. Styles will vary from round, to 90 degrees, to ornamental. Cost will vary depending on the material and design of the tile pieces.

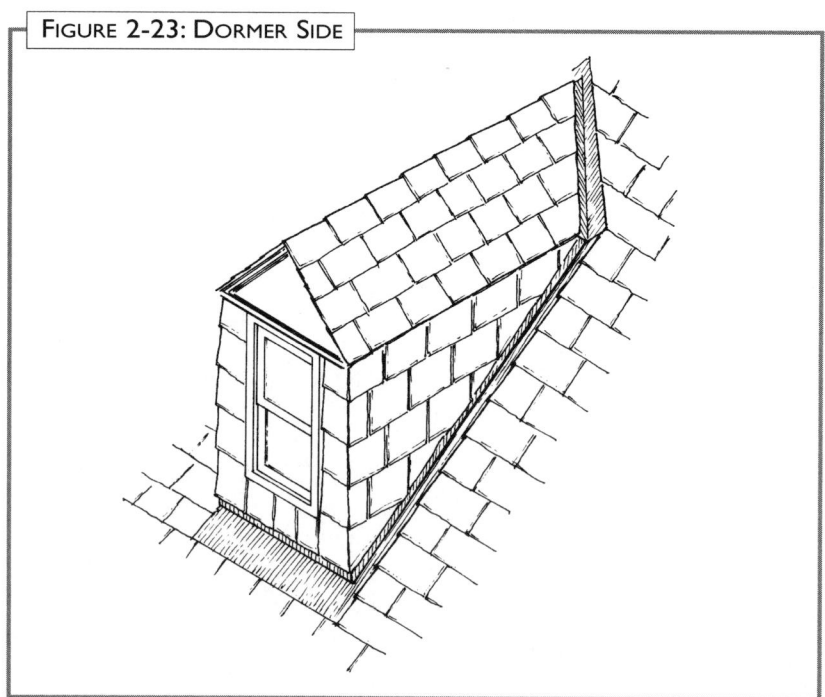

FIGURE 2-23: DORMER SIDE

Non-roof Surface (Figure 2-23)

In addition to choosing and specifying roof details, some designers may also specify the use of slate on non-roof areas, such as dormer sides, vertical walls, cupola sides, and chimney caps. Although this is not commonly done with other roofing materials, it is specified fairly regularly with slate.

Design Considerations

Flashings and Accessories

All roofs require the use of metal flashings at intersections where the roofing material does not create a watertight overlap. Because of the longevity of slate as a roofing material, selecting a flashing material that is equally as durable and long-lived becomes an issue. Copper is the most popular choice, but there are a variety of other metals used, such as lead-coated copper, zinc, lead, and stainless steel. Cost and compatibility with other roofing materials are important considerations when choosing flashings. Depending on the design, flashings may be visible in some areas, such as the valleys, hips, ridges, drip edge, chimneys, skylights, pluming vents, aprons, finials, gutters, dormer sides, and vents.

In addition to roof details, the slate roof design may incorporate any number of different accessories, as described below.

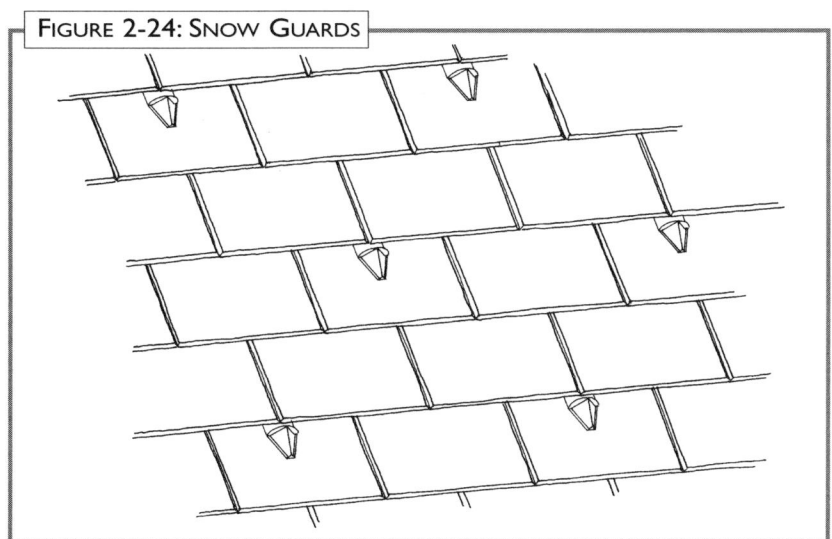

FIGURE 2-24: SNOW GUARDS

Snow Guards (Figure 2-24)

These are devices used to prevent snow and ice from avalanching off the roof. Like all smooth textured roof materials, slate will shed snow and ice from roof surfaces all at once, which may cause damage below. Snow guards are available in a variety of configurations and materials.

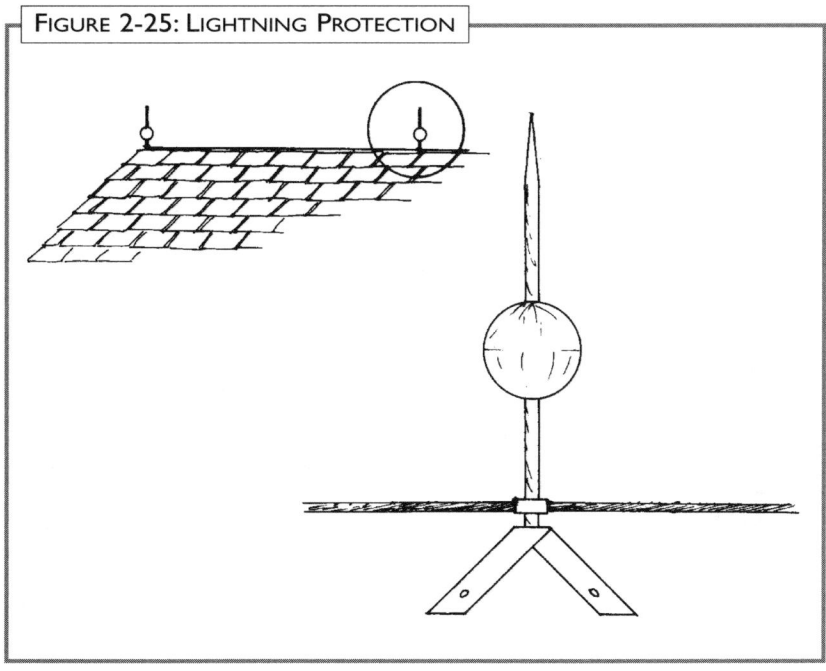

FIGURE 2-25: LIGHTNING PROTECTION

Lightning Protection (Figure 2-25)

Lightning rods are devices used to arrest lightning strikes. These devices are available in a variety of configurations and are typically placed on the highest points of the roof. They are typically copper or brass rods, and they are connected to a ground.

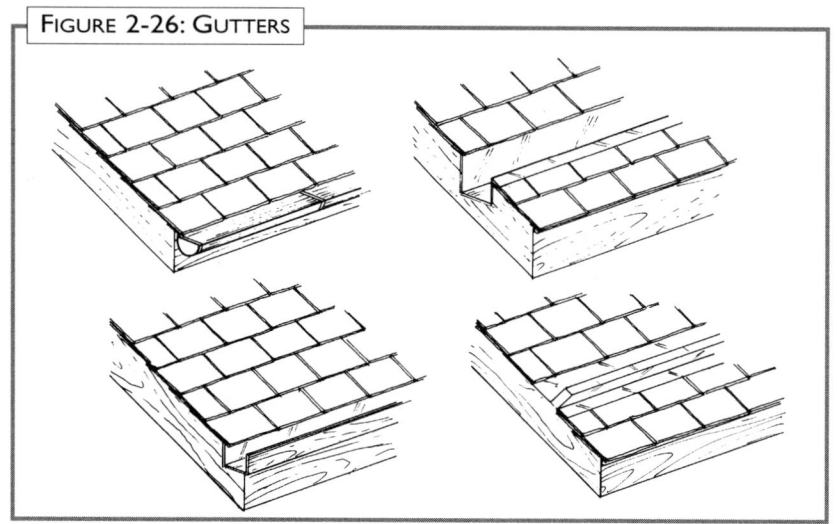

FIGURE 2-26: GUTTERS

Gutters (Figure 2-26)

Gutters collect water running off the roof and channel it into drain pipes. Gutters are available in a variety of materials and configurations, and they can be installed in a variety of ways in relation to a slate roof.

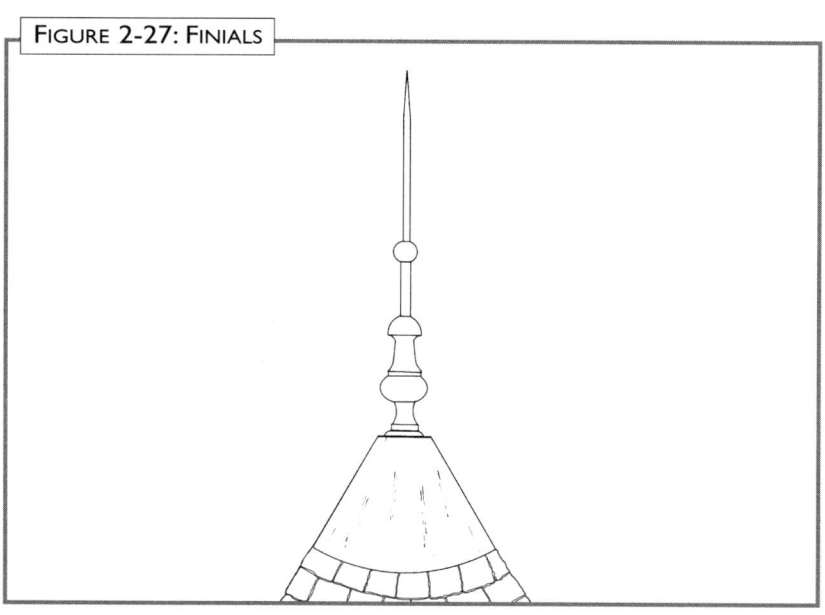

FIGURE 2-27: FINIALS

Finials (Figure 2-27)

Finials are fabricated decorative caps, often used at the peaks of hip roofs, cones, and towers. They are available in a variety of materials, including metal, clay, and wood, and may range from very simple to very ornate.

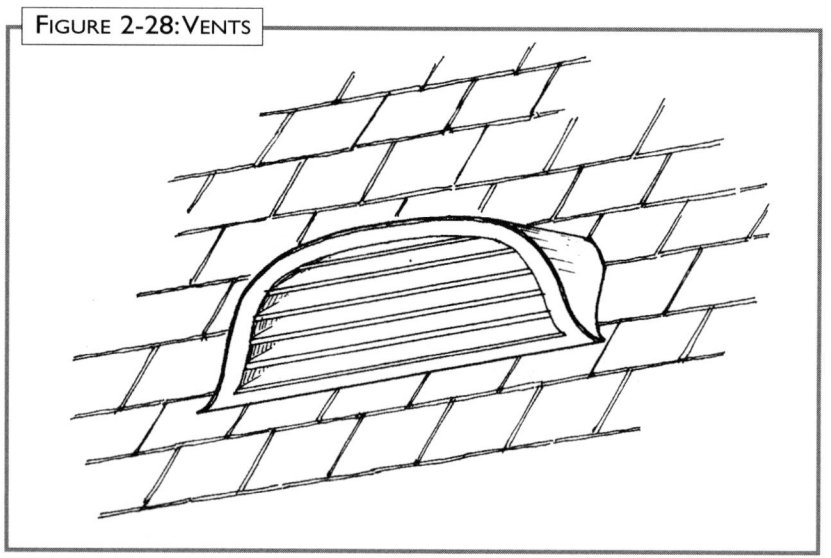

FIGURE 2-28: VENTS

Vents (Figure 2-28)

Vents are devices used to ventilate the cavity created by the roof line. Ventilation is required today by building code. These units come in a variety of materials and configurations including ridge venting.

Design Considerations

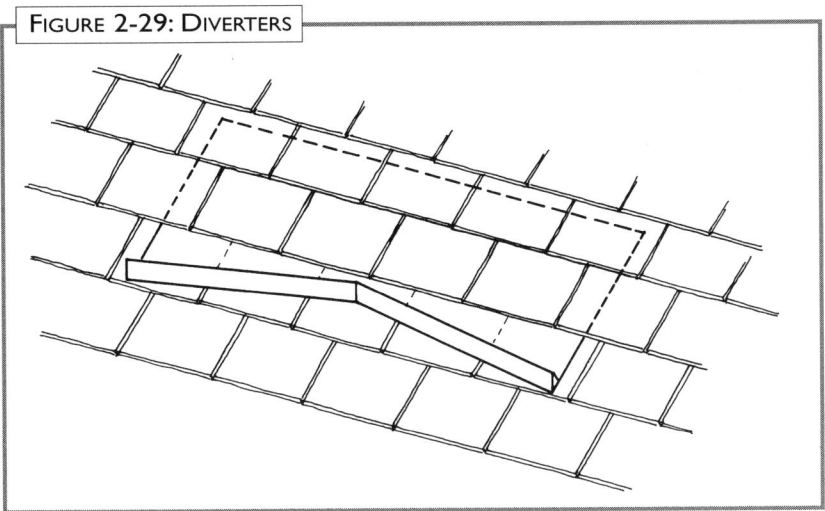

FIGURE 2-29: DIVERTERS

Diverters (Figure 2-29)

In some cases, owners or planners will request a rain diverter. The rain diverter is sometimes used above a doorway, for example, in lieu of a gutter. In areas where snow and ice are not a consideration, this detail is relatively straightforward. However, if snow and ice are considerations, this detail will have to be approached as though it were an above-the-eave gutter to avoid damage or failure.

The problem with a diverter is that it will put pressure on the slate below unless the original design provides for proper framing and flashing. In regions with heavy snow and ice, installing a diverter in a slate roof will probably cause unnecessary damage. If a rain diverter is necessary in a snow region, you should consider snow guards to protect the diverter.

FIGURE 2-30: SNOW SLIDES

Snow Slides (Figure 2-30)

Snow slides are used at the eaves of a roof to prevent snow and ice from accumulating and causing water to back up under the slate roof. A typical snow slide would be a standing-seam copper roof, three feet wide along all the eave areas. Although snow slides were commonly installed (before people understood of the need for venting), they are rarely used today. Modified bitumen installed under the slate at the eaves has eliminated the need for snow slides on many projects. (See Modified Bitumen in Chapter 8.)

Chapter 2

Establishing a Budget

In the process of designing a slate roof, naturally the owner/architect needs to be able to establish a realistic budget. The budget may be established before the design phase begins, it may be established step-by-step as the design phase proceeds, or the total cost may be calculated at the end of the design phase. In all likelihood, the budgeting process will include all of the above.

The object is to budget for the best possible installation, both in materials and labor, that will meet the design criteria. Decisions made throughout the design phase will certainly affect the cost of the project, and the owner/architect should be aware of the costs of alternatives throughout the project. There are alternatives that may sacrifice some design criteria without sacrificing integrity.

In general, establishing a budget should include the following steps:

1. Evaluating architectural style to see if slate fits the structure.

2. Determining whether or not slate meets all of the criteria for the project (such as fire resistance, quality, engineering).

3. Considering the structural design. Depending on the project, you may need to budget for the structural requirements of a slate roof (such as added structural members to support the weight of the slate).

4. Choosing the color or colors of slate to be used.

5. Choosing the size, texture, and thickness of slate.

6. Calculating the cost of materials with the help of the chosen supplier.

7. Estimating labor costs, scaffolding, consultant fees (if applicable), permits, architectural fees, and the cost of debris removal.

8. On the average a slate roofing mechanic can install one square of slate per day.*

9. Anticipate that a budget price for an installed slate roof of standard thickness and simple architectural design will be approximately $800–$1200 per square (labor and materials installed).

*This figure will vary. It is strictly a basic budget figure.

CHAPTER 3

Project Planning and Estimating

Having made some preliminary design decisions, you are now ready to proceed with project planning and estimating. The roof design and the budget both are still considered "preliminary" at this point since any decisions made during the planning phase could potentially affect the design and its costs.

This chapter provides an overview of the steps involved in planning a slate roof project and estimating costs. These steps, presented more or less in the order they will occur, include:

- Hiring Professionals
- Evaluating Structural Design
- Estimating Budget
- Estimating Labor
- Estimating materials
- Procuring Equipment
- Scheduling

Hiring Professionals

A slate roof project may involve an architect, a designer, an engineer, or other consultants. Any one of these professionals may be responsible for project planning. Although not every slate roof project will have a professional planner involved, many projects may require the services of one or more of these professionals. Some of these services are provided on an hourly basis, while others are based on a percentage of related cost, and in some situations two or more professionals will work together on a contract basis. Be sure to identify how this cost applies to your project.

Evaluating Structural Design

Whenever slate is to be installed on a roof, the roof's structure needs to be engineered by an architect or a structural engineer, particularly in relation to the structural members and the sheathing used to support the weight of a slate roof. This is especially true when thick slate will be used.

In the early days, slate was almost always laid over 1-inch thick, rough-cut boards with rafters of proper dimensions placed approximately 16 inches on center. Late in the 1800's, building codes in some of the metropolitan areas dictated the use of fire-retardant materials, such as lightweight concrete, for roof structures. In some situations, tongue and groove sheathing of varying thicknesses was used. The thickness of the sheathing would sometimes allow for varying the spacing of the rafters below. Throughout Europe slate was, and sometimes

35

Chapter 3

still is, installed over open strapping or battens.

When the revival of slate roofing began in the mid-1970's, a variety of new sheathing materials were available, including plywood, nail board with attached insulation, nailable concrete products for fire safety situations, as well as pressure-treated plywood and framing materials.

Taking into consideration all of the options now available, the structural engineer or the designer should specify structural members and sheathing that best meet the design criteria. The weight of the slate, proper fasteners, and the life cycle of the sheathing should all be considered, while keeping in mind what works in the southern United States may not necessarily work in the Rocky Mountain regions. Fortunately, there are building codes that dictate building design loads for any given geographical location. Given the testing information for various sheathing materials, national and regional building code requirements for a given geographical area, and the weight per square of the roofing material, an engineer can design the roof structure and sheathing to support the slate properly.

Table 3-1 shows the average weight per square for different thicknesses of roofing slate. This average weight includes the slate, 30# felt underlayment, and fasteners.

Estimating Budget

Budget figures for an average slate roof will vary considerably. Listed below are a number of contributing factors:

- Labor rates differ from one geographic region to the next.
- Insurance rates vary from 20% to 50% of payroll depending on location.
- Material costs vary.
- Unusual site conditions can contribute to material handling and scaffolding costs.
- Building codes may require more than minimum underlayment, fasteners, or other materials.

As a rule, the average cost per square of roof surface for a slate roof of standard thickness on a typical residential building, in today's dollars, is $800.00 – $1,200.00 per square, which includes labor and materials installed. However this figure can and does vary.

Estimating Labor

On average a slate roofing mechanic should be able to install one square of slate per day, but each job will vary. A large gable roof with no penetrations can certainly be slated faster than a hip roof with numerous valleys and dormers. For budget purposes, assume one square per installer, per day, on a project of average complexity. Adjustments should be made accordingly.

Table 3-1: Weight per Square

Slate Thickness in Inches	Average Weight of Slate in Pounds per Square (100 sq. ft) Sloping Roof (3-in. Headlap)
Standard 1/4"	850
3/8"	1,200
1/2"	1,800
3/4"	2,700
1"	3,600
1 1/4"	5,000
1 1/2"	6,000
1 3/4"	7,000
2"	8,000

Estimating Materials

Estimating the cost of all the materials required for a slate roof can be an involved process, depending on the complexity and size of the project. The most important materials to include in this estimate are the slate itself, flashings, materials for roof preparation, fasteners, and cant strips. The costs of accessories and other features should be added to these basic costs.

Estimating Field Slate

Slate is sold by the square, which is defined as the amount of roofing material required to cover a 100-square foot area (10 ft × 10 ft, for example) when properly installed. To begin estimating the amount of slate needed for the roof, determine the number of square feet of roof area using standard estimating procedures (multiplying the width by the length of the roof deck). Use standard geometric principles to determine actual area, and then divide the total number of square feet by 100 to determine the number of squares. Add 7-10% to the field area calculations (depending on the complexity) to cover repair, cutting loss in field areas, damage from handling, and future repair inventory.

The number of squares calculated in this manner provides a starting point for estimating materials and costs, but a number of other factors must also be considered. For example, unlike most other roofing materials, slate is available in a variety of lengths and widths, so the number of pieces of slate per square will vary depending on the size of the shingle. This will affect the installation time. Table 3-2 shows the most common slate sizes.

The cost of slate does vary with color. Weathering or semi-weathering colors are usually less expensive than the non-weathering colors. An order that calls for multi-color material will vary in cost depending upon the percentages of the colors chosen. A prospective buyer should be aware that he may be able to get the color he wants within budget if he has some flexibility about the percentages of colors chosen.

Choosing the proper size slate to fit the roof is critical in relation to aesthetic considerations, and size is also very important to overall job cost. Random-width slate is almost always less expensive than slate of one width. If one slate width is desired, it may be wise to ask for alternative pricing for random widths. Slate that is 14 inches or 12 inches long is sometimes less expensive because of its abundance at the quarry, but these lengths can cost more to install. Slate this small can also make the roof look busier and more textured, depending on the scale of the building.

As a rule, suppliers quote slate prices based on a standard 1/4-inch thickness. If a thicker slate is desired, it must be specified. Material thicker than the standard will cost more per square, cost more to ship, could add cost to the structural frame, and may increase labor costs.

Table 3-2: Standard Slate Sizes

Face Dimensions of Slate (in.)	Face Dimensions of Slate (in.)	Face Dimensions of Slate (in.)
10" × 6"	14" × 9"	18" × 12"
10" × 7"	14" × 10"	20" × 10"
10" × 8"	14" × 12"	20" × 11"
12" × 6"	16" × 8"	20" × 12"
12" × 7"	16" × 9"	20" × 14"
12" × 8"	16" × 10"	22" × 11"
12" × 9"	16" × 12"	22" × 12"
12" × 10"	18" × 9"	22" × 14"
14" × 7"	18" × 10"	24" × 12"
14" × 8"	18" × 11"	24" × 14"

Chapter 3

Headlap is the amount of the slate shingle that is lapped by the second course of shingles above. This headlap is what makes a slate roof watertight. In a standard slate roof installation, slate is applied with a 3-inch headlap although it can be 2-inch for roofs with a greater than 20:12 pitch or 4-inch for roofs with less than a 6:12 pitch. However, when discussing the purchase or installation of roofing slate, it is assumed that the slate is being applied with a 3-inch standard headlap. Quarries sell slate based on the number of pieces required per square for installation of slate using this standard 3-inch healap.

Exposure is the amount of slate that can be seen once the courses above have been laid. To calculate exposure (E), take the length of the slate (L), subtract the headlap (H), and divide by 2. The equation for figuring the exposure of a piece of slate is: $E = \frac{(L) - (H)}{2}$

The exposures for different lengths of slate with 3-inch headlap are shown in Table 3-3.

The number of pieces of slate per square can be found in Table 3-4. The number of slate pieces per square can also be calculated by dividing the number of square inches in a square of slate (14,400 sq. in.) by the number of square inches in the area of exposure of the slate (E_a). The area of exposure of a piece of slate is calculat-

Table 3-3: Slate Exposure

Slate Length	Slate Exposure
10"	3.5"
12"	4.5"
14"	5.5"
16"	6.5"
18"	7.5"
20"	8.5"
22"	9.5"
24"	10.5"

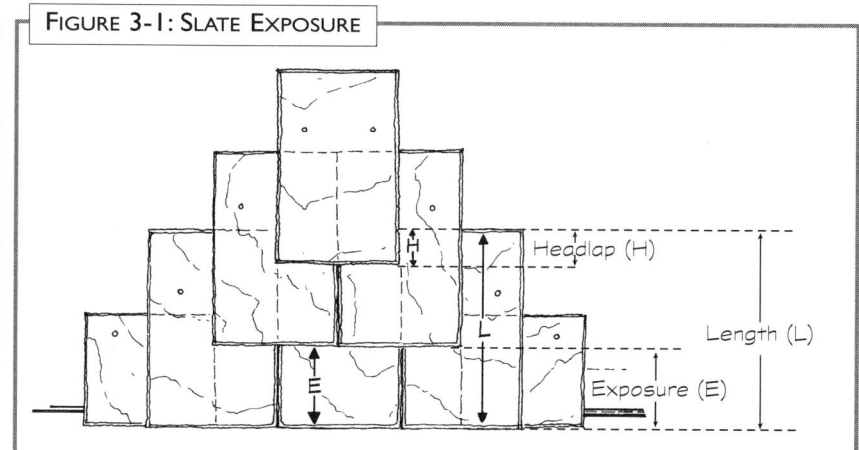

Figure 3-1: Slate Exposure

Table 3-4: Number of Slates per Square (3-in. Headlap)

Face Dimensions of Slate	Minimum Number per Square	Face Dimensions of Slate	Minimum Number per Square	Face Dimensions of Slate	Minimum Number per Square
10" × 6"	686	14" × 9"	290	18" × 12"	160
10" × 7"	588	14" × 10"	261	20" × 10"	169
10" × 8"	515	14" × 12"	218	20" × 11"	154
12" × 6"	533	16" × 8"	277	20" × 12"	141
12" × 7"	457	16" × 9"	246	20" × 14"	121
12" × 8"	400	16" × 10"	221	22" × 11"	138
12" × 9"	355	16" × 12"	185	22" × 12"	126
12" × 10"	320	18" × 9"	213	22" × 14"	109
14" × 7"	374	18" × 10"	192	24" × 12"	115
14" × 8"	327	18" × 11"	175	24" × 14"	98

ed by multiplying the exposure by the width of the piece of slate. Therefore, the equation for calculating the number of pieces of a given size of slate (N) in a square is:

$$N = \frac{14{,}400 \text{ in.}^2/\text{square}}{E_a \text{ in.}^2}$$

For example, for an 18-inch by 12-inch piece of slate, the area of exposure is (7 1/2 inches × 12 inches) = 90 square inches. Therefore, the number of slate pieces per square is: 14,400 ÷ 90 = 160 pieces of 18-inch by 12-inch slate per square.

Converting Quantity for Headlap

If the slate is laid with a 2-inch (20:12 pitch or greater) or 4-inch (6:12 pitch or less) headlap, then a square will cover more (2-inch) or less (4-inch) than the standard square with 3-inch headlap. If you are laying the slate with a 2-inch headlap, multiply the squares by the corresponding percentage in column 3 in Table 3-5. For example, if you are using 18-inch long slate, and the roof is 100 squares, multiply 100 by 94% = 94 squares. This means that you will only need to order 94 squares of 18-inch slate to cover 100 squares of roof area with a 2-inch headlap. If you are laying the slate with a 4-inch headlap, multiply the squares by the corresponding percentage in column 5. With 4-inch headlap, you will need more squares than the number of squares normally calculated for the roof. For example, using 18-inch long slate, multiply the number of squares on the roof by 107% to get the total number of squares needed.

Remember, slate should be ordered in terms of 3-inch headlap. Unless the roof slope is greater than 20:12 or less than 6:12, always use a 3-inch headlap. If you convert quantities for a different headlap, it is not necessary to order "100 squares of 18-inch slate laid with a 2-inch headlap."

All of the figures discussed thus far can be used in calculating the overall number of squares of slate needed for the project.

Estimating Slate for Valleys, Hips, and Ridges

Depending on the project, you may also need to calculate the slate needed for roof details. These quantities will need to be added to the field slate estimate.

To estimate slate for valleys, determine the lineal footage of the valley on the blueprints. Table 3-6 shows the actual length of valley or hip versus what can be measured off the blueprints. Add one square foot of material for cutting loss for each lineal foot of valley. Be sure to divide this quantity by 100 to convert it to squares and add this figure to your field slate estimate. This formula for calculating the cutting loss associated with valleys

TABLE 3-5: CONVERTING QUANTITY FOR HEADLAP

Slate Length (in.)	Slate Exposure 2-inch Headlap	% of 3-inch Headlap to cover at 2"	Slate Exposure 3-inch Headlap	% of 3-inch Headlap to cover at 4"	Slate Exposure 4-inch Headlap
10"	4"	88%	3.5"	117%	3"
12"	5"	90%	4.5"	113%	4"
14"	6"	92%	5.5"	110%	5"
16"	7"	93%	6.5"	108%	6"
18"	8"	94%	7.5"	107%	7"
20"	9"	94%	8.5"	106%	8"
22"	10"	95%	9.5"	106%	9"
24"	11"	95%	10.5"	105%	10"

is an estimate only. Actual quantities will depend upon roof pitch, slate exposure, and valley style.

For hip approach slate and mitered hips, determine the lineal footage of hip from the blueprints using Table 3-6. Add one square foot of material for each lineal foot of hip to account for cutting loss related to the approach slate. Convert this quantity to squares and add this quantity to your field slate estimate. In a mitered hip and its variations, the approach slate becomes the hip slate. Therefore no additional material will need to be calculated beyond the approach slate.

However, if you are using a saddle hip you will also need to calculate the material used for the hip caps. This is done by dividing the length of the hip (Lh) in inches by the exposure of the hip (Eh) in inches (the exposure is usually equal to the width of the hip cap) to find the number of pieces per side (Ph) and multiplying that by 2 to get the total number of hip pieces (Th). To calculate the number of squares of hip slate needed, divide the total number of hip pieces (Th) by the number of pieces of hip slate per square (Qs).

$Lh \div Eh = Ph$
$Ph \times 2 = Th$
$Th \div Qs =$ Number of Squares of Hip Slate

Add 10% to this number to account for loss. This material needs to be specifically ordered as hip slate so that it comes unpunched. Table 3-7 shows typical hip and ridge cap sizes, their exposure and the number of caps per lineal foot of hip or ridge.

For a strip or comb ridge, if finishing slate is being ordered from the quarry, (see finishing slate) add one square foot of material for each lineal foot of ridge to account for the capping slate. If the finishing slate is not being ordered, add 2 square feet of material for each lineal foot of ridge to account for the finishing course and the capping slate. Convert this quantity to squares and add it to your field slate estimate.

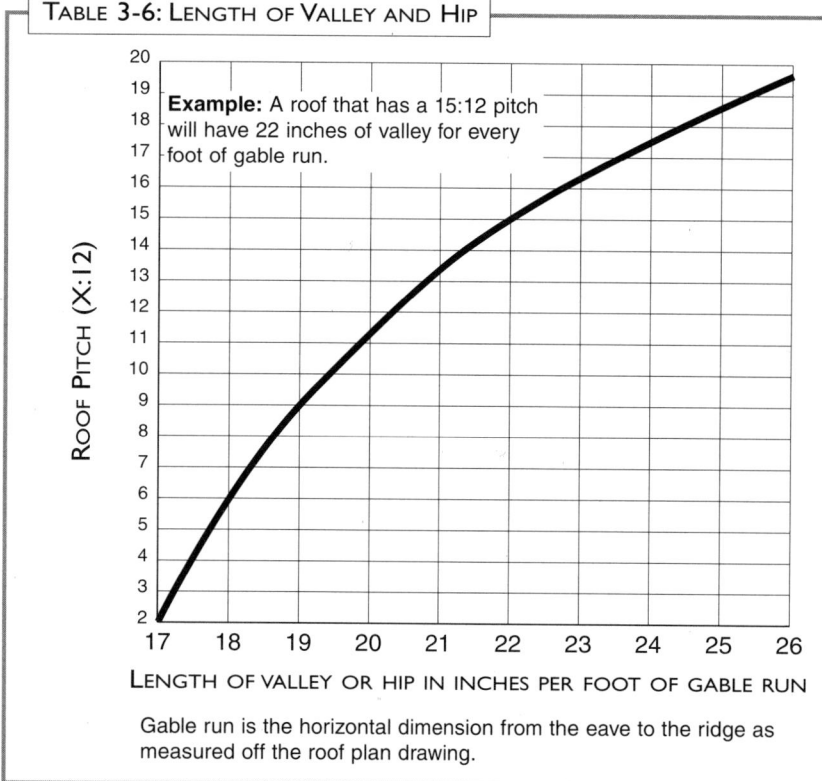

TABLE 3-6: LENGTH OF VALLEY AND HIP

Example: A roof that has a 15:12 pitch will have 22 inches of valley for every foot of gable run.

Gable run is the horizontal dimension from the eave to the ridge as measured off the roof plan drawing.

TABLE 3-7: SLATE SIZE FOR A SADDLE HIP OR RIDGE

Length of Slate	Hip and Ridge Cap Length × Width	Hip and Ridge Exposure	Number of Pieces of Cap per Lineal Foot of Hip or Ridge
12"	12" × 6"	6"	4
14"	12" × 6"	6"	4
16"	14" × 7"	7"	3.43
18"	16" × 8"	8"	3
20"	18" × 9"	9"	2.67
22"	20" × 10"	10"	2.4
24"	22" × 11"	11"	2.18

Project Planning and Estimating

TABLE 3-8: STARTER LENGTH

Slate Length (in.)	Starter Length (in./actual)	Starter Length (in./quarry run)
10"	6.5"	7"
12"	7.5"	8"
14"	8.5"	9"
16"	9.5"	10"
18"	10.5"	11"
20"	11.5"	12"
22"	12.5"	13"
24"	13.5"	14"

For a saddle ridge, determine the lineal footage of ridge in the same manner as you determined the lineal footage of saddle hip. Add 10% to this number to account for loss. This material, like saddle hips, needs to be specifically ordered as saddle ridge slate so that it comes unpunched.

Estimating Starter Slate and Finishing Course

To estimate the amount of starter slate you will need, calculate the lineal footage of the eaves. You will need to order this quantity of starter slate. Size will vary depending upon the length and exposure of the slate being used on the first course of the roof. The length of the starter slate is calculated by adding the exposure (E) and the headlap (H) of the first course of slate and rounding up to the next full inch:

$$(E) + (H) + 1/2 \text{ inch.}$$

For example, 18-inch long slate has an exposure of 7 1/2 inches: 7 1/2-inch Exposure (E) + 3-inch Headlap (H) + 1/2 inch = 11-inch long starter slate.

Table 3-8 above shows the starter slate length as it is calculated by the formula just described. If the quarry will be providing the starter slate, the starter length needs to be rounded up to the next full-inch dimension.

If you do not order starter slate cut to size by the quarry, each eave will have an extra material requirement of one square foot per lineal foot to account for starters.

Finishing slate can be ordered from the quarry if the layout procedure described later in this book is used. The finishing slate size found in Table 3-9 below is based on that procedure. If a vented ridge is used, this finishing slate size does not work (refer to the section on ridge vents in Chapter 8 for more information).

If you do not order finishing slate cut to size by the quarry, each ridge will have a material cutting loss factor of one square foot per lineal foot.

Estimating for Chimneys, Skylights, and Dormers

When estimating the slate needed for the entire roof, do not deduct the roof area taken up by a chimney, skylight, or dormer unless the area is larger than 50 square feet or 1/2 square. If slate is required on dormer cheeks, calculate these areas in addition to the roof area and add 10% for cutting loss.

Estimating for Flashing

A variety of metal flashing materials are available for roofing purposes, including galvanized sheet metal, painted steel, aluminum, copper, lead-coated copper, zinc, and lead. (Table 3-11 on the following page shows the weight and thick-

TABLE 3-9: FINISHING SLATE LENGTH

Length of Last Full Course of Slate at Ridge	Finishing Slate Length	Length of Last Full Course of Slate at Ridge	Finishing Slate Length
10"	none *	18"	12"
12"	8"	20"	14"
14"	8"	22"	16"
16"	10"	24"	18"

*See Chapter 5, Roof Layout, for explanation.

Chapter 3

ness of different metals used for flashing.) The flashing material chosen should be able to last as long as the slate, but it is safe to say that the flashings will likely wear out before the slate does.

Probably the two most common flashing materials used for slate roofing in this country are copper and lead-coated copper; zinc is more widely used in Europe. In residential construction, 16-oz. copper is probably the most common flashing. However, it may be wise to compare the cost of 16-oz. copper with that of 20-oz. copper. The slight difference in added expense is well worth the investment for the added service life.

One important thing to keep in mind is that copper will oxidize and form a patina. This oxidation can stain lower walls or structural details; however, staining is less likely if a good gutter system is used. Lead-coated copper will sometimes leave a gray milky residue as it oxidizes. Oxidation will continue until a protective coating (the patina) has developed, at which point it either stops or slows considerably.

Table 3-10 at right lists metals in order of their nobility (the quality of being chemically inactive, or inert). The higher the number, the more chemically stable the metal is; the lower the number, the less noble the metal. The greater the

TABLE 3-10: NOBILITY OF COMMON METALS
1 Zinc
2 Aluminum
3 Steel
4 Iron
5 Stainless Steel (Active)
6 Lead
7 Tin
8 Copper
9 Stainless Steel (Passive)

TABLE 3-11: Weight and Thickness of Flashing Materials

Thickness			Copper		Aluminum		Zinc		Stainless Steel		Galvanized Steel	
Gauge	Inches	Weight oz./ft	Thickness (in.)	Weight lb/sq ft	Thickness (in.)	Weight lb/sq ft	Thickness (in.)	Weight lb/sq ft	Thickness (in.)	Weight lb/sq ft	Thickness (in.)	Weight lb/sq ft
		32	.0431	2.00								
18			.0403	1.87	.0403	.563	.0400	1.50	.0500	2.10	.052	2.16
19			.0359	1.66	.0359	.507			.0438	1.84	.046	1.91
		24	.0323	1.50								
20			.0320	1.48	.0320	.451	.0310	1.20	.0375	1.58	.040	1.66
	1/32		.0313	1.45	.0313	.441						
21			.0285	1.32	.0285	.394			.0344	1.44	.037	1.53
		20	.0270	1.25			.0270	1.00				
22			.0253	1.17	.0253	.352			.0313	1.31	.034	1.41
23			.0226	1.05	.0226	.324			.0281	1.18	.031	1.28
		16	.0216	1.00			.0200	.750				
24			.0201	.932	.0201	.282			.0250	1.05	.028	1.16
25			.0179	.830	.0179	.253			.0219	.919	.025	1.03
		12	.0162	.750								
26			.0159	.737	.0159	.224			.0188	.788	.022	.906
	1/64		.0156	.725	.0156	.220						
27			.0142	.658	.0142	.200			.0172	.722	.020	.844
		10	.0135	.625								
28			.0126	.584	.0126	.177			.0156	.656	.019	.718

Project Planning and Estimating

distance between two metals on the list, the greater the electrolytic reaction when they come into contact. Therefore, if possible, do not mix metals. This electrolytic reaction will cause the more noble metals to break down the less noble ones. Eventually the reaction will eat a hole through the metal.

Drip Edge

When estimating flashing, begin with drip edge, which is installed around the perimeter of the building. Drip edge is not a required flashing, but it is recommended. The drip edge will help to prevent windblown snow and water from infiltrating the roof at the perimeter. This flashing can be manufactured in several different configurations. The detail you choose will depend on the shape of the perimeter material you are covering, the distance down from the roof line you wish to cover, and the appearance desired.

The material chosen for the drip edge should match the flashing on the rest of the building. However, some installers will use commercially available drip edge from a local building supplier. This will work fine as long as the roof flashing, if it is of a different metal, is isolated from the drip edge.

Drip edge is usually sold by the lineal foot. To estimate the quantity needed, calculate the entire footage of exterior roof perimeter. This will include the eaves and gables of the main structure, dormers, and any other penetrations designed to receive slate. If you are having this detail custom-fabricated, multiply the total lineal footage calculated by the width of the base flashing required prior to fabrication. Add fabrication cost to this material cost, and add installation cost to the labor bid.

Valley Flashing

To estimate valley flashing, determine the valley style needed for your project and apply the rules pertaining to the style.

For **open valley** flashing, the open valley should be exposed 2 inches wide on either side from the center of the valley at the top. The width of the open valley flashing should increase 1/2 inch per 8 feet on either side as you progress down the valley. Therefore, a valley that is 24 feet long would start with an exposure 2 inches wide at the top on either side and end with an exposure 3 1/2 inches wide at the bottom on either side.

> For a valley that is 24 feet long:
> 24 ft ÷ 8 ft = **3**
> (3 × 1/2") = 1 1/2"
> 2" + 1 1/2" = 3 1/2"

The valley slate needs to cover the valley flashing by not less than 4 inches, but you should design for a 5-inch minimum. Therefore, the valley flashing material will need to be a minimum of 16-inch wide stock prior to fabrication for the uppermost valley (2 inches open + 5 inches covered by slate + 1 inch hem = 8 inches × 2 for both sides of the valley = 16 inches). This same valley with a "W" configuration would need to be 18-inch wide stock in order to account for

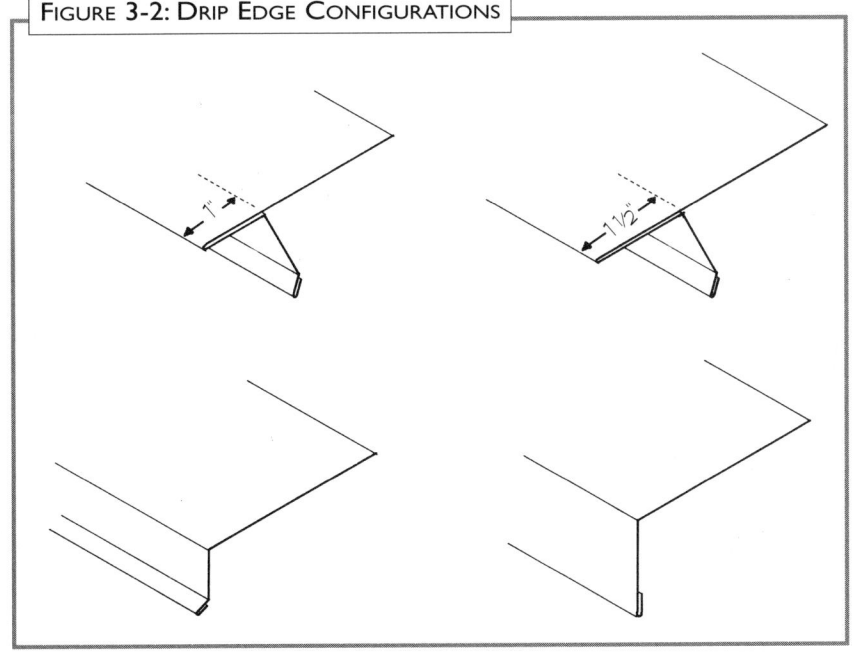

FIGURE 3-2: DRIP EDGE CONFIGURATIONS

Table 3-12: Open Valley Flashing Width

Length of Valley (lineal ft)	"V" Valley Style Stock Width	"W" Valley Style Stock Width
1 - 8 ft	16"	18"
9 - 24 ft	18"	20"
25 - 40 ft	20"	22"
41 - 56 ft	22"	24"
57 - 72 ft	24"	26"
73 - 88 ft	26"	28"
89 - 104 ft	28"	30"

the "W" bends. Table 3-12 shows some standard stock widths for open valley flashings.

The **closed valley** that uses a **continuous flashing** is the least desirable method for flashing a closed valley. It is very simple to calculate. Determine the valley length and multiply by 1 foot to determine the square footage of flashing needed.

Conventional stepped flashing for a closed valley is always 12 inches wide, bent at 6 inches to fit the angle of the valley. The length of this stepped flashing should match the length of the slate. This is the best method of flashing a closed valley where two different roof pitches intersect.

To determine the estimated square footage of materials needed to step flash a valley when the slate length is not yet determined, multiply the length of the valley (as determined from Table 3-6) by 1.75.

If the slate length is known, use Table 3-13. Whenever you have a valley formed by the intersection of two different roof pitches, use the shallower roof pitch to determine the valley length.

A closed valley using **stepped flashing bent point to point** will require the least amount of material. However, it can only be used if the intersecting roof surfaces are of the same pitch. This flashing is always cut square and bent from point to point. To determine the size of flashing required, use Table 3-14: Point-to-Point Flashing for a Closed Valley or Mitered Hip.

Hip flashings

To estimate hip flashings, determine the style of hip you are using and apply the rules for that style.

Mitered hips and **Boston hips** can be flashed with modified bitumen flashing, metal hip flashing, or both. However, these hips are very delicate, so avoid using excessive flashing. Modified bitumen flashing for mitered and Boston hips is calculated the same way as point-to-point flashing for a closed valley. Metal hip flashing for these hips are always cut square and bent from point to point. To determine the size of flashing required in either case, use Table 3-14: Point to Point Flashing for a Closed Valley or Mitered Hip.

With **saddle hip** flashing, the slate and wood nailer under a saddle hip will first be covered with modified bitumen as a base flashing. The quantity of modified bitumen needed depends on the size of the hip slate. Multiply the width of the hip cap slate (Wh) in inches by 2 to account for both sides of the hip, and then subtract 2 inches so that the modified bitumen is

Figure 3-13: Conventional Step Flashing for a Closed Valley

Length of Slate	Flashing Length and Width	Square Feet of Flashing per Piece of Flashing	Square Feet of Flashing per Lineal Foot of Rafter *
12"	12" × 12"	1 sq ft	2.66 sq ft
14"	14" × 12"	1.16 sq ft	2.53 sq ft
16"	16" × 12"	1.33 sq ft	2.45 sq ft
18"	18" × 12"	1.5 sq ft	2.40 sq ft
20"	20" × 12"	1.66 sq ft	2.34 sq ft
22"	22" × 12"	1.83 sq ft	2.31 sq ft
24"	24" × 12"	2 sq ft	2.28 sq ft

* Note that the factors provided are based on rafter length not valley length. On blueprints the rafter length can be directly measured, but the valley must be calculated using geometry.

Project Planning and Estimating

concealed by the slate cap. Multiply this by the length of the hip (Lh) in inches to get the total number of square inches of base flashing needed for the hip. Divide by 144 square inches to get the number of square feet of material needed for the base flashing.

Some saddle hips will require modified bitumen as a step flashing in addition to the required base flashing described above. To determine the square footage of hip flashing per lineal foot of saddle hip, refer to Table 3-15. Multiply the lineal footage of hip by the square footage of hip flashing required per lineal foot.

If the hip requires metal step flashing, the square footage is determined the same way you determine the square footage of modified bitumen hip flashing (use Table 3-15). Some projects will specify an exposed **metal hip** flashing cap. In these cases, size and style will vary, so consult the specifications and details to clarify quantities.

Ridge Flashings

To estimate ridge flashings, determine which style of ridge you will be using and apply the rules for that style.

The amount of flashing needed for a **saddle ridge, strip ridge,** or **comb ridge** is determined the same way as for a saddle hip both for the base and stepped flashing. Just as with a saddle hip, metal, modified bitumen, or both kinds of flashings may be required. Some projects will specify an exposed metal ridge cap. If this is the case, size and style will vary, so consult the specifications and details to clarify quantities.

TABLE 3-14: POINT-TO-POINT FLASHING FOR A CLOSED VALLEY OR MITERED HIP

Length of Slate	Size of Step Flashing	Square Feet of Flashing per Piece of Flashing	Square Feet of Flashing per Lineal Foot of Rafter *
12"	9" × 9"	0.56 sq ft	1.49 sq ft
14"	10" × 10"	0.69 sq ft	1.51 sq ft
16"	11" × 11"	0.84 sq ft	1.55 sq ft
18"	12" × 12"	1 sq ft	1.6 sq ft
20"	13" × 13"	1.17 sq ft	1.65 sq ft
22"	14" × 14"	1.36 sq ft	1.72 sq ft
24"	15" × 15"	1.56 sq ft	1.78 sq ft

* Note that the factors provided are based on rafter length not valley length. On blueprints the rafter length can be directly measured, but the valley must be calculated using geometry.

TABLE 3-15: ESTIMATING HIP FLASHING

Length of Slate	Hip and Ridge Cap Slate Length × Width	Hip and Ridge Exposure (Eh)	Hip Step Flashing Length and Width (Lhs) × (Wh)	Number of Pieces of Hip Flashing per Lineal Foot of Hip or Ridge (Qhs)	Square Feet of Hip Flashing per Lineal Foot of Hip or Ridge
12"	12" × 6"	6"	9" × 10"	2	1.25 sq ft
14"	12" × 6"	6"	9" × 10"	2	1.25 sq ft
16"	14" × 7"	7"	10" × 12"	1.72	1.43 sq ft
18"	16" × 8"	8"	11" × 14"	1.5	1.6 sq ft
20"	18" × 9"	9"	12" × 16"	1.34	1.78 sq ft
22"	20" × 10"	10"	13" × 18"	1.2	1.95 sq ft
24"	22" × 11"	11"	14" × 20"	1.09	2.11 sq ft

Chapter 3

Table 3-16: Estimating Stepped Flashing

Length of Slate	Stepped Flashing Length × Width	Square Feet per Lineal Foot of Rafter
12"	8" × 8"	1.18 sq ft
14"	9" × 8"	1.09 sq ft
16"	10" × 8"	1.02 sq ft
18"	11" × 8"	.97 sq ft
20"	12" × 8"	.94 sq ft
22"	13" × 8"	.91 sq ft
24"	14" × 8"	.88 sq ft

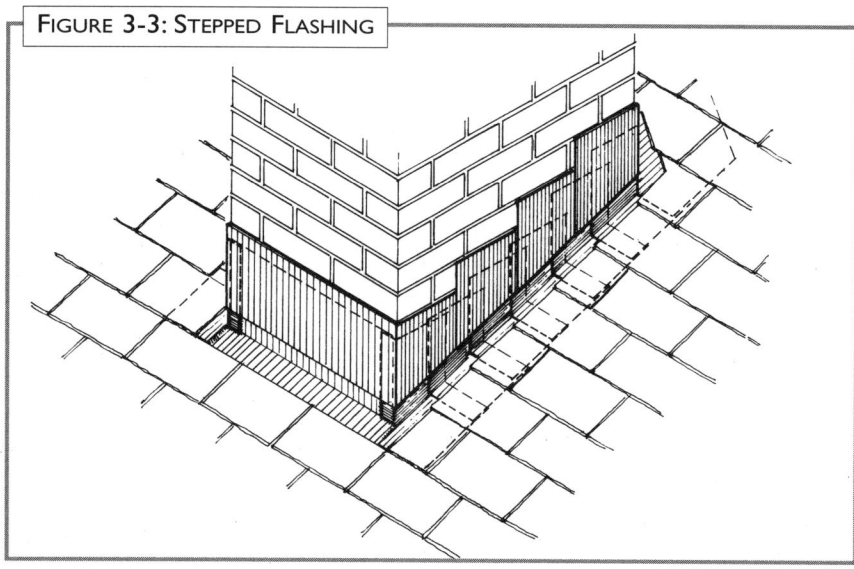

Figure 3-3: Stepped Flashing

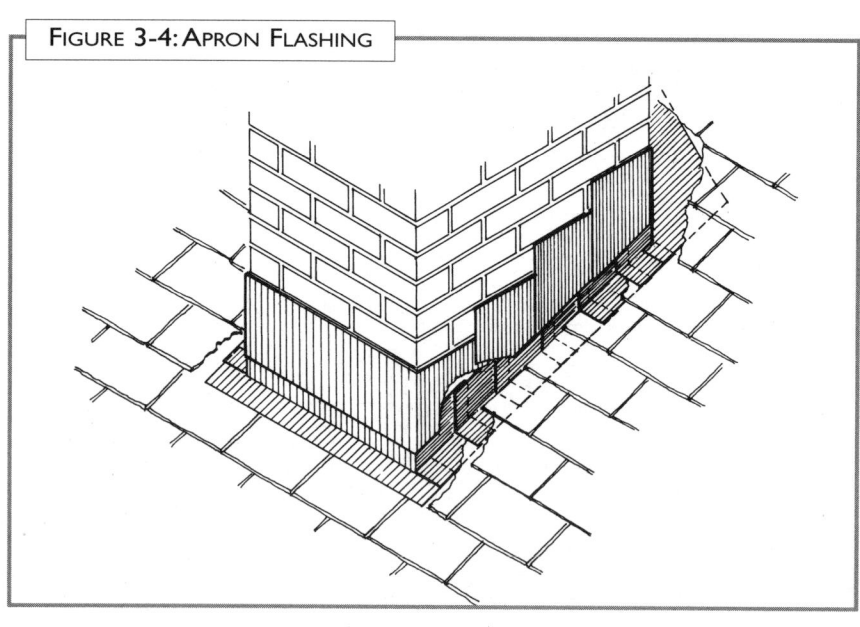

Figure 3-4: Apron Flashing

Stepped Flashing or Side Wall Flashing

Stepped flashing or side wall flashing is laced into each course of slate at a wall intersection (see Figure 3-3). To determine the length of the stepped flashing needed, add 1/2 inch to the length of the concealed portion of the slate. The width of the stepped flashing will be 8 inches for slate less than 1 inch thick and 12 inches for slate thicker than 1 inch. The flashing is then bent at a 90-degree angle so that 4 inches (6 inches) are on the roof surface and 4 inches (6 inches) are against the wall surface. To determine the quantity needed, use Table 3-16: Estimating Stepped Flashing.

Apron or Skirt Flashings

Apron or skirt flashings are installed at the face of all obstructions, such as walls, chimneys, skylights, and hatches. As a rule, these flashings will always extend up the wall face a minimum of 4 inches. They will extend out over the slate below, such that the bottom of the apron or skirt lines up with the butt of the course that covers the same slate course as the apron itself. To estimate the square footage of apron flashing, determine the length of area to be covered (Al) and multiply by 12 inches. In many cases, apron flashing will terminate within the roof surface. An example of this would be a dormer face within the roof. At these intersections, a corner will need to be soldered on.

Project Planning and Estimating

Counter Flashing

Stepped, cricket, valley, apron, and transition flashings are all examples of roof-related base flashings. However, when a masonry wall is encountered, the base flashings will need to be capped by "counter flashing" extending out of the masonry and over the base flashing. Masonry counter flashings should be installed by the mason, but this may not always be the case. Read the project specifications carefully to be sure who is responsible for this detail. (See Figure 3-5.)

On new projects, the counter flashings should be installed when the masonry is installed. This flashing then becomes through-wall flashing, and it is typically stepped into the courses of masonry. Size and quantity required will vary based upon the pitch of the roof. For estimating purposes, assume 2 square feet of flashing per lineal foot of counter flashing.

It is very common for lead sheet or soft copper to be specified for counter flashing instead of the more rigid roof base flashings. This allows the roofing installers to lift the counter flashing, place their roof base flashings behind it, and then lay the counter flashings back down. These softer metals are much easier to lay back down tight to the wall. If rigid counter flashing is used, it may be necessary to "tack solder" the pieces together

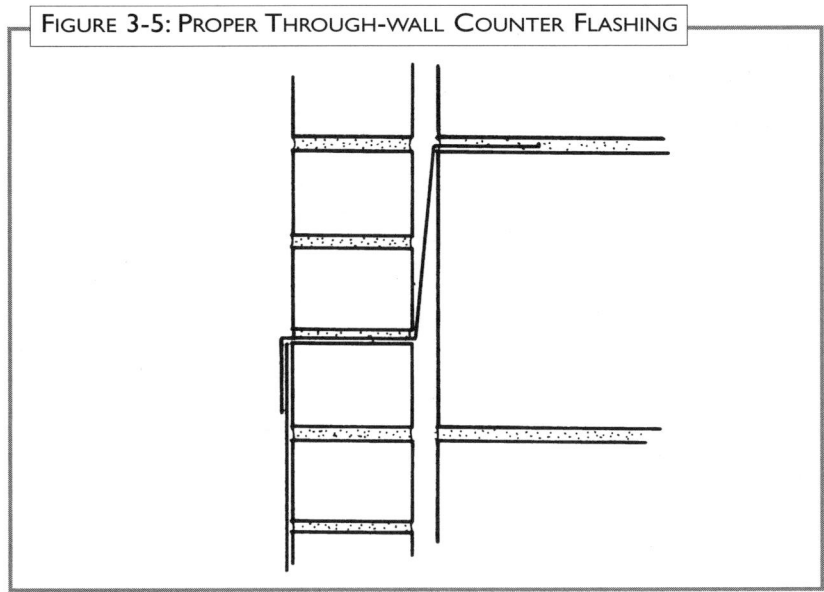

FIGURE 3-5: PROPER THROUGH-WALL COUNTER FLASHING

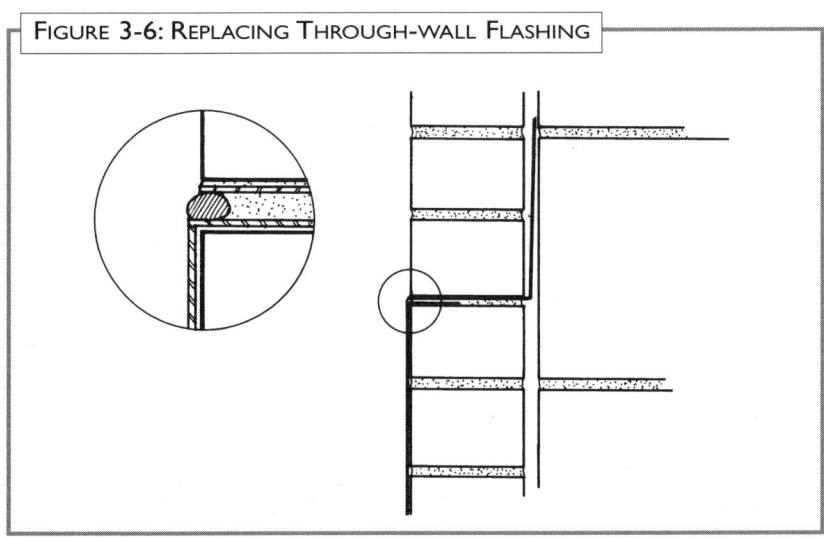

FIGURE 3-6: REPLACING THROUGH-WALL FLASHING

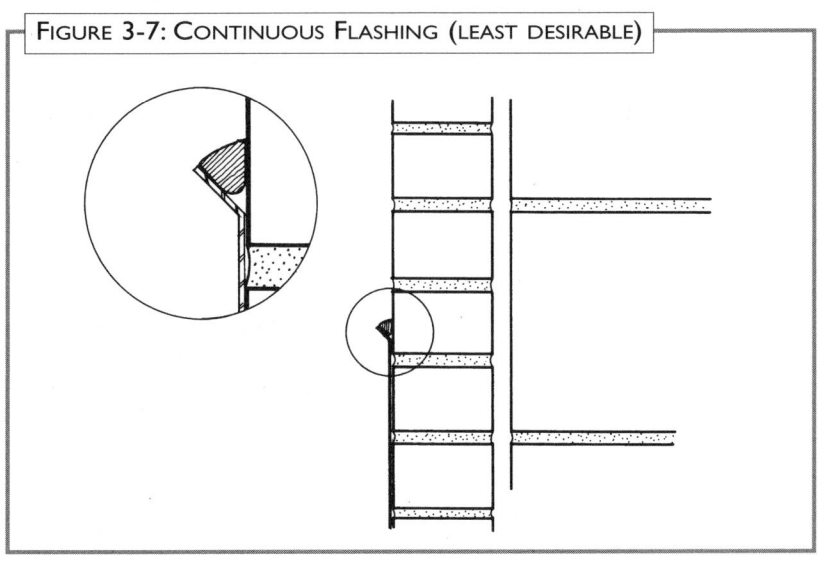

FIGURE 3-7: CONTINUOUS FLASHING (LEAST DESIRABLE)

Chapter 3

to hold them down after they have lifted.

On existing projects which require reroofing, the counter flashing will sometimes need to be replaced, (see Figure 3-6) and this is usually done by the roofing contractor. In most cases, the first step is to cut away the existing counter flashing. The installer then cuts a raggle into the masonry mortar joint with an abrasive saw. If possible, this raggle should be cut below the edge of the old counter flashing that is still left in the wall. The new flashing is then installed in the raggle just below the old through-wall flashing. This helps maintain the integrity of the old through-wall detail.

At one time it was fairly common for counter flashing details to be specified as one continuous piece (Figure 3-7), which was attached to the surface of the wall with masonry anchors and caulked to prevent leaks. This detail is probably fine on a composition shingle roof where the life cycle is 30 years or less. However, used on a slate roof, it will require maintenance long before the slate does.

Transition Flashing

A transition flashing may be required at any point where the roof changes pitch, such as where a shallow-pitch eave detail changes to a steeper pitch, or the opposite, such as on a gambrel roof.

This flashing is usually the same material being used as flashing on the rest of the roof. It is estimated in lineal feet. Cost will vary depending upon the specified installation technique, shape, and size. If there is no way of knowing how the detail will actually appear in the field, it is safe to assume that it will require approximately 1.5 square feet per lineal foot of transition flashing required.

Cap Flashing

On mansard roofs and slate roofs that intersect with a flat roof above, a cap flashing will be necessary. The size is determined by multiplying the length of the flashing needed by the width of the material needed. The width of the flashing will depend upon how simple or complex the shape.

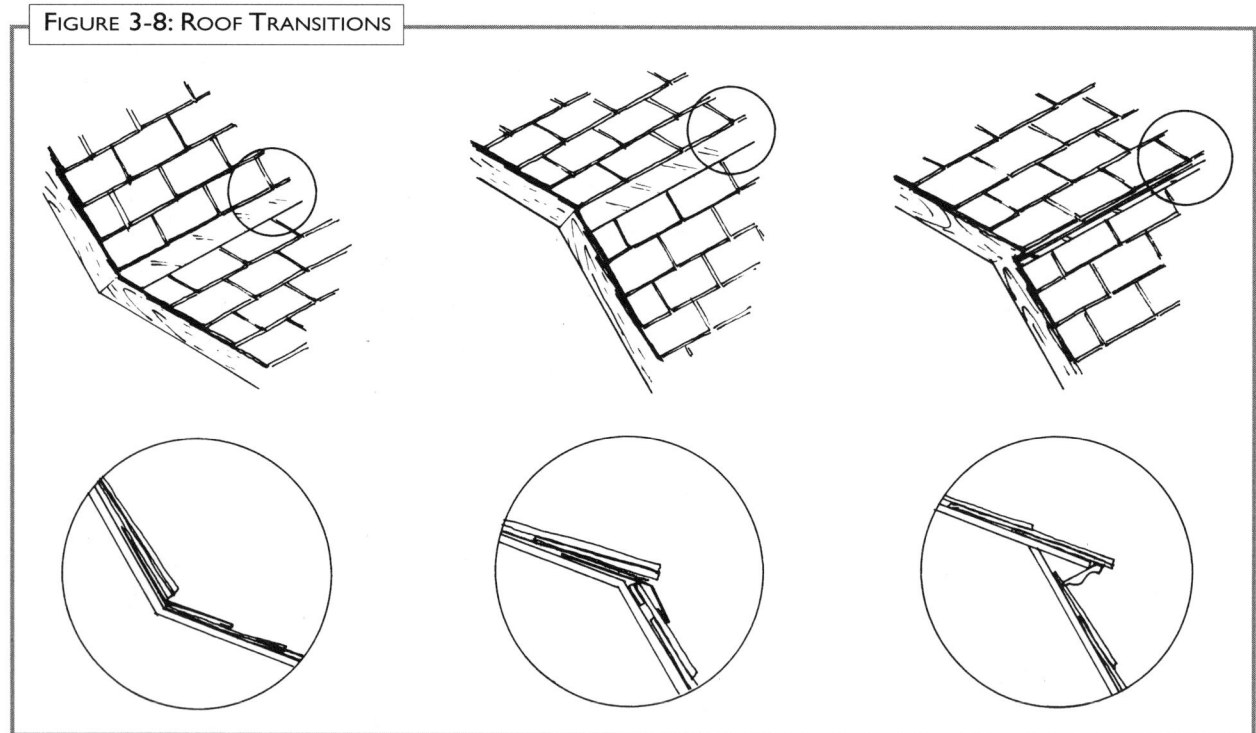

FIGURE 3-8: ROOF TRANSITIONS

Project Planning and Estimating

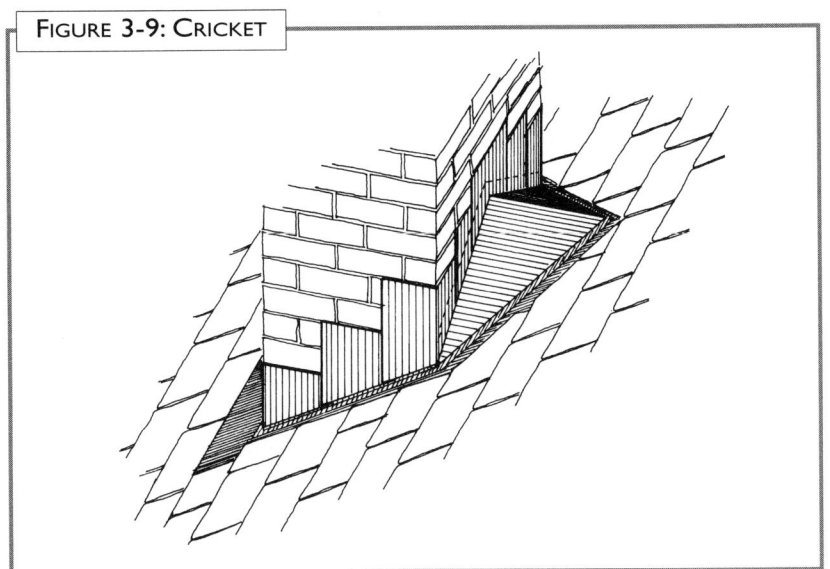

FIGURE 3-9: CRICKET

Cricket Flashing

Typically, a cricket is the roof flashing behind a chimney or other similar obstruction. Its purpose is to direct water, snow, and ice away from the back of the chimney or obstruction. This flashing detail will vary in cost based on the size and complexity of the cricket, and in most cases it will be installed by a mechanic skilled in working with sheet metal—either one of the slate mechanics or a separate subcontractor. Estimates will vary depending on the approach used.

Estimating for Gutters

Gutters will need to be estimated on a job-to-job basis. Standard square-foot estimating practices will apply. Keep in mind that built-in gutters can slow down the slate installation. The slate may need to be cut around the fastening brackets at the starter course level. The gutter installers will often need to share the scaffolding with the roofers, and the finished gutter will need to be protected and cleaned regularly. Gutters which are hung after the roof is completed will usually cost less.

Estimating for Roof Preparation

When estimating costs of materials for roof preparation, plan for a minimum 30# felt paper underlayment. This paper should be installed with galvanized roofing nails and roof buttons or discs to keep it from blowing off. Because a slate roof takes considerably more time to install than some other shingled roofs, this paper could potentially be exposed to the elements for a long period of time. Properly installed, the 30# felt paper will keep the structure watertight while the roof is being completed, allowing some limited work to proceed inside the structure.

When estimating for roof preparation, plan to cover the entire roof deck with 30# felt paper (unless local building codes specify otherwise). Install the felt in horizontal layers starting at the eave with a minimum horizontal overlap of 2 inches. Overlap all vertical joints by a minimum of 6 inches, and hips, valleys, and ridges by a minimum of 12 inches on each side. This is a minimum recommended underlayment. However, local building codes and job specifications vary and must be adhered to.

Like slate, felt paper is sold by the square. The 30# felt is usually sold in rolls, which are 3 feet wide. Be sure to confirm with the supplier the size (square footage) of the rolls being quoted. Add 10% to the estimated roof size to account for overlaps and waste.

Most new construction today incorporates the use of a modified bitumen around the perimeter of the roof, and in most cases up the entire length of each valley. This material should go on before the felt paper. Modified bitumen is commonly specified around the entire perimeter of the building from the eaves up to a point 24 inches within the exterior supporting wall. It is available from a variety of manufacturers. To determine quantities needed, check with the chosen manufacturer to find out the number of square feet per roll.

49

Chapter 3

Calculate the number of square feet needed based on the building specifications. Divide the total number of square feet required by the number of square feet per roll. If the project requires modified bitumen over the entire roof surface, follow the estimating procedure for the 30# underlayment, with the understanding that the material will come in various size rolls at varying cost. Keep in mind that installing this underlayment may also increase labor costs.

Estimating for Fasteners

Several different types of nails may be needed for a slate roof project. The underlayment is usually installed with galvanized roofing nails, although some areas require capping nails for this purpose (consult the building codes for your area). The slate will usually be installed with copper nails or stainless steel nails. Other kinds of nails, such as 16d framing nails, will be needed for some tasks, such as erecting roof scaffolding.

When choosing fasteners for the slate, keep in mind that the nail metal should be compatible with the metal being used for the flashing. The general rule is that dissimilar metals should not be used together because of the galvanic reaction or electrolysis that occurs between dissimilar metals. Copper nails should not be used with zinc flashing, for example, because the resulting electrolytic reaction will destroy the zinc.

Copper nails are the most commonly used nails and are compatible with copper, lead-coated copper, and lead. Stainless steel nails work with copper, lead-coated copper, zinc, and stainless steel flashings. A slate roof that is properly installed should outlast the flashings, so be sure to choose nail material that will last at least as long as the flashing.

Types of fasteners will also vary depending on the type of roof deck installed and the thickness of the slate. On many older buildings in metropolitan areas, cement-based planking was used as the roof decking. If you are reroofing over a dense surface such as a concrete product, copper nails will probably be too soft to penetrate the deck.

The length of the nails will be determined by the thickness of the slate. The equation used to figure nail length is:

> 2 × slate thickness (T) + 1 inch.

Table 3-17 gives some standard nail lengths for different slate thicknesses.

On a project with large, graduated slate, you may need nails 3 inches long or longer. These projects will also require 4 nails per piece of slate. Since slate is hung on the shaft of the nail, a stronger shaft will be needed for these projects. Long copper nails have a tendency to bend easily, so you may want to use 16d stainless steel nails instead.

In general, 10-gauge slater's nails are the most suitable for slate roofing. Because the slate is hung on the shaft of the nail, the thicker nail shafts are better, provided they aren't wider than the nail holes in the slate. Slater's nails are available in 12-gauge and are suitable for slate that is less than 1/4 inch thick. Table 3-18 lists the gauge, shank, and head diameter of standard copper and stainless steel nails.

To estimate how many pounds of nails you'll need for your field slate, follow these steps:

1) Determine the slate size you will be working with. If the slate is random width, estimate an average size. Then consult the Table 3-4 on page 38 to determine the number of pieces of slate per square of that size.

2) To determine the number of nails per square, multiply the number of pieces of slate per

TABLE 3-17: NAIL LENGTH FORMULA	
Slate Thickness	Nail Length
3/16"	1 3/8"
1/4"	1 1/2"
3/8"	1 3/4"
1/2"	2"
5/8"	2 1/4"
3/4"	2 1/2"
1"	3"
1 1/2"	4"

Project Planning and Estimating

square by the number of fastener holes per slate.

3) Next, determine the slate thickness, then consult Table 3-17 on page 50 to determine the proper length of nails required for that particular thickness. Use Table 3-18 to determine how many nails of that length and gauge there are per pound.

4) Divide the number of nails needed per square (calculated in step 2) by the number of nails per pound (calculated in step 3) to determine the number of pounds of nails needed per square.

5) Multiply the number of pounds of nails needed per square (calculated in step 4) by the estimated number of squares and add 15%.

The starter course, finish course, and valleys will all be fastened with the field slate nails. By adding 15% to the nail quantity for the field slate, you have already accounted for these nails.

Nails for hips and ridges need to be calculated separately. For mitered hips, add 10% to the quantity of field slate nails. If you are using a saddle hip, add 1/2 inch to the length of the nails required for the field slate. (Note: This is not necessary on a graduated slate roof. Unlike the field slate, the thickness of the hip slate is fairly uniform, usually between 1/4 inch and 3/8 inch thick, so the hip slate nails are usually 2 – 2 1/2 inches.)

TABLE 3-18: NAIL DIMENSIONS AND WEIGHT

			Copper		Stainless Steel	
Length	Gauge	Shank Diameter	Head Diameter	approx. # per lb.	Head Diameter	approx. # per lb.
1 1/2"	12	.109	21/64	212	5/16	218
1 3/4"	12	.109	21/64	183	5/16	189
2"	12	.109	21/64	161	5/16	170
3/4"	11	.120	3/8	295	11/32	323
1"	11	.120	3/8	234	11/32	244
1 1/4"	11	.120	3/8	203	11/32	199
1 1/2"	11	.120	3/8	169	11/32	170
1 3/4"	11	.120	3/8	148	11/32	149
2"	11	.120	3/8	131	11/32	136
1"	10	.134	7/16	204	3/8	212
1 1/4"	10	.134	7/16	190	3/8	166
1 1/2"	10	.134	7/16	135	3/8	139
1 3/4"	10	.134	7/16	124	3/8	126
2"	10	.134	7/16	102	3/8	105
2 1/2"	10	.134	7/16	85	3/8	91
3"	10	.134	7/16	68	3/8	78
3 1/2"	10	.134	7/16	59		
3"	9	.148	15/32	57	3/8	67
3 1/2"	9	.148	15/32	48		
4"	9	.148	15/32	43		
3"	8	.165			7/16	51
3 1/4"	8	.165			7/16	48
3 1/2"	8	.165			7/16	44
4"	6	.203			1/2	25

To determine the number of nails needed, multiply the length of the hip in feet by the number of hip slates per lineal foot (see Table 3-7 on page 40: Slate Size for a Saddle Hip or Ridge). This gives you the number of pieces of hip slate. Actual nail quantity is the total number of hip pieces times 2 (number of nails per piece) plus 10% for waste. To estimate nails for ridge slate, repeat the steps for determining nails for hip slate.

During the estimating process, the roofing contractor is sometimes surprised by how much money the fasteners cost. Typical galvanized roofing nails cost much less than copper or stainless steel nails. Some roofing contractors may suggest saving money by using non-ferrous fasteners with the flashings only, but this is not recommended. The fasteners actually represent a very small percentage of the cost of a slate roof. Based on an estimated installed slate roof cost of $800.00/square, and with copper nails costing approximately $4.00/pound, the cost of nails for a given project is only 1/2 of 1% of the total roof cost.

Estimating for Cant Strips

Another item to consider in estimating materials is the **cant strip,** which is the strip installed at the eaves to elevate the butt of the starter slate. (See Chapter 8: Installing Slate.) The most commonly used cant strips are made of wood, but they are sometimes specified as metal, fabricated to the required thickness. With rare exceptions, the cant strip thickness required is usually the thickness of the first course of slate.

Wood is the most commonly used cant strip. It can be used alone or in conjunction with metal edge details. The wood cant strip does not need to be more than 1 1/2 inches wide.

There can be a significant cost difference between a typical wood lath cant strip and a custom copper cant incorporated into the drip edge.

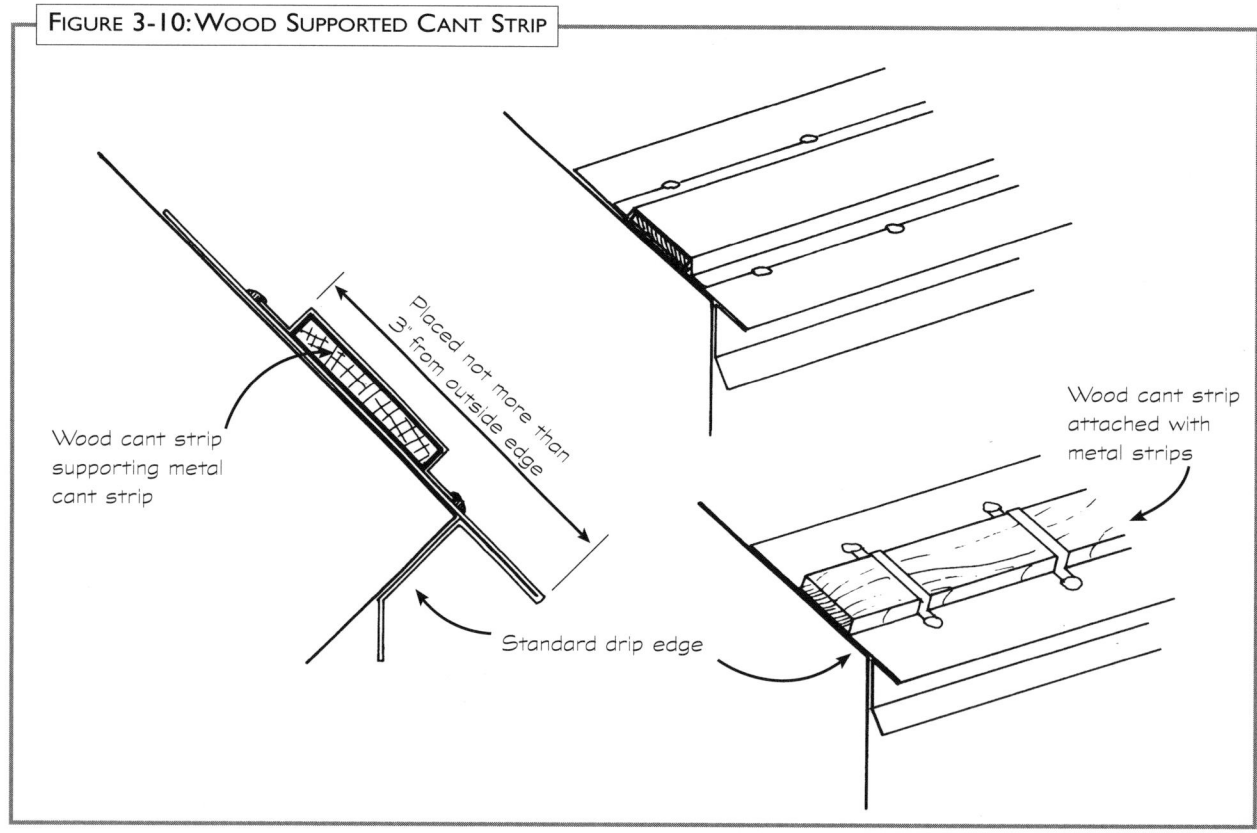

FIGURE 3-10: WOOD SUPPORTED CANT STRIP

Project Planning and Estimating

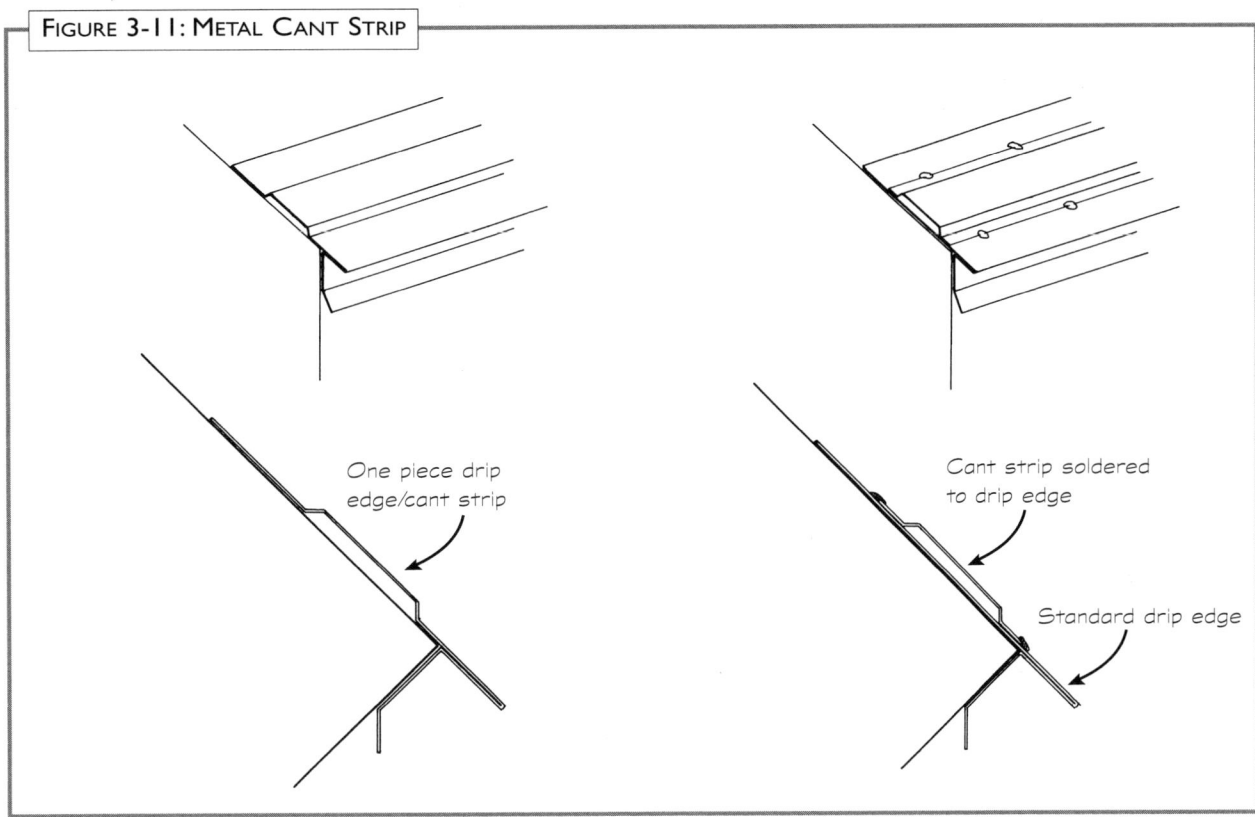

FIGURE 3-11: METAL CANT STRIP

If the cant strip is specified as fabricated metal, the estimator will need to multiply the length of the eaves by the width of cant strip material required. The elevated section of the metal cant strip does not need to be wider than 1 1/2 inches. This metal cant strip will sometimes be incorporated into the drip edge or gutter flange. It will sometimes be specified as a separate cant strip that is soldered to the drip edge or gutter flange.

When the slate exceeds 1/2-inch thickness, the metal cant strip will often need to be supported to prevent it from being crushed. This can be done by inserting wood under the metal cant strip, clipping wood to the metal base by soldering on clips, or increasing the gauge of the cant strip material so that it can support the thicker slate. Whenever possible attempt to fabricate the cant strip into the base metal. It will save a good deal of labor in the field.

Estimating for Accessories

Many roofs will have additional accessories, such as snow guards, finials, lightning rods, weathervanes, skylights, and access hatches. The estimator may need to account for these accessories, since they may require custom pieces. In some cases, these details cannot be bid on unless they are clearly specified with potential manufacturers' names and contact information.

The slate supplier may be able to provide more information and resources if needed.

Snow guards are a necessary accessory for slate roofs installed in snow and ice regions. To determine quantities, consult the specified manufacturer. Many snow guard manufacturers leave product placement to the discretion of the architect and the roofer. This approach has often caused a good deal of damage within the roof. Ask the manufacturer to provide test results for a given product with respect to load capability. Using the load capability for each snow guard and a safety margin, you can use the following formula and Table 3-19 to calculate the number of snow guards needed.

Chapter 3

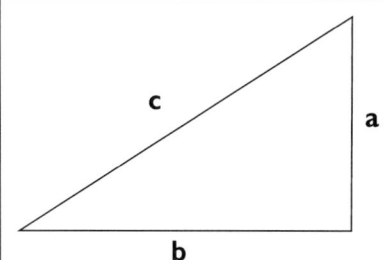

$y = [((\sin \angle a) \times \text{psf} \times 100 \text{ sq ft} \times b)) \div c] \div z$

- y = snow guards per square
- z = tested capability of the snow guard
- ∠ a = slope of the roof
- psf = building design load for the particular project
- b = horizontal distance from eave to ridge
- c = rafter length of eave to ridge
- 100 = number of square feet in a square

TABLE 3-19: ROOF PITCH VS. ANGLE OF INCLINE			
(Conversion of Slope to Degrees)			
Roof Pitch	Angle of Incline	Roof Pitch	Angle of Incline
1: 12	4.76°	13: 12	47.29°
2: 12	9.46°	14: 12	49.40°
3: 12	14.04°	15: 12	51.34°
4: 12	18.43°	16: 12	53.13°
5: 12	22.62°	17: 12	54.78°
6: 12	26.57°	18: 12	56.31°
7: 12	30.26°	19: 12	57.72°
8: 12	33.69°	20: 12	59.04°
9: 12	36.87°	21: 12	60.26°
10: 12	39.81°	22: 12	61.39°
11: 12	42.51°	23: 12	62.45°
12: 12	45°	24: 12	63.43°

For example, a building with a 6:12 slope, 77 psf design snow load, 62-foot horizontal distance, 64-foot rafter length, and a snow guard capability of 400 lbs/unit will require snow guards as follows:

6:12 slope = 26.57° and sin 26.57° = 0.44729; therefore, y = [(0.44729 × 77 × 100 × 62) ÷ 69.3] ÷ 400; y = 7.7 snow guards per square of roof area.

Placement of the snow guards will be determined after the proper quantity needed for your geographical area has been calculated.

This accessory, if not carefully estimated and executed, can become a major maintenance item, which is an especially important consideration when it comes to warranties. Ask the distributor about labor cost for installation of each unit. Most costs will be estimated relative to the cost of installing one piece of slate.

Roof Vents

Slate roofs can be vented with some of the commercially available roof vents. Like all other accessories, the vent used should have a service life that matches that of the rest of the roof. There are many roof vent styles available, some of them in copper. Quantities required will be dictated by building codes. Depending on the complexity of the vent, each vent can take from 1/2 hour to an hour to install and slate around.

Vented ridges can be installed on a slate roof and capped using slate or metal. Check the details to determine whether or not a ridge vent is specified. If it is, the layout of the roof, as well as the time to cap the ridge, can be affected.

Vented ridge caps have become somewhat of a necessity in modern construction. Unlike buildings constructed at the turn of the century that were poorly insulated, today's structures are virtually airtight. Without proper ventilation, interior moisture could cause the building structure to rot from the inside out.

It is possible to install dormer vents, traditional modular vents, and a variety of power vent fans. However, most owners and planners agree that they prefer to avoid installing vents that look non-traditional.

Project Planning and Estimating

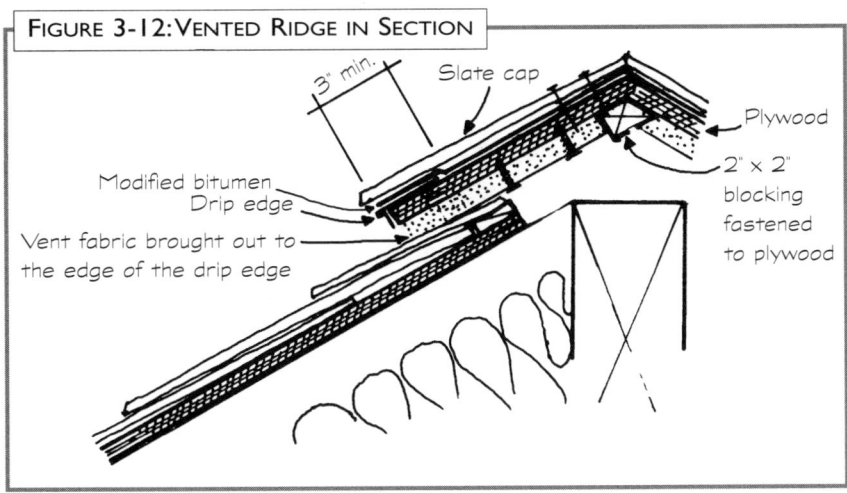

FIGURE 3-12: VENTED RIDGE IN SECTION

For composite shingles and wood shingles, there are a number of ridge vent systems available which allow the installer to fasten the ridge cap shingles through the vent material. The ridge cap shingles are then elevated above the field material by 1 inch to 1 1/2 inches (the thickness of the vent material). Although it is obvious that the ridge is elevated to allow for ventilation, it isn't obtrusive.

Many designers desire a similar appearance for a slate ridge cap. This can be done; however, special provisions will have to be made for it in layout, roof preparation, and installation. For more information, refer to discussions of these specific areas in Chapter 6: Layout and Chapter 8: Installation. Another thing to consider is that slate shingles are much heavier than composition shingles. The weight of the individual slate caps, when supported by nails long enough to penetrate slate, vent, and deck, will cause the cap slate to sag. The result is a vented ridge detail that is not likely to last very long and that may cause leaks.

For planning and estimating, determine the types of vents to be used, the costs of the materials, and the amount of time required both to install the vents and to slate around them.

Plumbing Vents

Plumbing vents should be made from the same flashing as the rest of the project. A slate roof should have the vent type shown in Figure 3-13A.

Some plumbing vent types rely on a neoprene or lead gasket to clasp the protruding plumbing vent tube. (See Figure 3-13B.) The life and quality of this type gasket will determine the service life of the plumbing vent. Because the slate roof will last as long as the flashings

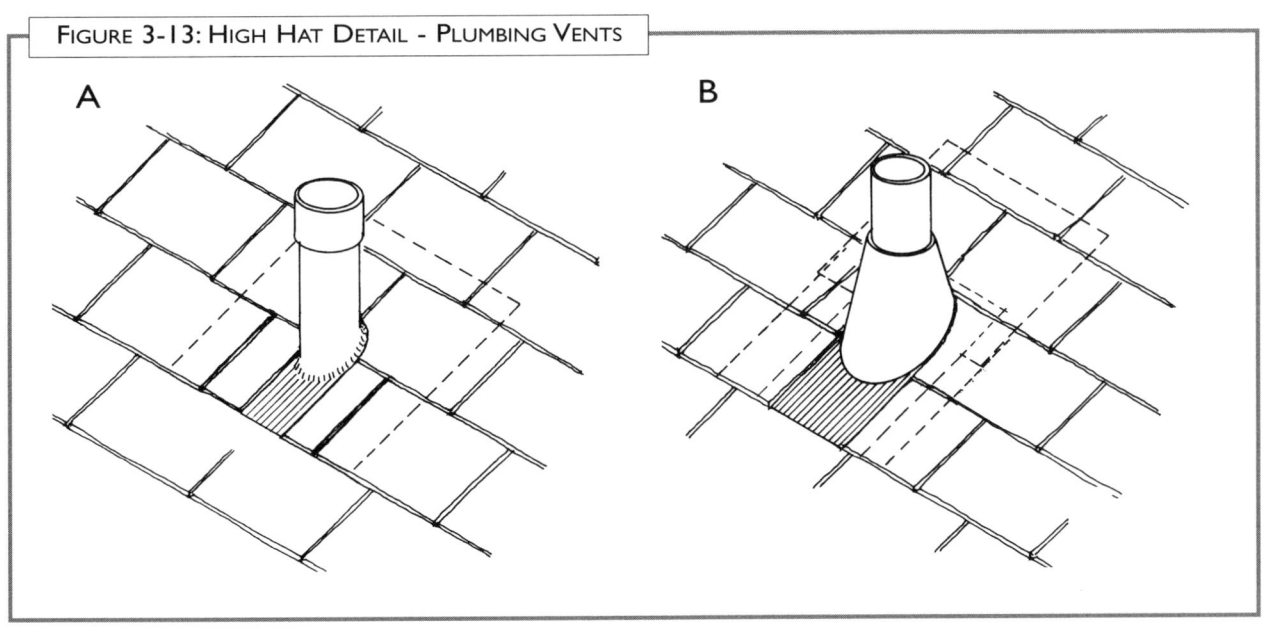

FIGURE 3-13: HIGH HAT DETAIL - PLUMBING VENTS

or longer, the plumbing vent, too, should last that long. Ask your local building materials supplier or slate quarry to recommend a good source for this accessory.

The average plumbing vent requires approximately 1/2 hour of additional installation labor. The cost of the vent itself is estimated separately.

Estimating Equipment Needs

In planning and estimating a slate roof project, consider the cost of equipment needed for the job and the logistics involved. In particular, the roofing project will require ground-to-eave scaffolding, roof scaffolding, material handling equipment, and safety equipment.

Scaffolding

The roofer usually provides his own scaffolding. However, on large projects where several contractors need to use scaffolding, it is better for the owner or general contractor to provide it. Pipe scaffolding up to the eave level is the best scaffolding for a slate project. Hydraulic lifts may also be an option for projects where terrain permits them.

Careful consideration should be given to the scaffolding needs for a slate roofing project. In all cases, OSHA regulations govern proper scaffolding and safety requirements. When possible, use pipe scaffolding, which provides several benefits. When starting the slate roof, you often need starter slate, drip edge, gutters, cant strip, field slate, nails, tools, and mechanics. If pipe scaffolding is used and is fully planked, the platform usually allows for all of the starting materials and provides a good eave-to-roof staging area. Pipe scaffolding also allows for proper safety rails and screens, providing safety for both the mechanics on the scaffold and for any tradespeople working near it on the ground. Do not overload this scaffolding. Never drop a full pallet of slate in one location on this or any other scaffolding that hasn't been properly engineered.

Roof Scaffolding

The roof scaffolding refers to the slate brackets and planking that the roofer installs as the slate progresses up slope. These brackets and planks are needed every 4 or 5 feet up slope. The maximum horizontal spacing between brackets should not exceed 7 to 8 feet. If the brackets are too widely spaced, the planks will flex between brackets and rest on the finished slate roof. Be sure to use scaffolding-quality planks as approved by OSHA regulations.

The brackets are left in place as the mechanics proceed up slope. When a particular roof surface has been completed to the peak with all related details, the installer will begin to work back down the roof, removing those planks and brackets as he goes.

It is not uncommon for the roofer to leave some of these brackets in place for other contractors to use. Be sure to estimate the cost of the brackets that will remain as well as an extra cost for their removal and the repair of any slate damaged by other contractors using the scaffolding.

Material Handling Equipment

When slate is delivered to the job, it comes on pallets that weigh approximately 4,000 pounds each. Unloading the truck and distributing pallets around the site may require some form of forklift. As a rule, anything over 5 squares should be unloaded with a forklift. Five squares represent an estimated 2 pallets or less of material. Anything less than 2 pallets can be unloaded by hand, if necessary.

On new projects, these pallets should not be placed close to the building where they may be in the way. Consequently, the roofer must move the slate from some obscure location to the scaffolding. If the entire pallet is not used, it must then be moved back to storage or to the next roof section.

There are several points to consider when planning for unloading:

1. The slate usually arrives on a commercial carrier. If loaded with slate only, these trucks tow open flatbed trailers. For small

Project Planning and Estimating

FIGURE 3-14: ROOF SCAFFOLDING BRACKET

orders, the slate sometimes arrives on a closed trailer.

2. Most slate companies do not ship their material on their own trucks. They hire commercial carriers. These subcontracted shipping companies usually don't have portable unloading equipment. It becomes the owner's or roofer's responsibility to arrange for unloading.

3. At the quarry, the slate is loaded onto the trucks with a forklift. The greatest amount of damage occurs later at the building site during the handling and stacking of the slate. Having handling equipment available at the site can ultimately save you money both in labor and in materials.

4. Because of their size, delivery vehicles usually are unloaded at or near the entrance to the project. Rarely can slate be left where it is initially unloaded. For this reason, a forklift can be invaluable.

5. If laborers who would otherwise be working on the roof are used to unload a truck, then the cost of labor must be weighed against the cost of a forklift.

In some cases it may be wise to consult a local roofing material supplier. Some of the larger local suppliers own delivery vehicles ranging from trucks with detachable forklifts, to boom trucks, to mechanical elevators, to cranes. Some suppliers will agree to have the slate delivered to their commercial location, where it then can be unloaded with a forklift and safely stored. The local supplier later delivers the slate to the site on the appropriate delivery vehicle. There is obviously a cost associated with this service, but it may be worthwhile in the long run. Some local suppliers, however, may provide this service at no additional cost if the slate is purchased through them.

On many projects the roofer will need to move the slate pallets from the unloading area to a storage or staging area. The staging area on many sites is in a safe, but out-of-the-way place. A forklift may be required on site for the duration of the slate installation to retrieve this stored material. On some projects, the forklift is shared by all contactors. If this is the case, the roofer is also expected to share the cost of this machine.

Project size should be considered when selecting the type of forklift, particularly for getting the slate from the ground to the roof. On larger projects, a forklift that reaches the eaves can save on material-handling costs. On smaller projects or difficult access sites, a ladder hoist is the best equipment for getting slate to the eave. Other options include a crane, roof-mounted cable hoists, rope and wheel, or portable ladder hoists similar to those used throughout Europe.

Purchasing from a supplier who has boom or pallet delivery capability can eliminate the need for the forklift. However, the slate will then have to be moved around the site by hand. It is important for all parties to understand that increased labor costs and damage are directly related to the amount of handling the slate endures before being installed. The owner should be prepared for this cost and should understand the importance of having a forklift on site.

Scheduling

Any work from the roof up that needs to be done by other contractors should be completed before the slate is installed. There are two reasons for this. First, it can be difficult to go back later and fill in roof sections that were left open for other work to be completed, and second, you should avoid walking on a slate roof once it is installed.

Coordinating the proper installation sequence with other contractors working on the project is essential. Stop and start-up cost will naturally upset the progress and attitude of the installer. Job meetings are critical since so many people are more accustomed to working around other roof types that can easily be repaired and accessed later. A pre-job start up meeting with all the involved parties present can help prevent possible roof damage and frustration later on.

Planning for Clean-up

A large slate roof project involves significant effort during the final clean-up phase, and the costs of clean-up should be considered in the project estimate. Obviously, the clean-up should be completed by the roofer, but the responsibilities and logistics should be discussed. For example, leftover stone might be able to be left on the site, or it may have to be hauled away at additional cost. All parties should know from the start who will be responsible for the clean-up, the costs of removing debris, and so on.

The Bidding Process

When project planning and estimating have been completed, it is time to solicit bids. This section discusses:
1. the procedure for finding qualified contractors,
2. how to provide contractors with a detailed specification that describes exactly the materials and details for bidding purposes, and
3. how to evaluate bids.

Finding a Contractor

There are a few methods you can use to find qualified bidders to install a slate roof. The first and best method is is to contact the slate quarries and ask for recommedations. Because roofing contractors normally purchase slate directly from the quarry, the quarries will generally know the slate roofing contractors in any given area and be familiar with their work. Often, owners will inform the quarries when an installation has gone well. However, the quarries are also usually the first to be contacted when an installation is going poorly. Because there are so few slate installers around, a quarry will often have more than one installation by which to judge a roofer.

A second method is to follow the same procedure described above with your local roofing supply company. These local suppliers receive feedback that is similar to that received by the quarry, and usually have a broader knowledge of the local contractor base. If slate is rarely installed in your region, the local roofing supply house might be able to point out the best roofers in your area.

A third alternative is to ask your architect, local property owners, or local building contractors for a recommendation.

It is not uncommon for slate roofing contractors to travel throughout a region or even throughout the country. Consider a combination of the approaches mentioned above.

Qualifying the Bidders

You should now have a list of potential bidders. Ask each one for references of prior installations that are as similar in scope to your project as possible. The local slate repair handyman may never have actually installed an entire slate roof. It is preferable that each installer have a minimum of five years prior experience installing slate.

Project Planning and Estimating

Unfortunately, it is not always possible to find an installer with the proper qualifications. This situation is hardly unique to the slate roofing industry. As a result, Registered Roof Consultants (RRC) are gaining popularity as project advisors.

Mock-ups as discussed on page 24 are another good method for evaluating a potential bidder's ability. However, mock-ups rarely include every detail of the job and should not be used as the only criteria for evaluation.

Bid Package

It is best to get at least three bids. Provide each bidder with a complete set of plans, specifications, and a detail checklist. (See Appendices B and C.) The plans and specifications ensure that all parties are bidding on the same thing. Adding a "detail" checklist helps clarify the preferred style for each detail. You should also provide all bidders with a standard bid form. This standard form allows you to compare the numbers from the bid with the numbers you generated in your preliminary estimate.

Evaluating Bids

Bids for a slate roof can vary widely from contractor to contractor. Before awarding the contract, take the time to go over the bid with the prospective contractor. It is in everyone's best interest, before any work begins on the roof, to determine whether or not the contractor can, in fact, complete the work for the bid price. If, after work has begun, it is discovered the contractor has made a mistake in his bid, it will be nearly impossible to find another contractor to take over.

Awarding the Contract

Slate roofing contracts vary in complexity, depending on the project. A good source for contract forms is the American Institute of Architects (AIA). Although not all of the details in an AIA contract may apply to your project, you can usually custom-tailor the contract to fit your individual project needs. Contact your state or local AIA chapter or your architect for more information.

NOTES

CHAPTER 4

Ordering and Handling Slate

When the project has been awarded, it is time to order the slate and have it delivered to the site. This chapter discusses procedures for ordering, receiving, storing, and handling slate.

Ordering Slate

In terms of both time and labor, the production of roofing slate is an intensive process, so quarries prefer to become involved with the project as early as possible. By tracking pending projects, quarries can anticipate upcoming needs.

When you are ready to place an order, contact the quarry or a sales representative. Using the order form checklist in Appendix B, communicate to the salespeople exactly what it is you wish to purchase. Request that this form be signed and returned to you along with the quarry's quote confirming the price and availability of the slate you desire. Any changes from the original request, including approved changes in size or thickness, should be made in writing.

The lead time for slate delivery will vary on almost every project. As a rule, semi-weathering slates are the most readily available. However, availability of the desired material should be addressed as early as possible in the planning process. The fact that slate is quarried from the ground can pose some problems with availability, for any of several reasons.

1) In some cases, geological faults in the material may render the slate useless as roofing material. The quarry may spend days blasting and removing slate that cannot be used just to get to useable material.

2) Winter conditions can slow or stop the quarrying process. Most larger quarries have erected large milling buildings to allow them to process materials year-round. However, heavy snow and severe cold will prevent them from mining raw material during the worst winter months. Be sure to plan the placement of your order to avoid seasonal delays.

3) Special orders of thickness, cut butts, color mixes, or large sizes can cause delays. Any size slate is available, given enough lead time. However, for many projects, time is of the essence, and there are compromises that can be made without adversely affecting the specification. Minor changes in size or other characteristics can change the lead time by weeks or months. Identify your preferred size and, if time is critical, identify possible alternative sizes. This flexibility will allow the quarry to fill your order more quickly and, in some cases, may reduce the material cost. The thicker material will be more expensive because it is harder to trim, punch, handle, and ship.

4) Desired colors seem to run in cycles of popularity. In one year the greatest percentage of orders will be black, the next year semi-weathering green, and the following year purple. Don't assume that because it was easy to obtain unfading green on a project two years ago, the same will be true this year.

Placing your order as early in the process as possible will help guarantee that you get the material size, color, and texture you want without having to compromise.

Payment Terms

Historically, for projects of any size, roofing slate has always been purchased directly from the quarry, although it can now be ordered from a local supplier. Once an order has been filled, the quarry will want to ship it and be paid promptly. If special arrangements need to be made for storing the slate at the quarry or handling the slate on site, be sure to clarify this up front.

Some quarries will extend payment terms to qualified customers. However, most will ask for C.O.D. payments, usually made by certified check. This is standard procedure for an order of standard-thickness slate. Terms may vary for custom orders. The buyer needs to understand that graduated, special butt cuts, and sometimes orders of special sizes may often require a non-refundable deposit up front. When asked to produce a special order, the quarry needs to be assured that the order will not be canceled later. The time lost on a special order (while other standard orders could be filled) can be costly for the quarry. Terms and conditions of these special orders will vary, so be sure to clarify payment terms when estimating the job. The project owner should know ahead of time about any special payment conditions.

As a result of these special payment conditions, local building supply distributors are becoming more involved in this market. As product manufacturers, most quarries prefer to sell to repeat customers with a good payment history. Consequently, they will consider selling their slate to local distributors at a discounted rate. The distributor then accepts the burden of establishing payment terms and conditions with the local purchaser.

During the bidding process, be sure to consider the terms and conditions of payment with your bid. The owner should know up front if a deposit is required or if a shipment needs to be paid C.O.D. If extended payment terms are required and not available through the quarry, you (the bidder) may have to purchase the material through a local supplier with whom you have credit.

Receiving Slate

Once the slate has been fabricated and placed on slate pallets, delivery arrangements can be made. Slate is usually delivered on open flatbed trucks. Most of these trucks are capable of carrying approximately 44,000 pounds of material. If you are using standard-thickness slate that weighs an average of 850 pounds per square, this translates into approximately 50 squares per truckload. Most quarries do not own and operate their own delivery trucks, so even small

Ordering and Handling Slate

quantities of slate are generally shipped by common carrier.

Make sure that shipping costs are defined up front when bidding a project. Some quarries will quote their material on a cost per square basis F.O.B. (freight on board) their yard. In other words, the shipping is not included in the price. Others will quote a delivered price. In order to compare quotes properly, you need to clarify this shipping cost.

Table 3-1: Weight per Square on page 36 shows the average weight per square of material based upon its thickness. If 1/2-inch to 5/8-inch material is being used, the shipping cost can be twice that of a standard thickness 1/4-inch to 3/8-inch material. While this may not represent a significant cost difference close to the quarry, if material is being shipped cross-country, it can increase the cost dramatically.

Finished slate is stacked on edge on heavy hardwood "stone" pallets. Material is stacked 2-3 tiers high on each pallet depending on the size of the slate. The pallets are then usually banded with steel bands to prevent them from shifting and breaking during the loading, shipping, and unloading process. Each pallet weighs approximately 4,000 pounds, so there are rarely more than fifteen pallets per truckload. If thick slate is being used, not only will more trucks be needed, but more pallets will also be required. This is an important consideration since additional pallets will require more storage space. Also, some quarries charge a pallet fee for the pallets. On a large project, you may be able to return these pallets for a refund, but that is rarely the case. Be sure to verify this pallet fee when bidding. Some quarries will add this cost later (usually $20.00 per pallet), so what may appear up front as a better material price, may actually be higher than the price from a supplier who includes all of the costs in the per-square-delivered price.

Rarely are the slate pallets themselves stacked. When slate is shipped from foreign countries, shippers do use crates that allow for stacking. However, these crates are packed tightly into overseas containers that prevent them from shifting side to side. Most trucks will reach their weight limitations before the pallets need to be stacked.

The quarry numbers or labels each pallet and provides the shipping company a bill of lading and a packing slip, identifying each pallet number and the number of pieces of each slate size per pallet. These counts are usually accurate but can vary slightly.

When the slate has been loaded and is ready to ship, the quarry usually calls the purchaser to let him know that the slate is on its way and to provide an estimated time of arrival. As discussed in Chapter 3, you will likely need to have a forklift available on the site for unloading the truck and moving the slate.

When the slate truck arrives, the driver will scan the horizon for a forklift, then present you with a copy of the bill of lading and the quarry's packing slip. Depending on the payment terms established, the driver may request payment for shipping and/or materials prior to unloading. This is common practice. Make sure you establish the payment terms and conditions with the supplier prior to shipment.

(While you are unloading the truck, the driver might ask to borrow your cell phone to call a dispatcher to find out where his next pick-up is; this final step may indeed prove to be the most expensive part of the project.)

While the slate is on the truck, it is generally the responsibility of the shipping firm. There will often be some breakage during shipping, but usually no more than 3-4 pieces per pallet. Any breakage or significant damage to pallets or slate should be noted in writing (referenced by pallet num-

Chapter 4

ber) and agreed upon by both you and the driver. This documentation will provide you and the quarry some recourse later should the shipper refuse to pay for materials damaged in shipping.

The buyer is responsible for counting the slate on several or all pallets to ensure that the quantities match those on the packing slip. Quantities will sometimes vary slightly. (When counting thousands of pieces of slate, it is easy to make mistakes.) If quantities do vary from those shown on the shipping document, the quantities and pallet numbers should be noted and confirmed with the truck driver. Any discrepancies should be reported immediately to the seller. Explain what you discovered during your inspection and, prior to unloading the slate, determine how any variations will be rectified.

Be prepared to take photos of your shipment. The photographs may prove helpful during discussions with the supplier.

After you finish inspecting the shipment, ask the driver to sign your page of notes, confirming your inspection. Some drivers may be reluctant to do this since you are in essence asking the driver to witness and agree with your inspection. If sizable variations are detected, the driver may ask to contact his dispatcher as well as the quarry before signing off.

In some areas, unloading the slate from the truck prior to inspection indicates that you have accepted the material as is. If nothing else, it relieves the trucking company of any responsibility for damages that may have occurred while the slate was on board. Any damage can then be blamed on the unloading process.

A wise truck driver will want to have the slate inspected prior to unloading since damage often does occur during the unloading and storage process. Ultimately, an initial inspection will serve to protect all parties. Once the slate has been shipped, received, paid for, unloaded, and stored on a job site, it will be very difficult for a supplier to accept at a later date that the buyer is beginning to notice problems.

Whenever possible, try to have all of the slate shipped to the job site before starting the roof installation. This is not always possible, but the installer should mix the slate from all of the pallets even if it is a one-color job. This will ensure that slate from each delivery is mixed throughout the roof.

Storing Slate

Once the slate has been delivered, it will have to be stored until it is needed. As a rule, try to store the slate out of sight and at a reasonable distance away from the building. Having a forklift on site will provide you tremendous flexibility in establishing and utilizing a safe storage area. There are a number of things you should consider when selecting a storage location.

- The weight of each pallet of slate and the use of a forklift can damage finished landscaping and can crush subsurface drainage pipes and septic systems.
- Slate pallets are constructed of hardwood and will usually last for a year or more. However, if they are stored in a wet area for a long period of time, they can pull apart when moved. In cold climates, the pallets will sometimes freeze to the ground. If you anticipate storing the pallets for a long period of time, try to elevate them using 2 X 4's or the like. This will help keep the pallets from freezing to the ground or getting bogged down in mud.
- The slate is always stacked on edge, which allows moisture to pass through the stacks. However, in cold climates, water from melting snow and ice will often freeze between the pieces. Try to keep the pallets

that you will need during winter conditions covered. If the slate is being stored indefinitely, it does not need to be covered unless you expect to use it during the winter months.

- When slate is stacked flat, capillary action pulls water in between the pieces. In cold climates, this water will freeze and expand. The compressive weight of the flat stack will often break slate as the ice creates movement. Once breakage begins, it will continue.
- Oils of any kind, including, hydraulic fluid, motor oil, and suntan oil, will permanently stain slate.
- Avoid storing slate too close to the job site's general delivery area. Pallets hit by trucks and other vehicles do not move easily; instead, the slate breaks.
- If storing the slate near the building, make sure that it is not in anyone's way. Since the pallets are not easily moved, tradesmen will work around, over, and on top of them, which usually results in broken slate.

Finally, protect your slate from theft and from "casual use" as the job location dictates. Most people realize that slate is expensive, and it is not uncommon for someone to help themselves to a couple of pieces to paint or write on (although it is difficult to steal large quantities because of its weight). Also, curiosity about slate is common, so people may want to touch it and handle it. The pallets make great workbenches and good seats, too, so try to store your slate where it is not easily accessible to other tradespeople and passersby.

Handling Slate

In many ways, roofing slate is similar to glass. Like glass, slate must be handled gingerly and safely. The edges are very sharp. When the slate is trimmed for size at the quarry, the finished edge resembles a serrated knife edge. Therefore, it is best to wear gloves when handling this material and never allow the slate to slide through your hands. This sliding action will quickly illustrate the sharpness of the "knife edge." The corners of a piece of slate are sharp enough to stick through a piece of 1/2-inch plywood if dropped from the roof. On the job site, always exercise caution. Falling slate is both sharp and heavy. Stay out from underneath the area where the roof is being installed unless you have proper safety equipment (e.g., hard hat, scaffold tunnel).

Slate is also similar to glass in that you can gauge its quality by ringing or sounding each piece. This can be accomplished by gently but firmly tapping the slate with a hammer. Imagine doing the same thing with a piece of glass. If the glass is solid and free from structural defect, the glass will ring when tapped. If the glass is fractured or defective, the tapping will produce a rattling or thudding sound. The same is true for roofing slate.

One of the single most important habits for the slate roofing mechanic to develop is the habit of sounding each piece of slate before it is installed. This simple habit will eliminate the need for possible repair after the roof has been installed and will help insure the highest possible quality. Ideally, the slate mechanic will be the last person to touch each piece for the life of the roof; "ringing" the slate is among the last, but most important steps, in quality control.

Another similarity between slate and glass is that a person is much more likely to get cut on a broken piece than on a whole piece. Yes, the serrated edges are sharp, but the rough edge of a broken piece can be as sharp as a razor. When sounding a piece of slate, hold it away from your body. If, in fact, the piece is bad, it will sometimes break in your hands. To

Chapter 4

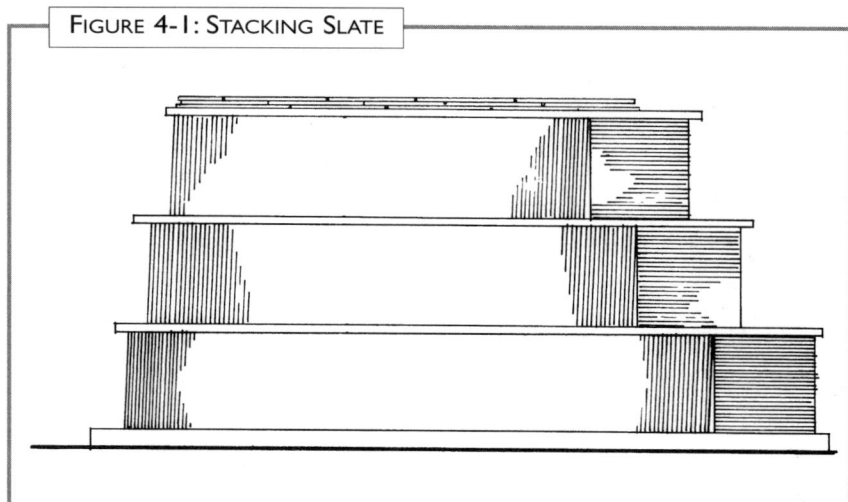

Figure 4-1: Stacking Slate

avoid injury hold the slate in one hand, in front of your body, rather than cradled in your arm.

During the roof process, it is likely that some of the pieces sounded will be rejected by the installer as unusable material. These pieces, along with the fall-off from valley, hip, and miscellaneous cuts, should be stored neatly in one area of the scaffolding. Any unusable material should then be removed each time someone climbs down from the roof.

When slate is removed from the pallets and stacked elsewhere, it needs to be stacked on edge. Use flat end stacks to support the rest of the slate stacked on edge. These end stacks only need to be placed on one end of a row of slate. Never stack the slate more than three tiers high, and always place a wooden spacer between tiers. It is not necessary to cover this material except under winter conditions.

Getting Slate to the Eaves

As the roof installer works, slate will need to be moved from the storage area to the eaves. Even on existing structures where the roofing contractor is the only contractor on the site, the building occupants have routine and building access needs that must be considered. In most cases, the most logical storage area for the slate will probably be the least logical location for the roofer. As mentioned earlier, the best means of getting slate to the staging platform will likely be a forklift or a ladder hoist. The less the slate is handled by the mechanics and laborers prior to installation, the less breakage there will be. In addition, the roof will go on much more quickly if the slate can be brought to the mechanics using the most efficient method that project conditions will allow.

Lifting equipment becomes very important on a project that calls for graduated length slate. Because the length of the slate gets progressively shorter as you proceed up slope, there is a need to change lengths often. If you happen to be working with 18-inch long slate and the roof layout dictates that you graduate to 16-inch long slate, ideally you will take any remaining 18-inch material and load it back on a pallet with other 18-inch long material which is then placed back in storage. Therefore, it is very helpful if the apparatus being used to bring the slate to the roof can also bring unneeded material back to the ground. Any material brought back to the ground will need to be returned to the storage area until it is needed again on a different section.

Regardless of what means of lifting is employed, be sure to distribute the material brought to the scaffolding or roof as evenly as possible. Never place an entire pallet of slate which has been lifted by crane or forklift directly on a scaffolding platform. This is not a safe practice.

CHAPTER 5

Roof Construction and Preparation

The roof framing and roof deck must be properly constructed to support a slate roof. The roof must be completed and properly prepared before actual slate installation can begin. For most slate roof projects, the roofer is responsible for inspecting the roof sheathing, trim, and masonry details before beginning his work. Any incomplete or unacceptable work should be noted and presented to the owner or building contractor in writing. Corrections to any of these problem areas should be made before the roofer begins installing the underlayment.

On many projects, the builder will ask the roofer to start a section of roof that has had all of the details completed, while the builder readies additional roof areas. This is a common and reasonable request. Inspect each new roof section before progressing, and note any trouble areas as stated above.

This chapter provides information on roof construction and the materials and procedures to be used in roof preparation.

Roof Construction

As described earlier in Chapter 2, slate roofs in the early days were almost always installed on 1-inch thick, rough-cut boards supported by rafters of varied size and spacing. Today, with the implementation of building codes and the availability of new construction materials, roofs can be constructed in a variety of ways and with many different materials, such as plywood, nail board with attached insulation, gypsum products for fire safety situations, and pressure-treated wood products. The structure on which a slate roof is to be installed should be professionally engineered by an architect or a structural engineer to meet specified building design loads for a given geographical location and to support the weight of the slate.

Framing techniques for all projects will be dictated by the architect's specifications or the building design. However, some

Chapter 5

aspects of the roof construction related to a slate roof will vary somewhat from other roof types. For proper slate roof installation, the roof construction must meet the following requirements.

1) All sheathing must be securely nailed to prevent bouncing. If the sheathing is bouncy, the slate already attached can break or the vibration can pull up the nail heads. Bouncy sheathing is not acceptable.

2) Sheathing should extend as far as trim details permit. When elaborate trim details are specified, it is common for the framing contractor to end the sheathing well short of the finished trim. The result is a void between the finished trim and the sheathing, which creates a problem both at the eaves and on gable-end details. At the eaves it can prevent the attachment of the cant strip at the proper proximity to the eaves. At the gables it can prevent the installer from being able to nail the slate close to the exterior edge of the gable. As a result, the slate can sag or be lifted by wind.

3) Valley framing and sheathing must be cut and fit to create a straight valley line. In some framing details, several valley rafters will be laminated together to form a structural beam. Unless the jack rafters tying into these valleys are set properly, a flat spot can occur in the valley. Unlike some flexible roofing materials, slate will not lay well through this flat spot.

4) Hip sheathing must be cut and fit to follow a straight line up the hip. This sheathing does not need to be miter-cut, but it must be installed straight. This is especially important if the project requires a mitered slate hip detail. The appearance of the mitered hip will be affected by the quality of the framing and sheathing installation.

5) For both hips and valleys, the closer the sheathing is fit, the better the roofer will be able to fit and secure the slate.

6) For a standard ridge detail, the sheathing should be cut and fit as close to the peak as possible. This can be difficult if a thick beam is required at the ridge for structural purposes. The framer will have to provide a true peak so that the nailers that the ridge slate are attached to can be set as close to the peak as possible.

7) For a vented ridge, the sheathing will be set back from the ridge as specified by the architect. This space needs to be kept as small as the governing building codes permit. Should thick structural ridge beams become a problem, a different ventilation method may become necessary. Otherwise, the vented ridge detail could become extremely large and difficult to attach (see vented ridge sections in Chapters 6 and 8).

Roof Sheathing

There are many types of materials that can be used for roof sheathing under a slate roof. Plywood is probably the one most commonly used in today's construction. If the roof is sheathed with plywood, the plywood should be at least 3/4 inch thick. However, proper thickness should be determined by the engineer. Plywood is easy to cover with underlayment and, when properly engineered, supports slate well.

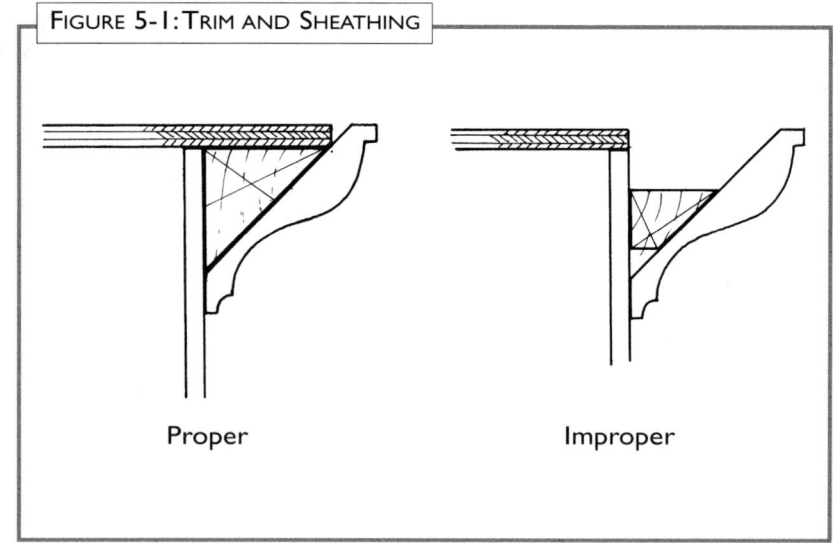

FIGURE 5-1: TRIM AND SHEATHING

Proper   Improper

Roof Construction and Preparation

Years ago one-inch thick rough-cut lumber and sometimes tongue-and-groove boards were the most common sheathing used. These boards were and still are easy to cover with underlayment, and they provide an excellent nailing surface. If one-inch boards are used, be sure to install the underlayment as soon as possible to prevent warping.

During slate application, the installer will sometimes hit a joint between boards while nailing. If this happens, the slate must be punched to provide proper fastening.

Some projects will require fire-resistant sheathing materials, such as lightweight and aerated nailable concrete products. These materials are fairly common with commercial work. If the roof sheathing is made of these materials, be sure to use a heavy-gauge nail to avoid bending the nails during installation.

Slate has been installed over open batten systems throughout Europe for years. In theory, this technique is an excellent approach to roof sheathing for slate. The slate, being elevated on battens, allows air to circulate and provides for natural ventilation. However, in some locations, windblown snow, silt, and sand will filter in behind the slate. This situation is more common with thicker slate. To avoid this problem, a variety of fabrics have been developed that can be installed over the rafters before installation of the horizontal battens. The fabric is intended to replace the underlayment and prevent moisture and debris from infiltrating the structure.

If a batten system is desired, install solid roof sheathing first wherever possible. Then install the specified underlayment over the sheathing. Next, install vertical counter battens directly above the roof rafters and, finally, install horizontal nailer battens over the vertical battens. In essence, this will create a double roof sheathing situation. If you decide to use this approach, be sure to evaluate adjoining trim and structural details. The batten and counter batten system will elevate the nailing surface and is of particular concern in a reroofing situation.

These batten systems can be constructed from either wood or metal. With a metal batten system, be sure that the flashings and fasteners chosen for the slate are compatible with the metal battens. Thickness of the chosen batten material will be dictated by the weight and thickness of the slate, but typically, wooden battens are 1 inch by 4 inches.

With steel becoming more and more popular in modern construction, the question of whether or not slate can be installed directly to a steel deck often arises. Technically it can; however, underlayment is still required over the steel before installation of the slate, and the fasteners will probably need to be stainless steel screws. The logistics of this approach often make it cost-prohibitive.

Trim

Keep in mind that the cant strip at the eaves will elevate the starter and first course of slate above the finished trim detail. Drip edge will usually cover the resulting void at the eaves. However, some contractors will elect to install their finish trim after the slate has been installed. By doing so they can fill any voids created by the cant strip. If you choose this method, the installer must allow the slate to overhang a sufficient amount to account for the thickness of the trim not yet installed.

In most cases, building trim details are installed prior to the slate. In areas where the trim intersects the roof (such as a dormer soffit detail), the trim should be held off the roof surface three times the thickness of the slate. In addition, these areas will require that flashing be slid behind the trim. The trim installer should allow a nail-free area equal to the slate thickness plus 4 inches. This way the roofer can install the proper size flashing without interference from the nails holding the trim in place.

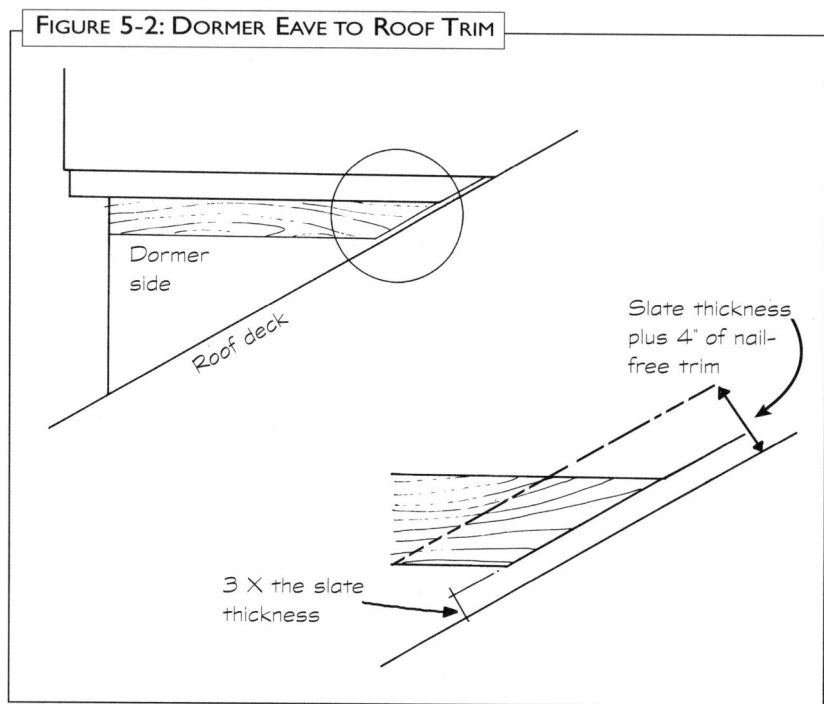

FIGURE 5-2: DORMER EAVE TO ROOF TRIM

In some cases, the trim may need to be installed after the slate, but try to avoid this situation whenever possible.

Roof scaffolding used by the roofer will need to be left up for the trim contractor. This can significantly delay roof completion and also damage the finished roof. Most roofers provide for these contingencies by stipulating in their bids that they be paid extra for these types of situations.

Masonry

All masonry, with the exception of some stucco details, must be completed and cleaned prior to the installation of the slate. Circumstances may dictate exceptions to this rule on occasion, but anytime masonry work is done above a finished slate roof, slate damage and/or staining is likely to occur.

When possible, counter flashings related to masonry should be through-wall flashings. The masonry contractor should install the counter flashings when he does the masonry work. Although some specifiers will ask that the roofer cut a raggle into the masonry and insert the counter flashing as part of the roofer's work, this is not the proper approach. (See Chapter 3, page 47.)

Unless the masonry is chemically sealed, it can allow water to seep in behind it. A leak can then develop which appears to be roof-related but is actually masonry-related. On new projects, always have the counter flashing installed with the masonry as through-wall flashing.

Counter flashing must be placed high enough to allow for the typical 4-inch height of the roof base flashings. On steep roof surfaces it will sometimes be necessary to install more and smaller pieces of counter flashing than normal to avoid using extremely large pieces.

If stucco is being used, then the stucco becomes the counter flashing. The base flashings that are installed with the slate are placed tight to the wall structure, and the stucco is then installed over the top of the base flashings.

The stucco installer must be careful to isolate the slate base flashings from the wire mesh that holds the stucco. In addition, he must not nail through the base flashings, and he must protect the slate from damage.

Whenever possible, install the base flashings before the stucco goes on, and then install the slate after the stucco work has been completed. If the roof is properly laid out in advance, most base flashings can be installed, supported by a spacer equal to the slate thickness, and the slate can then be installed later. (See Chapter 12: Advanced Situations.)

Stone chimneys seem to be particularly susceptible to leaks. This may be due to irregular mortar joints and structural contours. Be sure that these chimneys are through-wall flashed. The through-wall flashing will shed water that may get behind the masonry back out over the counter flashings.

Roof Construction and Preparation

Roof Preparation

Once roof construction has been completed and approved, then roof preparation can begin. Roof preparation involves the installation of underlayment, cant strips, nailers, and flashings.

The minimum recommended pitch for slate roof application is 6:12. (Slate can be installed on a pitch as shallow as 4:12 if the headlap is increased to 4 inches.) Slate is considered a "steep slope" roofing material and application. This "steep slope" designation implies many things, but particularly the need to exercise caution and safety. Consult OSHA regulations for the guidelines governing safety on a "steep slope" roof project.

As a first step in roof preparation, it is most efficient to set up scaffolding (a platform between knee- and waist-high) at the eaves of the building. The type of eave scaffolding used is up the the roofer and OSHA. In most cases, site conditions will dictate the scaffolding chosen.

Keep in mind that standard-thickness slate weighs approximately 850 pounds per square — one square of slate and two men will weigh well over 1,000 pounds. Pipe scaffolding with a fully planked platform, complete with proper stair access, safety railings, and security fencing is the most comfortable for slate roofing.

Proper eave scaffolding becomes a staging area for the roof surface being worked on. Slate is installed at an average of one square per installer per day, so the eave scaffolding may be in place for a while. Slate does not install quickly. For those who haven't worked with slate before, this is probably one of the most difficult aspects of slate roofing to understand.

As well as functioning as an eave staging area, the platform provides protection for other contractors as the roofers progress up the slope. A piece of slate sliding off a roof can be very dangerous. The scaffolding platform helps protect everyone who needs to be on the site while the slate is being installed.

Underlayment

During the construction process, the framework goes up and the roof sheathing goes on. Depending upon the logistics and complexities of the trim details and roof penetrations, the roof sheathing could go on and be exposed to the elements for an extended period of time. For this reason, the sheathing needs to be covered, or "dried in," to protect it from damage sustained from the elements. Roof sheathing can be protected temporarily with tarps, but tarps are notorious for blowing off and should be considered temporary at best. The long-term effect of exposure is deterioration of the sheathing; the short-term effects are often warping and/or bubbling (if plywood is used), which can create problems for slate installation. An uneven deck will not allow the slate to lie flat. This can cause the slate to break after the roof goes on.

A 50-square slate roof may require as many as 50-man days to install, so the roof sheathing must be protected for a fairly long period. Installing a water-resistant underlayment over the entire roof surface should be the first step in roof preparation, thereby protecting the sheathing and structure from the elements during the roofing process.

Some people believe that the underlayment serves as a cushion for the slate to lie on. When using 90-pound felt paper, this is no doubt true. However, it is not necessary with natural slate to provide a cushion. Some of the synthetic materials may require a cushion, but natural slate does not. In fact, in many European applications, the slate is installed on batten systems. The result is that the field slate is actually unsupported over much of its length.

There are a number of older buildings that once had cedar shingles on them that were replaced with slate shingles. When the cedar was removed (as is typical of barns), the new slate was installed directly over the existing battens.

Chapter 5

After the slate goes on, the underlayment serves as back-up waterproofing. It's important to stress that the underlayment becomes a back-up waterproofing, because the slate and flashings serve as the primary waterproofing. Unlike many synthetic materials, properly applied slate with a 3-inch headlap is waterproof. However, because slate is essentially a loose-laid material, wind-blown snow can filter into the roof behind it. (See Figure 5-3.)

Standard thickness slate is approximately 1/4 inch thick. The edges of the slate are chamfered during the trimming process, so there will be a triangular-shaped gap created by the chamfer where the progressive courses overlap the preceding vertical joints. Obviously, this gap will increase as the thickness of the slate increases.

Because of the nature of the slate overlap combined with the headlap, these gaps do not allow water infiltration. In fact, they create natural roof ventilation that probably adds life to the slate itself. To further illustrate this point, on buildings with roofs that are installed over open battens, on a sunny day, you can see streaks of light from inside the building. This can be an issue when using textured and/or thick slate. Bees and sometimes bats can crawl into the gaps between the slate. If the slate isn't laid over a solid roof deck with an underlayment, unwanted pests could appear. Felt paper underlayment is notorious for wearing out. It dries out, becomes brittle, cracks, and eventually becomes somewhat powder-like. This process usually takes from 50 to 75 years. Obviously, there are slate roofs over 75 years old that are not leaking as a result of worn-out underlayment.

The primary purposes of underlayment are to prevent wind-driven snow or silt from infiltrating the structure and to keep the roof watertight while the slate is being installed. Heavier types of underlayment will add some insulation value. A modified bitumen or membrane product, for example, will definitely prohibit air and moisture infiltration, but it will also prevent the escape of air and moisture from within the structure. If the underlayment completely seals the roof, be sure to provide for proper ventilation of the sealed space.

There are a number of different types of materials suitable for use as underlayment, as described below.

The most common underlayment material is asphalt-saturated, organic felt paper, commonly referred to simply as "felt paper." Sold in rolls that are usually 3 feet wide, felt paper is available in a variety of thicknesses that are defined by their average weight per square.

Fifteen-pound felt paper is most commonly used with composition shingles. It can be used with slate but is the least desirable of the felt papers because it is the thinnest. Fifteen-pound felt does not hold up well if it is not roofed over relatively quickly. It is very susceptible to being torn or blown off by wind, and it is very difficult to walk on or move across without tearing. Because it is recom-

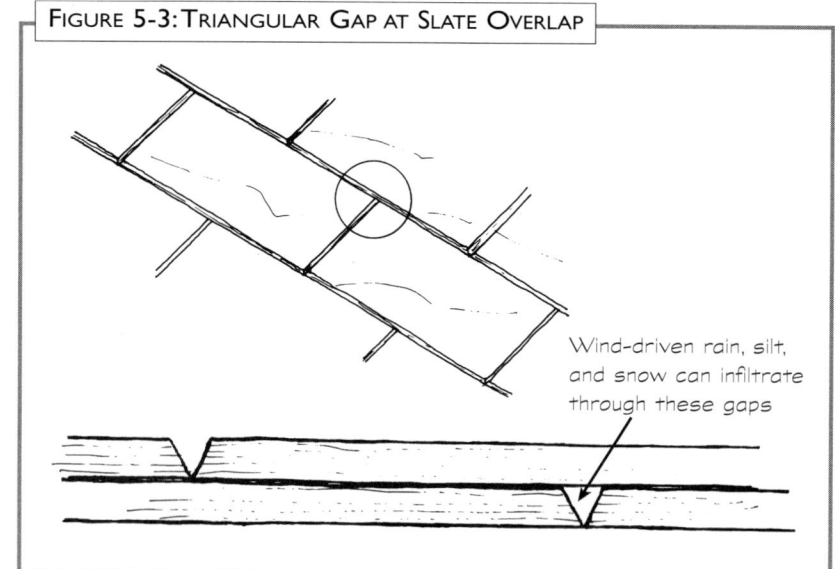

FIGURE 5-3: TRIANGULAR GAP AT SLATE OVERLAP

Wind-driven rain, silt, and snow can infiltrate through these gaps

mended that the roofer lay the roof out prior to starting the slate installation, it becomes necessary to access the entire surface prior to installing the slate. Fifteen-pound felt thus becomes very dangerous because it can tear under the roofer's feet. Finally, all felt paper has a tendency to bubble when exposed to the elements. This bubbling is minimal when using 30# felt or greater, but the 15# felt tends to bubble so much that it can crack and break when the slate is applied.

Thirty-pound felt paper is twice as thick as 15# felt. This is the most commonly used felt paper for slate roofing. In the first half of the 1900's, 45# felt was readily available and was specified a great deal. However, it is difficult to find and is rarely used today.

Ninety-pound felt is three times as thick as 30# felt. Some specifiers believe that this is the best underlayment material for slate roofing. They believe either that slate needs a cushion to sit on, or that the slate itself will not keep the roof watertight. Neither point is valid.

Modified bitumen is a relatively new roofing product. This product can be described as raw rubber in a roll. One of the most important qualities of this material is that when the bitumen is penetrated by a fastener, it forms a gasket around the penetration, thereby preventing any leaks. Typically, modified bitumen comes in 3-foot wide rolls. One roll will usually cover one square of roof area. One side of this material is sticky and will adhere to the sheathing. To prevent the roll from sticking together, a removable paper backing covers the sticky side and must be removed during installation.

Although modified bitumen has been around since the mid-70's, its use as an underlayment has just recently been accepted. Initially, modified bitumen was used as a perimeter waterproofing membrane. Since some of these membranes can be nailed through without creating a leak, this is a logical approach to perimeter underlayment. Because it appears to have worked quite well as a perimeter waterproofing, many specifiers have adapted the use of this material as the underlayment for the entire roof. In theory, this is not a bad idea, but it can cause a number of problems:

a) The attic space/roof sheathing will not be able to vent. Without proper ventilation, the substructure could develop serious moisture problems.

b) A waterproof roof that does not leak even when nails penetrate it allows for a very large margin of error for the roof installer. The goal that the specifier is trying to achieve is a watertight roof, regardless of the most adverse conditions. However, if the person or people overseeing the project are not familiar with proper slate roofing techniques, the slate can be and sometimes is installed incorrectly with little or no immediate consequence.

c) The above-mentioned situation opens up the opportunity for installers who may not be qualified as slate roofing installers to bid the project. Most roofers will realize that with a waterproof underlayment, even if they don't do the job properly, no one will know.

d) Some of these products are made for below-grade applications and can melt and run, which will cause staining. What starts out as a good idea could ultimately fail.

e) This material will shrink if exposed to the sun for extended periods. If the area covered with modified bitumen will be exposed for a long period, it may need to be covered by 30# felt.

Membrane sheet roofing materials, such as EPDM (typically resembling car tire inner tube material), were designed to be used as exposed flat roofing materials. They are sometimes used if the roof is going to go without slate for an extended period of time. An example of this might be a reroofing project where structural changes require that the slate stay off indefinitely.

In years past, hot tar (BUR) was often applied at the eave to serve the same purpose as today's

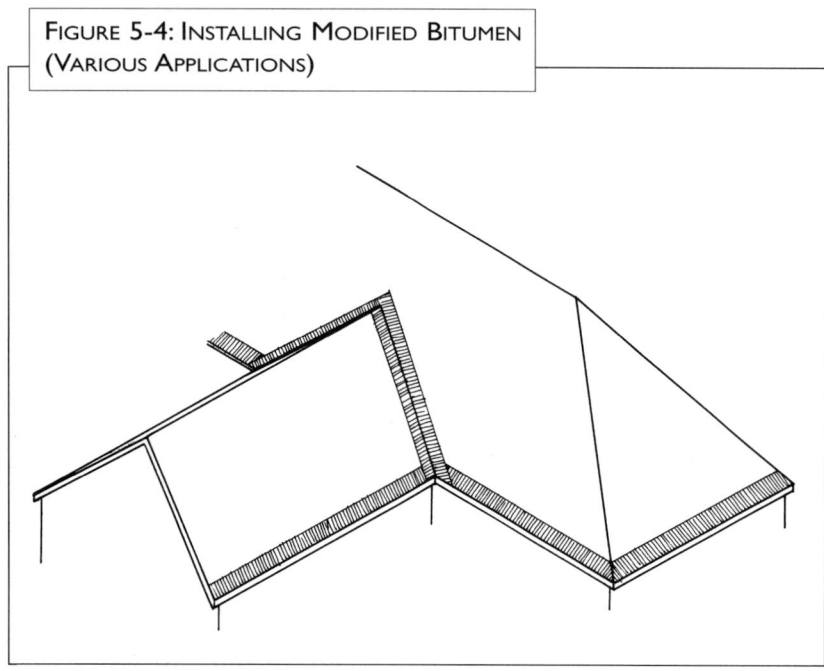

FIGURE 5-4: INSTALLING MODIFIED BITUMEN (VARIOUS APPLICATIONS)

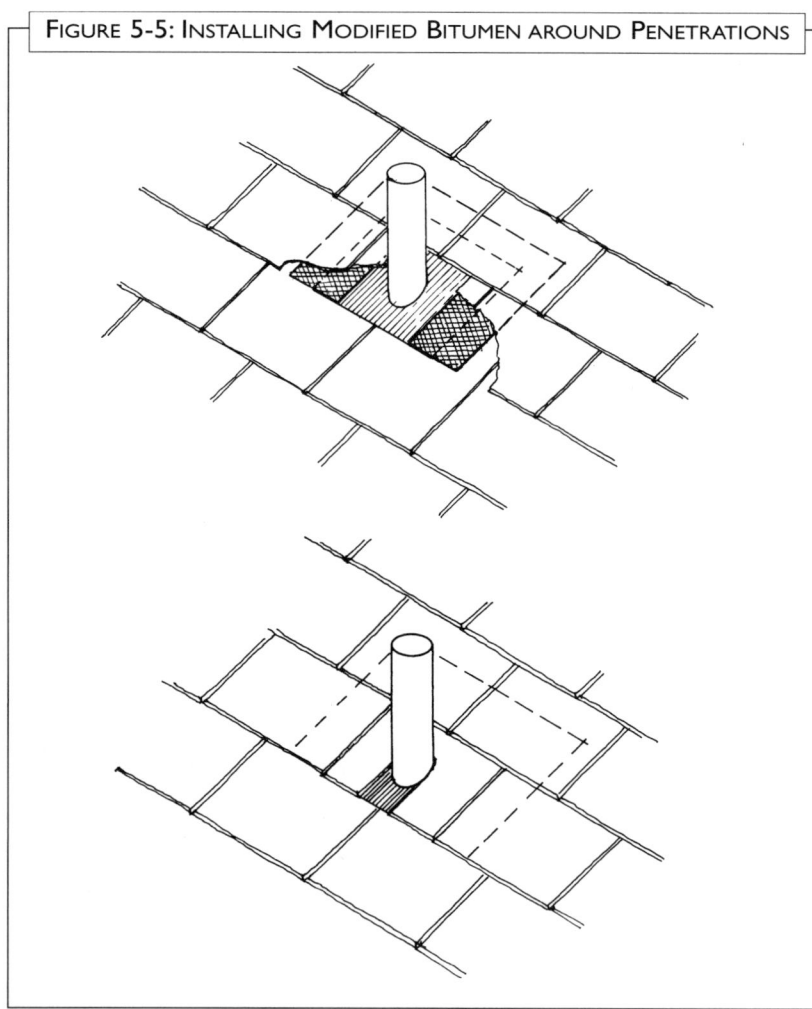

FIGURE 5-5: INSTALLING MODIFIED BITUMEN AROUND PENETRATIONS

modified bitumen. This material is rarely used anymore in slate roofing applications.

Rosin paper (rosin-sized paper or slip sheet) is typically associated with flashing details. It was occasionally used in the past as underlayment, but this is not recommended.

In Europe, a number of water-resistant breathable fabrics are being used. In most of these cases, the fabric is applied over a batten roof system. The result is that the roof can then "breathe." If one of these fabrics is desired, make sure that it will not leak if it comes in contact with the substructure. Just as with some water resistant tent fabrics, some of these fabrics will leak where they come in contact with a structural member. If a breathable fabric of a type that can leak is installed over plywood, it may actually retain moisture at the roof deck.

There may be additional underlayment materials available, but the most important thing to keep in mind when specifying these materials is to follow local building codes. A roof in Florida, for example, will certainly not experience the ice damming situations that modified bitumen was originally designed to protect against, but the same roof in Florida may experience hurricane conditions with wind-driven rain.

Today, proper underlayment is governed by local building codes, and proper application of the underlayment will vary from region

Roof Construction and Preparation

to region. Consult the project specifications for particulars. The installer may also need to consult the manufacturer if a product other than felt paper is desired. As a rule of thumb, assume that modified bitumen will be used around the entire perimeter of the building's roof and along the valleys; 30# felt paper will be laid over the bitumen to cover the remaining roof surface.

Sometimes the sequence in which the underlayment needs to be installed may not appear logical. In an attempt to "dry in a roof," the roofer or builder may install the modified bitumen in the valleys and along the perimeter and then apply the 30# felt paper over it. However, in new construction, the roof may need to be "dried in" before the perimeter trim is completed. Because the modified bitumen should extend beyond the finish trim, it may not be practical to install it at this point. If that is the case, you will need to install the eave-modified bitumen at the time that you are preparing to install the slate. Make sure that the successive course of underlayment properly overlaps the modified bitumen. Do not simply lay the eave-modified bitumen on top of the existing 30# felt.

Unless otherwise specified, be sure to use cap nails or roofing nails with roof tins to secure the underlayment to the roof deck. As stated previously, this underlayment will be in place longer than most builders and roofers are accustomed to. If not well secured, it is very likely to get blown off the roof. On new construction, before windows and doors are installed, air pressure from inside the building can blow the paper off. Repairing a felt paper underlayment that has been torn off or ripped is very time-consuming.

Underlayment is not what makes a slate roof watertight. However, the roof is likely to perform longer if the installer repairs or replaces any damaged underlayment prior to installing the slate. Figure 5-7 shows a typical minimum underlayment detail. Keep in mind that local building codes should prevail, assuming they meet this minimum.

Cant Strip

In roof preparation, the single most-neglected detail for a proper slate roof installation is the cant strip. After properly installing the specified underlayment material, careful consideration needs to be given to the cant strip detail. The purpose of the cant strip is to elevate the starter course into the same plane as the rest of the slate on the roof. Without a cant strip, the slate will not lie properly. Essentially, the cant strip replaces the head or top of the course that

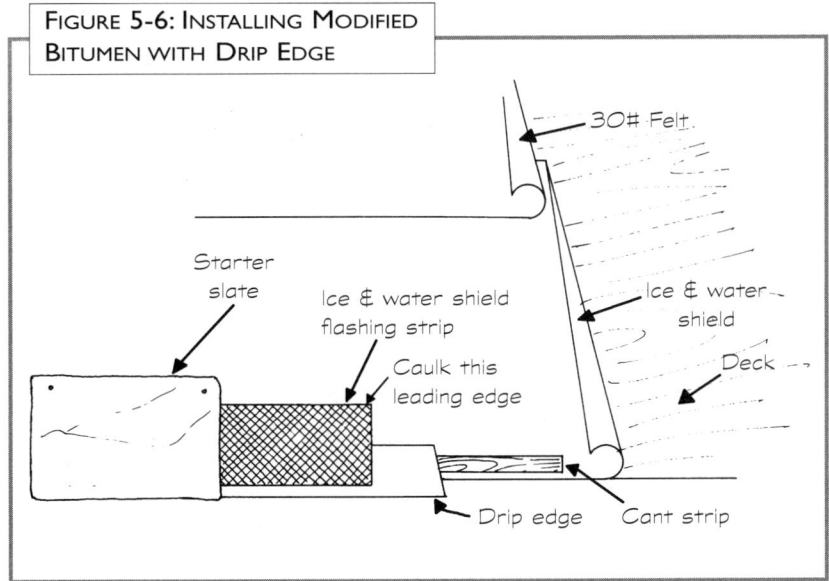

FIGURE 5-6: INSTALLING MODIFIED BITUMEN WITH DRIP EDGE

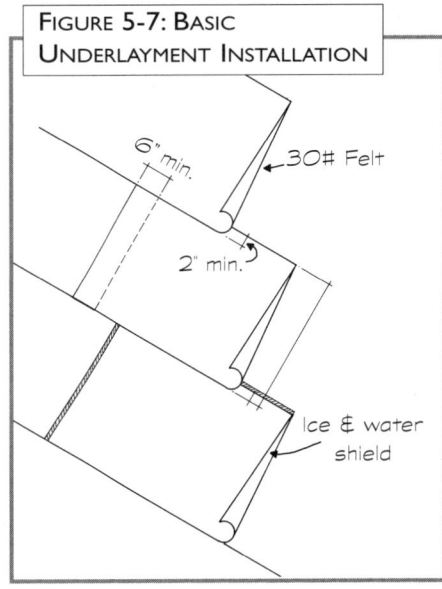

FIGURE 5-7: BASIC UNDERLAYMENT INSTALLATION

Chapter 5

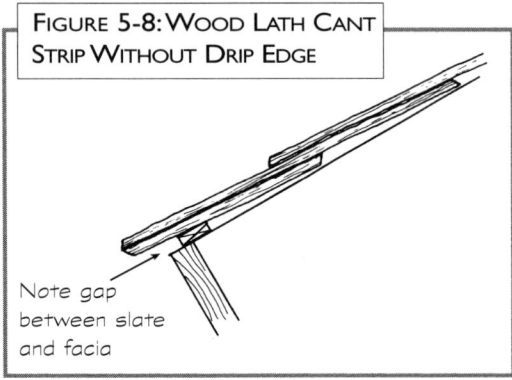
Figure 5-8: Wood Lath Cant Strip Without Drip Edge
Note gap between slate and facia

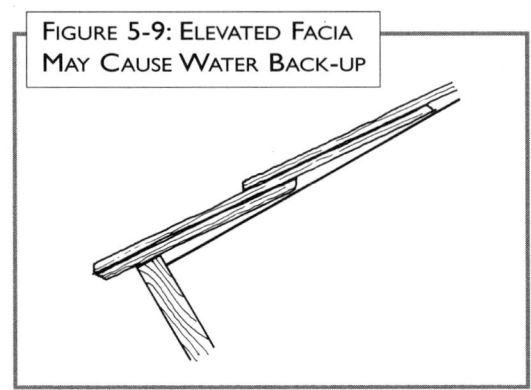
Figure 5-9: Elevated Facia May Cause Water Back-up

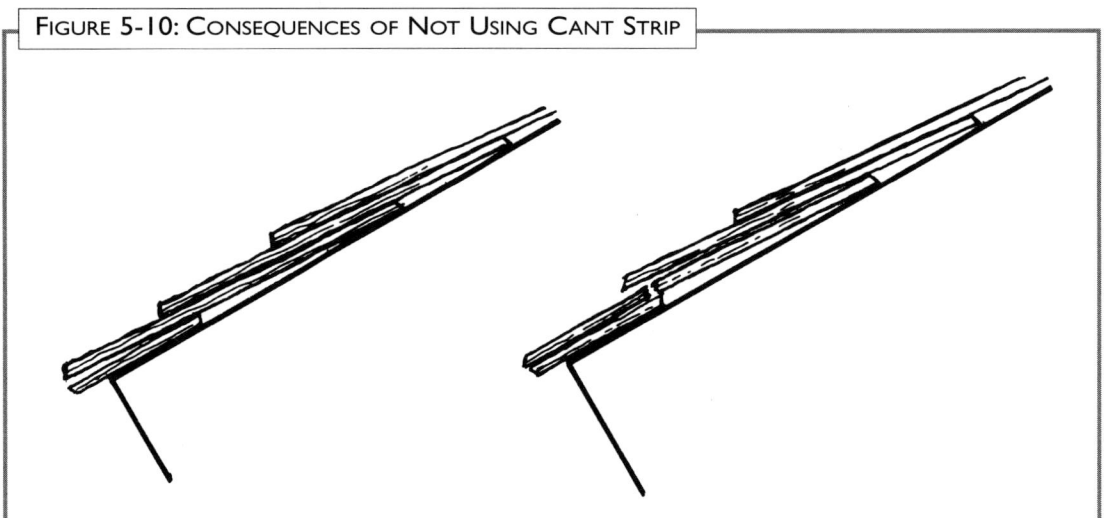
Figure 5-10: Consequences of Not Using Cant Strip

is missing, as illustrated in Figures 5-8 and 5-9.

Slate is a rigid material and, if not properly supported, can eventually break. If you are working with standard-thickness slate that is approximately 1/4 inch thick, then the starter slate length is at least 3 inches longer than the exposure length of the first full course. If the starter slate were to be nailed flat directly to the roof deck, with the first full course nailed flat directly on top of the starter, one of two conditions would occur:

1) Either the top of the first full course would be elevated off the roof deck 1/4 inch in order to have the butts of the first course and the starter course lie tight together; or

2) If the first full course is nailed tightly enough to draw the head of the first course down to the roof deck, a gap will develop at the base between the first course and the starter course. Eventually this will cause the first full course to break.

To determine cant strip thickness, simply match the thickness of the slate that you are working with. Therefore, standard-thickness slate (1/4 inch) will require a 1/4-inch thick cant strip under normal conditions. However, eave finish details can vary. If the installer is not certain, he should lay out a piece of starter slate in its proper location at the eaves and then lay a piece of field slate on top of the starter. Elevate the butt of the starter until the head of the field slate is touching the roof deck and is fully supported by the starter slate. Measure the gap from the eave to the back side of the starter slate to determine the thickness of the cant strip. This technique will almost always be used to determine the cant strip thickness for thicker slate.

The cant strip should be installed as close to the eave as

Roof Construction and Preparation

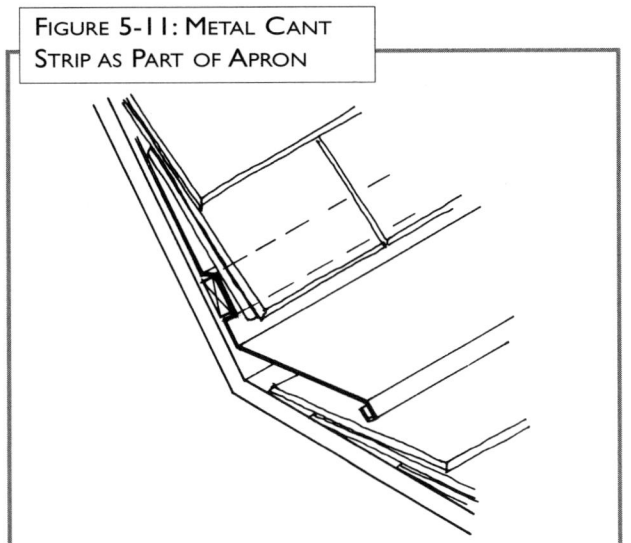

FIGURE 5-11: METAL CANT STRIP AS PART OF APRON

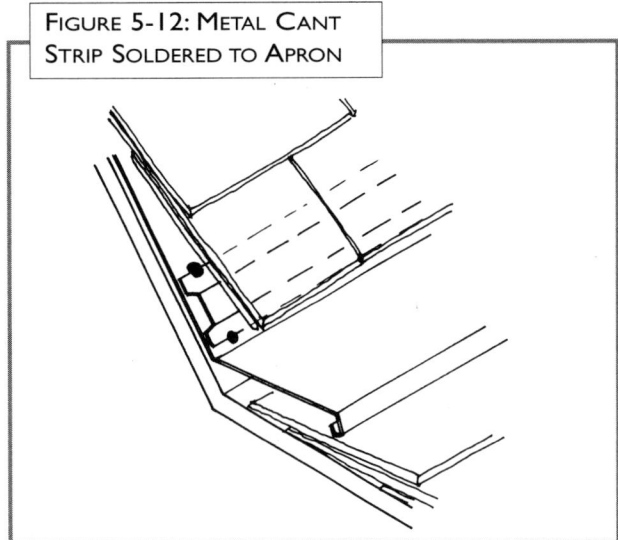

FIGURE 5-12: METAL CANT STRIP SOLDERED TO APRON

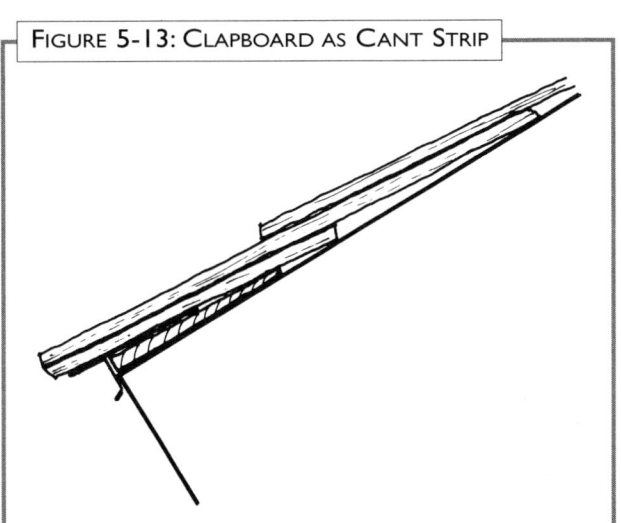

FIGURE 5-13: CLAPBOARD AS CANT STRIP

possible to do its job properly. If the eave detail does not permit placement of the cant strip close to the edge, the starter strip could actually be elevated too high; but this is rarely, if ever, the case.

As a rule, the top of the cant strip should not be more than the thickness of the field slate at a point 3 inches up from the bottom of the starter and first course. Keep in mind that its function is to substitute for the head of another piece of slate.

Wooden Cant Strip. Cant strips are most often wooden and are nailed to the edge of the first roof board into each rafter. The wooden cant strip is often a mason's wood lath, which is 1/4 inch thick, 4 feet long, and 1 1/2 inches wide. Most large lumber yards carry wood laths in bundles. They are referred to by several different names (such as wood lath, mason's lath, and furring strips).

Other materials used for wooden cant strips are clapboards (usually seconds) and cedar shim shingles. (See Figure 5-13.) The taper of the clapboards and shim shingles provide good support for the starter. However, because of their long taper, they will sometimes elevate standard-thickness slate too high out of the proper roof plane.

Metal Cant Strip. Cant strips can also be fabricated out of metal. This is commonly done with the same type of metal as that which is used on the rest of the project.

Chapter 5

The metal cant strip is a nice detail in terms of quality and longevity, and it is especially helpful when working with built-in gutter details or elaborate eave flashing details. However, if thick and heavy slate is being used, the weight of the slate could crush an unsupported metal cant strip. If possible, provide blocking underneath, or use heavy flashing and bracing.

The metal cant strip can be installed several different ways. One way is to fabricate it into the drip edge detail, which is then nailed into place. A second method is to fabricate the cant strip in a configuration that will allow it to be nailed in place. The third method is to solder the cant strip onto the eave flashing detail.

Slate Cant Strip. Some projects will specify a slate cant strip. The thickness of the stone cant strip is calculated the same way as the wood and metal cant strips. This piece of slate should be no more than 3 inches wide. If the slate cant strip is more than 3 inches wide, it will begin to elevate the starter slate out of the proper roof plane.

The slate cant strips are sometimes cut with a saw to avoid breakage. This is a very difficult detail to install properly at the eaves. The location of the nail holes can become a problem in relation to proper eave overhang.

Nailers

In some projects, roof preparation includes the installation of wooden nailers. **Hip nailers** are only required when using a saddle, metal, or tile hip cap detail. Their function is to provide support and a nailable surface for the hip cap. The nailer should be three times the thickness of the slate being used, and it should be installed on either side of the hip and as close to the hip line as possible. (See Figure 5-14.)

Finishing course nailers are needed when the finishing course is not fully supported by the last full course of field slate. (See Figure 5-15.) Twelve-inch and 14-inch long slate will not require finishing course nailers for standard application. (See Figure 5-16.) If a vented ridge is being used, a finishing course nailer will not be required unless the slate is 20 inches long or longer. (See layout instructions in Chapter 6 for further explanation.)

If a finishing course nailer is needed, it should be the thickness of the slate. The finishing course nailer is installed at the head of the last full piece of field slate. This nailer is not installed until the roof layout has been completed. Some contractors will wait until the ridge area is being slated to install the finishing course nailer.

Most ridges require a **ridge nailer**. The only exception is when using very small slate where the finishing course functions as a strip ridge cap. The function of the ridge nailer is to provide support and a nailable surface for the ridge cap.

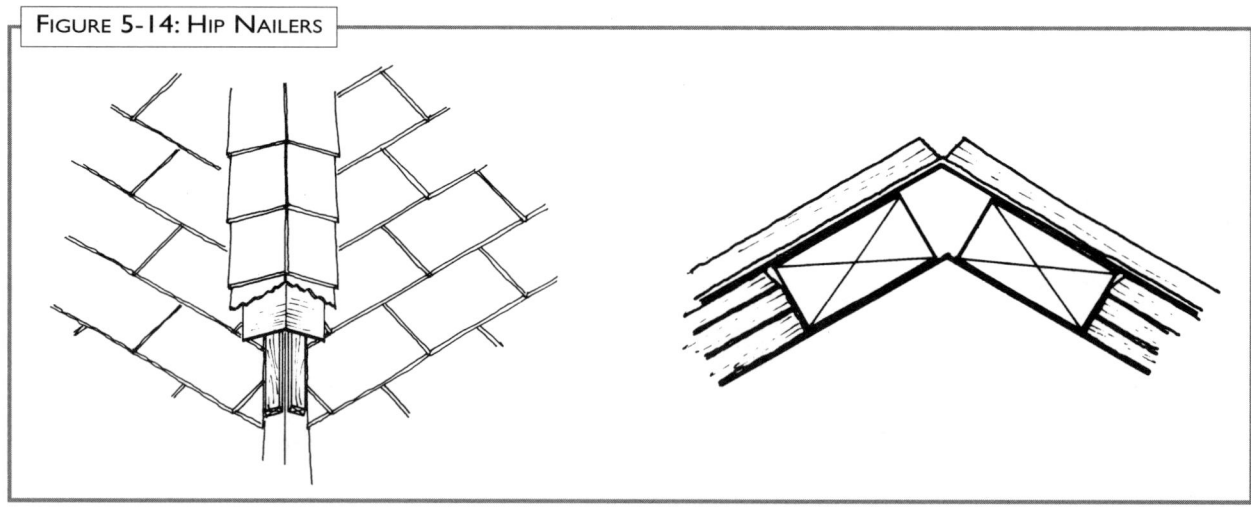

FIGURE 5-14: HIP NAILERS

Roof Construction and Preparation

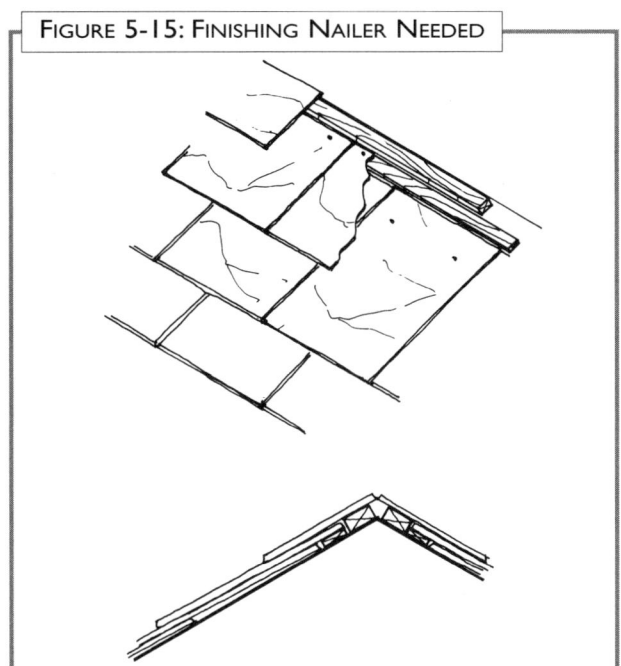

FIGURE 5-15: FINISHING NAILER NEEDED

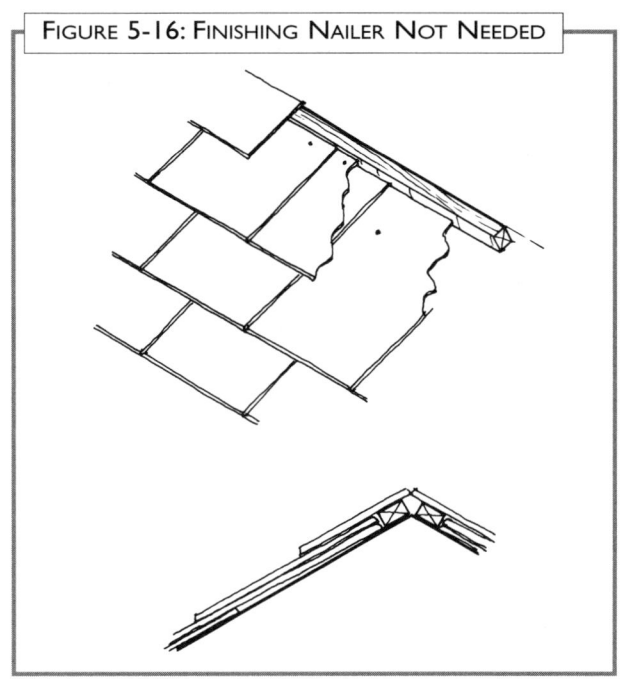

FIGURE 5-16: FINISHING NAILER NOT NEEDED

This ridge nailer should be two times the thickness of the slate being used and should be installed on either side of the ridge and as close to the ridge line as possible. Like the finishing course nailer, the ridge nailer is not installed until the roof layout has been completed. Some contractors will wait until the ridge area is being slated to install the ridge nailer. (See Figure 5-15 and Figure 5-16.)

Base Flashings

One of the most important steps in roof preparation is the installation of base flashings. Base flashings include drip edge, open valley flashing, and some gutter details. A proper slate roof specification and installation should include drip edge. For standard-thickness slate, the drip edge can be installed over the cant strip. It is then stripped in with modified bitumen. (See Figures 5-6 and 5-17.)

The top of the drip edge should not extend beyond the nail holes of the starter slate. This should be noted at all eave details. It is common to see specifications that call for base flashings to extend farther up the roof than the starter itself, but it makes no sense to go up farther than the first penetrations.

Open valleys should be installed before the roof layout begins. It will be helpful to snap a valley chalk line to lay the valley to. The cant strip should end at the outside edge of the valley flashing. This will prevent the cant strip from elevating the valley flashing above the roof deck.

The cant strip fasteners should never penetrate the valley flashing. Cant strips are sometimes specified up the entire length of open valleys, but they are not necessary. (For more information, see "Open Valleys" in Chapter 8.)

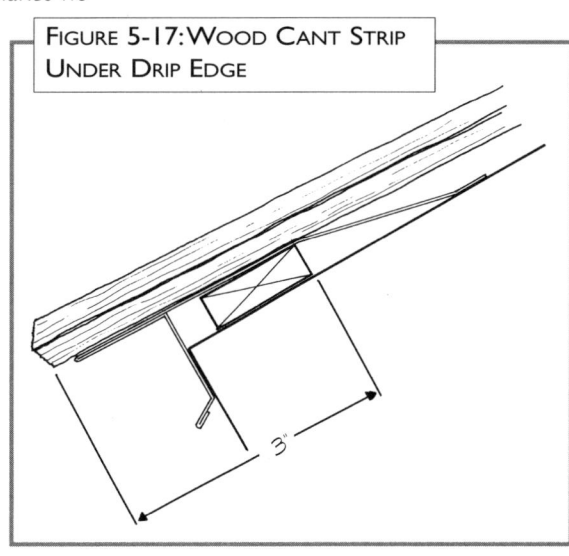

FIGURE 5-17: WOOD CANT STRIP UNDER DRIP EDGE

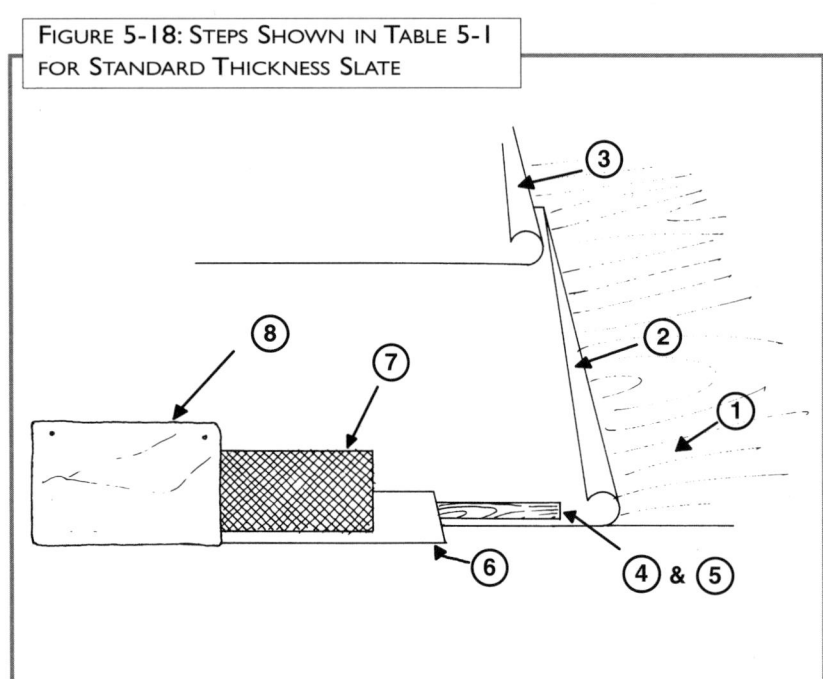

FIGURE 5-18: STEPS SHOWN IN TABLE 5-1 FOR STANDARD THICKNESS SLATE

Roof Preparation Summary

For every slate roof project, roof preparation begins with the inspection of the roof deck and installation of underlayment. However, the sequence of steps following the installation of underlayment may vary. Table 5-1 shows the proper sequence for roof preparation with no drip edge specified, with standard-thicness slate, and with heavy slate.

TABLE 5-1: ROOF PREPARATION SEQUENCE

No Drip Edge Specified	Standard-Thickness Slate	Heavy Slate
1. Inspect roof deck.	1. Inspect roof deck.	1. Inspect roof deck.
2. Install ice & water shield around perimeter and up valleys.	2. Install ice & water shield around perimeter and up valleys.	2. Install ice & water shield around perimeter and up valleys.
3. Install underlayment.	3. Install underlayment.	3. Install underlayment.
4. Install the cant strip.	4. Install the cant strip.	4. Install base flashings.
5. Install nailers where accessible.	5. Install nailers where accessible.	5. Install the cant strip.
6. Install base flashings.	6. Install the drip edge.	6. Install drip edge.
	7. Strip in the drip edge with ice & water shield.	7. Strip in the drip edge with ice & water shield.
	8. Install starter slate.	8. Install starter slate.

PART 2
INSTALLATION

The second part of this book has seven chapters about installation and repair techniques for slate roofing.

CHAPTER 6   **ROOF LAYOUT: PAGE 83**

An explanation of how to lay out a slate roof before beginning to install the slate

CHAPTER 7   **SLATE TOOLS: PAGE 95**

A brief description of the hand tools needed for installing slate

CHAPTER 8   **INSTALLING SLATE: PAGE 101**

Procedures and techniques for installing slate

CHAPTER 9   **FLASHINGS: PAGE 139**

Procedures and techniques for installing flashings

CHAPTER 10   **REROOFING: PAGE 151**

Special considerations for reroofing projects

CHAPTER 11   **REPAIR, MAINTENANCE, AND TROUBLESHOOTING: PAGE 153**

Procedures for repairing and maintaining slate roofs

CHAPTER 12   **ADVANCED SITUATIONS: PAGE 169**

Techniques for handling challenging roofing situations

CHAPTER 6
Roof Layout

The installation of slate should not begin until the entire roof has been laid out to the greatest degree possible. On surfaces which are not accessible without roof scaffolding, only the very bottom and the very top of the roof can be laid out initially. Successive layout will progress as roof scaffolding is installed. This is a critical step in slate roofing because of the nature of the material and how it is installed. Slate shingles are not as flexible or adjustable as other materials, such as composition shingles. This chapter explains the steps for completing the roof layout.

Before you begin, assume that the roof structure is not perfectly square and that the valleys and hips are not perfectly straight. This is by no means intended to be a criticism of the carpenters and framers who built the building. In many cases, slate is installed on existing buildings that need to be reroofed. Most existing buildings will have moved and settled somewhat over the years, creating a situation where the roof lines are no longer perfect. As a rule, it is safest to assume that no building, new or existing, is perfectly square.

The roof layout involves snapping horizontal chalk lines to align the top of each course of slate shingles. Vertical lines are not always required. All lines to be followed at a later date should be laid out using a chalk that will not wash away. Red, yellow, and some orange chalks are said to be permanent, but check the label on the side of the bottle before starting. Blue chalk is not a permanent color and should generally not be used for the roof layout; however, since it will wash away, it may be appropriate when snapping lines on the surface of the slate itself.

Slate roof shingles vary in length, so there will almost always be slight variations in the alignment of the slate butts. These variations are usually minimal, they are rarely seen when looking at the roof from the ground, and the uneven texture of the slate surface usually hides whatever variations may occur. However, some inexperienced project management people may panic when they observe this variation from the roof level.

Chapter 6

As a rule, when laid to chalk lines at the heads of the slate, the butts should not vary more than 1/4 inch. If the owner is going to see this roof from an unusual angle, he may specify that the installer snap chalk lines to align the butts of the slate instead of the heads. This method will be much more time-consuming and therefore more costly.

The steps for roof layout presented below are most appropriate for random-width slate. Layout procedures for single-width and graduated slate roofs vary somewhat, as explained later in this chapter.

Step 1: Starter Slate

Most slate roofs are installed with a 1 1/2-inch overhang at the eaves and at the gable ends. To establish this 1 1/2-inch overhang at the eaves, a starter slate chalk line needs to be laid out. The length of the starter slate is determined by adding the exposure (E) of the field slate and the headlap (H) of the first course of slate and rounding up to the next full inch:

(E) + (H) + 1/2 inch

For example, if you are using 18-inch long field slate, the exposed surface area (E) will have a reveal of 7 1/2 inches, with (H) being 3 inches; therefore, the starter slate will be 11 inches long.

Most quarries will provide you with starter slate upon request. (See Chapter 3 for more information about starter slate.)

Once the starter slate size has been determined, the starter layout line can be snapped on the underlayment. Assuming that the slate is 18 inches long, the proper starter size is 11 inches. Use one of the following methods to lay out the starter course line.

Starter Slate Layout without Drip Edge

1) Take the length of the starter course and subtract the overhang (usually 1 1/2 inches): thus, 11-inch starter − 1 1/2-inch overhang = 9 1/2 inches. This is the distance to measure up from the eaves to mark the head of the starter course.

Scratch a reference point in the underlayment at both ends of the roof surface at this point. This will create a layout line parallel to the eave. Scratching reference points in the underlayment may help later if mistakes were made. The 11-inch long starter slate will then hang 1 1/2 inches over the trim detail.

If you are working on a project where the eave trim details have not yet been completed, be certain you are measuring 1 1/2 inches past the actual eave edge. This location can be specified by the carpenter. Communication with the carpenters is critical to good layout.

2) A simple way to lay out this 1 1/2-inch overhang is to allow your tape measure to overhang the eave 1 1/2 inches and mark the underlayment at the 11-inch mark on the tape.

Starter Slate Layout with Drip Edge

The slate should extend 1 1/2 inches beyond the eave regardless of drip edge configuration. Some drip edge configurations will have this 1 1/2-inch overhang fabricated into them. If this is the case, it is not necessary to overhang the 1 1/2-inch drip edge overhang by an additional 1 1/2 inches. Starter slate installed with this drip edge configuration will usually be lined up flush with the drip edge overhang.

If the drip edge does not have a 1 1/2-inch overhang configuration, then the starter slate will be installed with a 1 1/2-inch overhang as if there is no drip edge.

Step 2: First Full Course of Field Slate at Eave

The next step after laying out and snapping the starter course line is to lay out and snap the line for the first full course of field slate. The butt end of the starter slate and the butt end of the first full course of field slate should align. To determine the layout of the first full course of field slate, measure up from the butt of the starter course the length of the field slate to be used. Be sure to account for the overhang of the slate when measuring for your layout.

Roof Layout

An alternate method for determining the placement of the layout line for the first full course of field slate at the eave is to measure up from the layout line for the starter slate. The distance to measure up from the starter slate layout line is determined by subtracting the length of the starter course (Ls) from the length of the first full course of field slate (L). *For example, 18-inch slate has a starter course length (Ls) of 11 inches, so 18 inches − 11 inches = 7 inches. The first full course layout line will be 7 inches above the starter course layout line.*

Step 3: Last Full Course of Field Slate at Ridge

The next step is to lay out a line that is parallel to the ridge. This is the layout line for the last full course of field slate. To determine the point at which to lay out the last full course, take the length of the slate exposure (E) and subtract 4 inches. This equals the distance (Dr) to measure down from the actual ridge. The actual ridge is the point at which the two roof planes intersect, not necessarily where the plywood ends. (See Figure 6-1.)

$$Dr = E - 4 \text{ inches}$$

Measure this distance (Dr) down from the actual ridge on both sides and snap the last full course layout line. This line will be parallel to the ridge. Using this formula will prevent an unusually large or small exposure at the ridge. For example, 18-inch long slate = 7 1/2-inch reveal − 4 inches = 3 1/2 inches, so you will want to strike a line parallel to the peak at a point 3 1/2 inches down from the actual peak.

You will now have chalk lines snapped at the eaves for the starter course and first full course, which are parallel to the eaves, and a chalk line at the peak for the last full course, which is parallel to the ridge.

Step 4: Valley and Hip Lines

While you are on the roof laying out the starter and last full course lines, take the time to snap lines up the length of the hips and valleys. This will give you a straight reference line to follow and in

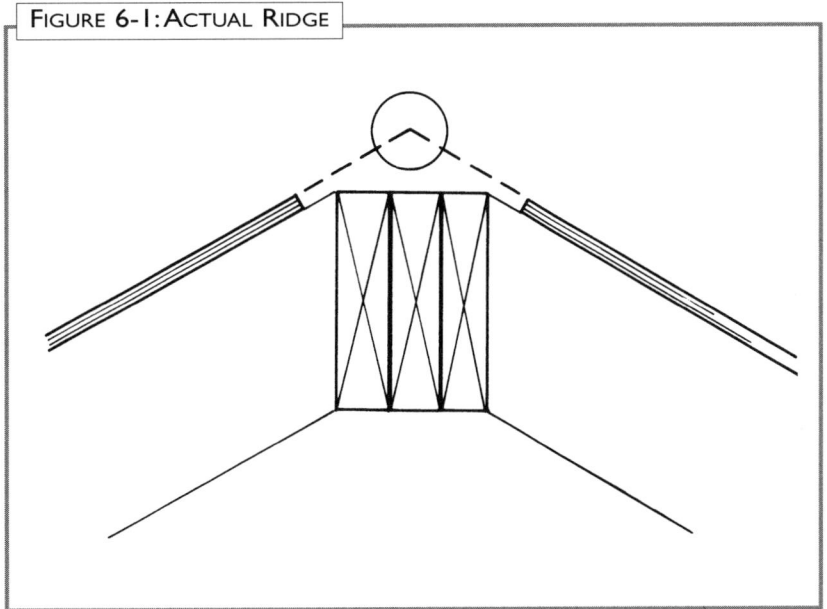

FIGURE 6-1: ACTUAL RIDGE

TABLE 6-1: FINISH SLATE LENGTH

Length of Last Full Course of Slate at Ridge	Finishing Slate Length	Exposure
10"	none	
12"	8"	4 1/2"
14"	8"	5 1/2"
16"	10"	6 1/2"
18"	12"	7 1/2"
20"	14"	8 1/2"
20"	16"	9 1/2"
24"	18"	10 1/2"

Chapter 6

some cases will assist in the layout of the field course lines. If you are working on a roof that has all the same roof pitch, horizontal reference lines snapped on adjacent roof surfaces should intersect at the valley and hip lines. This is especially helpful to know if you are laying out a roof with closed valleys and mitered hips.

This will not apply on a roof with open valleys. If the valleys are open, the valley flashing should already have been installed.

Snap valley reference lines on top of the open valley flashing to which the slate will be cut. (See *Open Valleys* in Chapter 3 and Chapter 8.) These lines in the valley will provide a reference point that the horizontal field lines should be snapped to or past. Remember these lines are closer to the center of the valley as you get closer to the ridge.

Step 5: Laying out the Field of the Roof for One-Length Slate

At each end of the roof, measure the distance between the first full course layout line and the last full course layout line. This distance can then be divided by the exposure of the slate being used to determine the number of courses to be laid out. Rarely will the roof layout divide equally into even increments.

When the roof layout does not work out to even course increments, the course layout must be adjusted to account for the extra area. In order to make this adjustment, the installer should always reduce the exposure, which will add one course to the roof layout. Never stretch beyond the recommended exposure.

For example, consider a roof that on the blueprints is 20 feet or 240 inches from eave to peak. You are using 18-inch long slate, so the starter slate is 11 inches.

Snap a starter course line at a point 9 1/2 inches up from the eaves to allow for a 1 1/2-inch overhang. Then snap a first full course line 7 inches up from the starter course line. Next snap a last full course line at a point 3 1/2 inches down from the actual peak. The area that still needs to be laid out is 220 inches. If you divide the remaining 220 inches of field distance by 7 1/2 inches (the proper exposure for 18-inch slate at 3-inch headlap), the result is 29.33 courses, or 29 full courses with 2 1/2 inches of field area remaining. The exposure will have to be reduced to convert this 2 1/2-inch discrepancy into a full 7 1/2-inch course. It is not necessary to divide this discrepancy into every course; the reduction can be accomplished by reducing the exposure of the first 10 courses to 7 inches. The remaining courses will then all lay out to 7 1/2 inches.

This example illustrates the importance of laying out the entire roof prior to installing the slate. Try to make course adjustments near the lower portion of the roof where the additional headlap is most beneficial.

A further complication in laying out field slate is that most roofs are not square. In the field, the same roof with a 220-inch field length on the left side may actually have a 218 1/2-inch field length on the right side. This discrepancy will require that 13 courses on the right side be laid out at 7 inches to compensate for the difference. This slight course adjustment is not noticeable.

While the exercise above is great in theory, it may not always be that easy. Steep roof pitch may make layout very difficult. If this is the case, snap as many lines as can be reached safely at both the eaves and the peak. Adjustments can then be made accordingly as you progress up slope and more accurate measurements can be obtained.

Step 6: Reference Lines, Reference Lines, Reference Lines

When an obstacle such as a dormer or chimney is encountered, reference lines need to be snapped. These lines have several uses.

The first and last full course lines are both reference lines. They provide reference points with which to tie all the roof surfaces together. When you encounter an

Roof Layout

obstruction such as a dormer, you cannot continue snapping lines. In this case, you must snap regular reference lines as close as possible to the bottom and the top of the obstruction. The layout on either side of the obstruction is then divided evenly between these top and bottom reference lines.

For example, consider a roof with a dormer. Using the proper layout for field exposure, snap as many continuous lines as possible up to the base of the dormer. Then go to the top of the roof and lay out the roof from the last full course reference line down to the last continuous line before reaching the top of the dormer. On each side of the dormer, you will need to lay the same number of slate courses. However, depending upon how square the roof is, the spacing between courses may not match exactly.

The important point is that you must tie into the same reference line as you progress over the top of the obstruction. On a roof with a large gable addition extending out from the middle of an elevation, you may need to lay reference lines down from the peak only.

There can never be too many reference lines. Unlike composition shingles where reference lines are sometimes laid every 5 courses (if at all), with slate you need to have lines to follow. Should you run into a situation where you need to

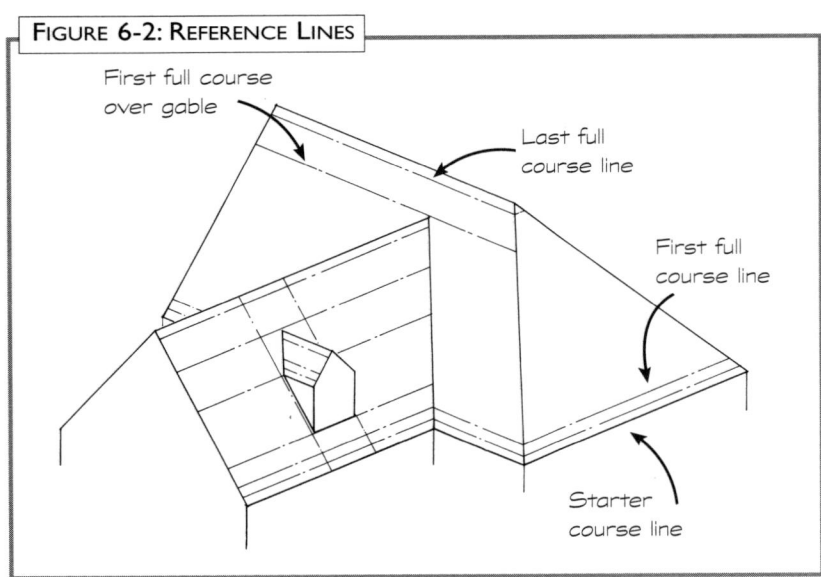

FIGURE 6-2: REFERENCE LINES

lay a temporary line, use a different color chalk.

It is common to make mistakes during the layout process. One common mistake is to scratch a reference point on the underlayment that is off by an inch in either direction. You will need to snap another line to correct the reference point. To make sure that the installers use the correct line, either scratch a mark through the wrong line with a nail or re-snap the correct line with a new heavy chalked line. It helps to scratch as many progressive reference points as possible from a given location. Then, before snapping any new layout reference lines, measure to the nearest reference line at the upper or lower end of the roof to verify that each side is still progressing as planned. Making and correcting reference lines is much easier and cheaper than removing slate because courses do not line up or are crooked.

Top Lines vs. Bottom Lines

A slate roof is laid out to top lines (aligning the tops or heads of slate in each course). This allows the roofer to lay out the entire roof prior to installing slate. This helps eliminate any confusion with layout as the roof progresses. Another advantage of having all of the lines laid out is that the installer can work on more than one course at a time. Typically, each mechanic will work three to four courses at a time across the roof.

If the installer chooses to snap bottom lines (to align the bottoms or butts of slate in each course), then an entire horizontal course must be completed before the next horizontal course line can be snapped. Some projects will be specified such that every fourth or fifth course must be laid to a bottom line. The logic behind this approach seems to have something to do with keeping the butt ends

87

Chapter 6

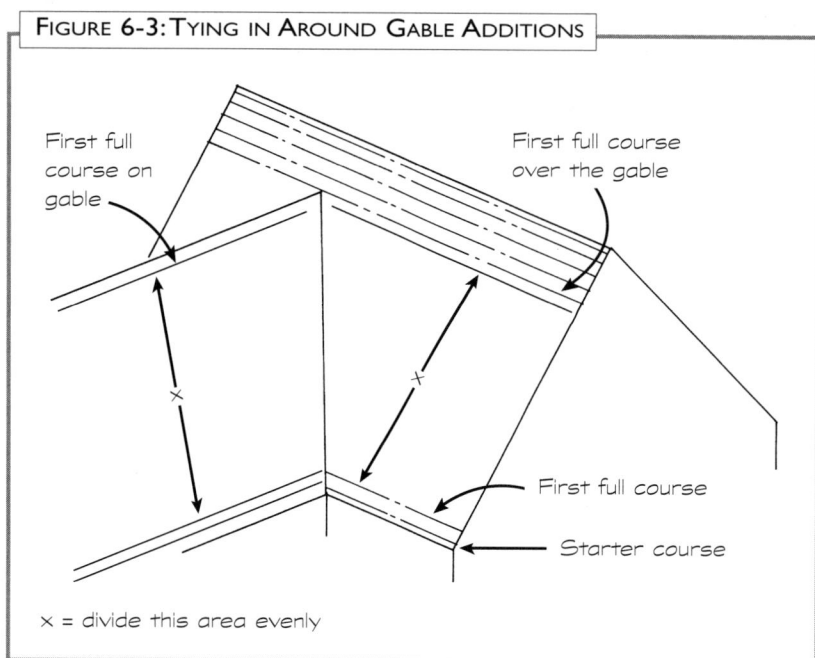

FIGURE 6-3: TYING IN AROUND GABLE ADDITIONS

of the slate in a straight line. The slate will vary in length from piece to piece sometimes as much as the thickness of the material. This is understandable when considering how the material is manufactured.

If a slate roof can be observed from a perspective which allows the observer to look down the length of a course, such as from a dormer window or scaffolding, the variations in butt alignment will be obvious. These variations are to be expected, but they should not exceed the thickness of the slate. This appearance adds to the texture of the roof and its appearance of authenticity.

Laying slate to bottom lines will slow down installation. However, if this approach is required, be sure to use a blue chalk or chalk of another color that will wash off.

Single-Width Slate Roof Layout

All of the layout described above applies to random-width slate roofs. Unless the roof is also graduated in length, the installer can now begin laying slate.

For a slate roof of shingles with one length and one width, you will also need to snap vertical reference lines. If you are working on a roof with gable ends, start from the gable and work in. Again, you will be working with a 1 1/2-inch gable overhang (same as the eave overhang). Determine the width of the slate supplied. It is best to measure several pieces to get an average width. Assuming that you are working with slate that is 18 inches by 10 inches, you will need to snap a vertical reference line that will offset each progressive course by 5 inches. Keep in mind that you will need a vertical reference line at the gable end so that the 5-inch piece of slate will overhang the gable 1 1/2 inches. In this case, you would measure in from the gable 3 1/2 inches at top and bottom, scratch your reference point, and snap a vertical line to those points. Drip edge profiles could affect vertical line placement. Be sure to take this into consideration.

Some installers insist upon using an exterior string line as a reference line up the gable end. This is done by attaching a string to a block at either end of the gable. The block should be 1 1/2 inches thick so as to hold the string the proper distance away from the gable trim. This usually works fine and will provide for a very straight reference line up the gable end. However, if expensive gable trim is being used, this is not recommended. An alternative to this string-and-block method is to use a drip edge profile that the slate is laid to. In most cases, the gable trim will be straight enough that the installer can mark 1 1/2 inches on the face of each gable piece of slate and set the slate consistently up the gable.

The gable reference line is just that, a reference line. Its purpose is to keep you from getting too far off course with your vertical joint alignment. You will want to snap vertical reference lines approximately every 5 feet across the

Roof Layout

roof surface and in even increments based upon the width of the material that you are using,. Vertical reference lines should also be snapped on either side of roof obstructions. If you begin to stray off these vertical lines by more than the thickness of the slate, you will need to cut the slate lengthwise to bring it back in line. Hip and valley cuts will be made accordingly as the evenly staggered courses approach these areas. This is different from the procedure using random-width slate.

Try to avoid snapping all of the vertical lines until you are sure that the layout is correct.

When single-width slate is used on a roof with obstructions, vertical lines will have to be laid out on either side of the obstruction. First, lay out a vertical line on one side of the obstruction. Then measure over in even increments of the slate being used and create a parallel vertical line on the opposite side of the obstruction. This will assure that vertical joints in the slate will be symmetrical as they join together over the top of the obstruction. The layout here is critical. Be sure both sides are tying into the same horizontal line level. (See installation techniques in Chapter 8.)

Perpendicular Line Layout

Unlike random-width slate, slate of one width involves more layout and imposes more restrictions. Vertical reference lines needed for single-width layout are relatively straightforward for a roof with a gable end to start from. However, if you are working on a hip roof or a roof with an obstruction in the middle, such as a dormer or a large gable, laying out the vertical lines will be more difficult but absolutely necessary. Use the following method to establish a vertical line on a hip roof section.

Find a point somewhere near the middle of the roof section that you are starting. The horizontal layout lines should have already been snapped.

1) Scratch a reference point on the starter course line at the approximate center of the roof.

2) Now measure out approximately 3 feet on each side of the center reference point. If you don't have enough distance to reach 3 feet, that's okay. However, it is critical that the distance measured on either side of the center reference point be exactly the same and on the same reference line as the center mark.

3) Now, find a piece of wood approximately 2 times as long as the distance measured out from the center reference point and drive a nail into each end so that it protrudes through the wood.

4) Drive the nail on one end of the wood into one of the reference points to the side of the center reference point. Now, with the nail sticking through the other end of the wood, scribe an arc on the underlayment above the center reference point.

5) Follow the same procedure on the wide reference point on the opposite side. The arc scribed in the underlayment from the second reference point will create an X directly above the center reference (at the starter course).

6) Snap a line from the cen-

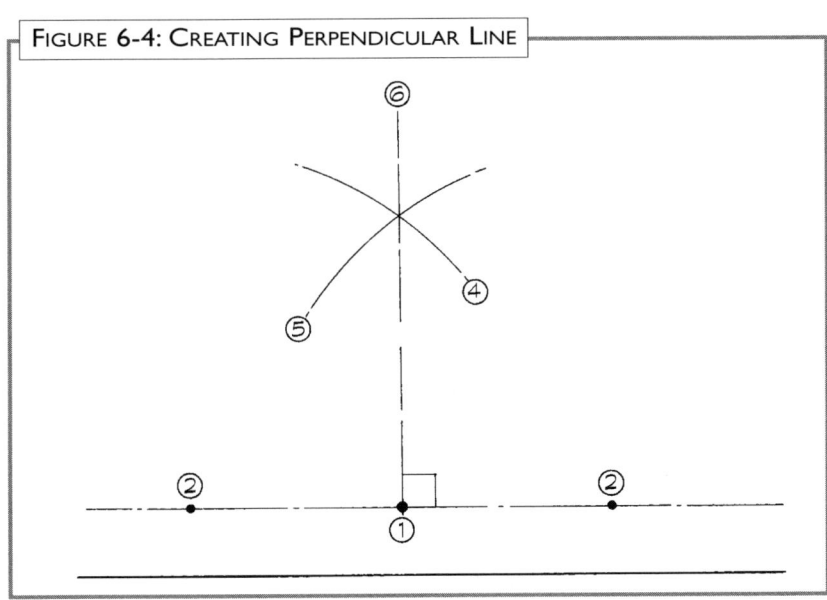
FIGURE 6-4: CREATING PERPENDICULAR LINE

Chapter 6

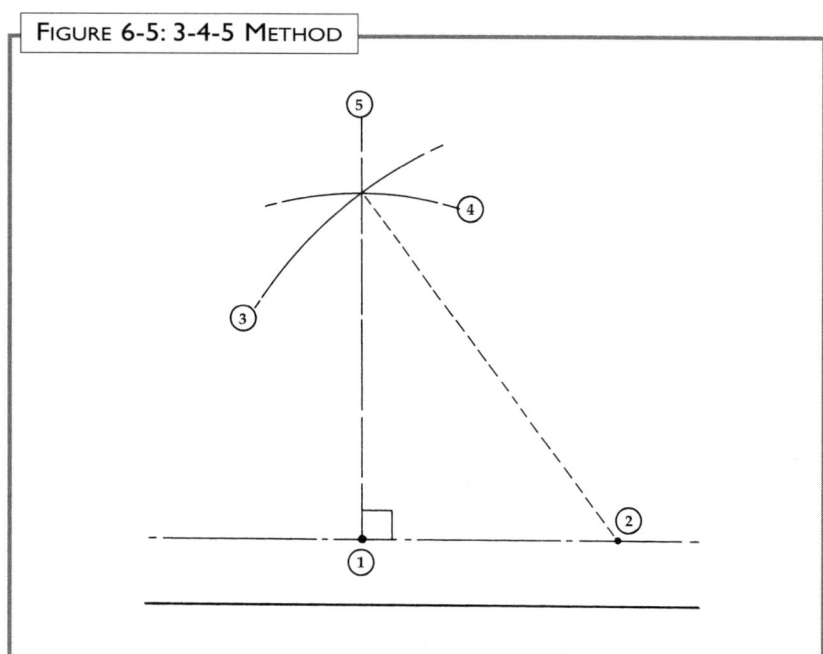

FIGURE 6-5: 3-4-5 METHOD

ter reference point near the eaves to the ridge that connects the center reference point through the X above. This line, if done with reasonable care, will be perpendicular to the starter course line.

Once you have established the first perpendicular point, follow the layout procedure as described for the gable layout. Approximately every 5 feet (based on an even increment of the slate width that you are working with), snap vertical reference lines over the entire roof surface (Figure 6-4).

Some installers prefer to use the **3-4-5 method**, which uses right-triangle geometry to create a perpendicular line. With this method you: **1)** pick a center point on the starter line (or ridge line); **2)** measure to one side a distance of 3 feet, and scratch a reference mark. **3)** From that point scribe an arc at 5 feet (in the area approximately above the center reference line below). **4)** Finally, scribe an arc 4 feet up from your center point. Where the 5-ft line and the 4-ft line cross, an X will be created. **5)** By snapping a line through your center point and the X created by the arc intersection, you will create a line perpendicular to the starter line (Figure 6-5).

Graduated Slate Roof Layout

In most cases, a graduated slate roof will be more difficult to lay out than a roof using random-width or single-width slate. Most quarries will provide a roof layout for graduated slate, but don't expect the quarry to provide this layout before the order is placed. (Quarries that can provide this information will not want to provide such revealing plans to those who can't.) In addition, don't expect the roof to lay out precisely as the quarry's layout shows. There are always changes on site that either don't show on the plans, weren't anticipated, or were simply overlooked. To carry out a graduated slate roof project properly, all parties must understand that it is more costly, difficult, and time-consuming than other projects.

Usually with a graduated roof, there are 3 to 5 courses of the longest slate at the eave area and 7 to 9 courses of the shortest slate at the ridge area. Using the principles of layout provided above, lay out the starter course, first full course of field slate, and last full course of field slate. Next, lay out all the courses for the longest slate at the eave and the shortest slate at the ridge. Be sure to check the roof to see how far out of square it is by measuring between these reference lines. These lines will not help you determine the number of actual courses yet, but they will tell you how far off the lines are from being parallel so that you can begin to make adjustments.

For example, assuming you have 9 courses of 12-inch material to lay out at the ridge, deduct one for the finishing course at the ridge. Then lay out the last full course at 1/2 inch from the peak (exposure 4 1/2 inches − 4 inches = 1/2 inch), followed down the

Roof Layout

roof slope by seven additional courses at 4 1/2 inches. Assuming that there are 3 courses of 24-inch long slate at the eaves and using the 3-inch headlap formula, these courses will have an exposure of 10 1/2 inches.

After these three courses are laid out, you will graduate down in size from 24 inches long with a 10 1/2-inch exposure to 22 inches long with a 9 1/2-inch exposure. You will need to snap the first line for the 22-inch long material at 8 1/2 inches above the last 10 1/2-inch reference line (see Figure 6-6). This will allow the first 22-inch long course to lay over the last 24-inch long course in a manner that still shows a 10 1/2-inch exposure on the last row of 24-inch long slate while maintaining the 3-inch headlap. Improper layout of these transition courses will result in improper headlap.

The next reference point will then be scratched at 9 1/2 inches, which is standard for laying 22-inch slate with a 3-inch headlap. Let's say there are to be 5 courses of 22-inch slate, one at 8 1/2 inches followed by 4 at 9 1/2 inches. The next graduation will be to 20-inch long material, which lays at 8 1/2 inches. Therefore, the first course of 20-inch slate will follow the last course of 22-inch slate at a reference point of 7 1/2 inches. Continue laying out the balance of the 20-inch long material at 8 1/2 inches. The first course of 18-inch material will lay over the 20-inch material at 6 1/2 inches. Stop there for the layout at the lower end of the roof.

Go back to the top of the roof. You will now have to work the layout in reverse from the eave. The slate size previous to the 12-inch long material is 14 inches long. The 14-inch long material is laid at 5 1/2 inches. Therefore, in order to graduate from 5 1/2-inch exposure to 4 1/2-inch exposure, you will need to lay out a course at 3 1/2 inches long as you work back down from the ridge. Then, assuming that there are 9 courses of 14-inch long slate, you would mark down from the 3 1/2-inch reference point 8 courses of 5 1/2 inches followed by the ninth course at 4 1/2 inches. The 4 1/2-inch increments allow for proper graduation into the 16-inch material.

While working on the layout, frequently measure between upper and lower reference lines and make adjustments for any parallel variations. Even though there are two slate lengths left to lay out, the reference lines should already have been corrected to make them parallel.

Determine which of the last two sizes you have the most of and lay out that length material last. Assuming that there is more 18-inch material than 16-inch material, finish laying out the 16-inch material by measuring down from the 4 1/2-inch graduation point in 6 1/2-inch increments as specified.

The last graduation reference point will be a 5 1/2-inch increment to allow proper graduation to 18-inch slate. Since you already have the reference line laid out for the first course of 18-inch long slate and you have worked down the roof to the last course of 18-inch long slate, simply divide the remaining roof area by 7 1/2 inches to determine the final layout.

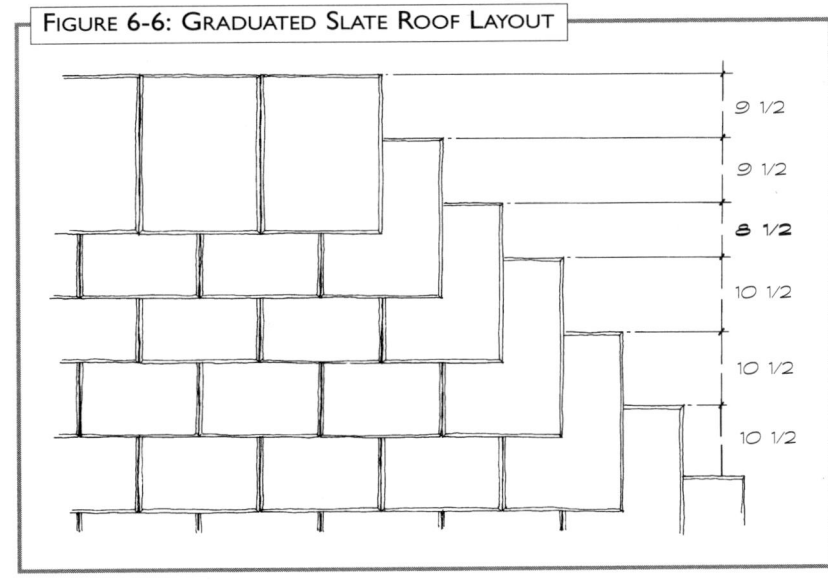

Figure 6-6: Graduated Slate Roof Layout

Chapter 6

Obviously, course adjustments will need to be made within the last slate size applied. It will probably not lay at exactly 7 1/2 inches. Remember to shrink and not stretch these courses. Try to lay the roof out so that the slate size of which you have the most material is laid out last. This will allow for the most adjustment for exposure. If there is not enough area within this one size, it is okay to adjust other sizes as necessary.

One of the most difficult aspects of laying out a graduated roof is in adjusting the course layout. If the last slate size laid out is such that the slate exposure has to be reduced significantly, it may be easier to use an additional course of either the smaller slate above or the larger slate below. Undoubtedly, adjustments will have to be made to the layout that neither the framer, architect, quarry draftsman, nor installer can anticipate.

Laying out a graduated slate roof can be confusing. It is highly recommended that reference marks be established and checked before snapping all of the lines. This type of roof should be laid out by an experienced installer.

Change of Pitch

In many situations, the roof will change pitch. The classic example of this is a gambrel-style roof. The layout for a roof with a change of pitch should be done in a manner that will minimize the exposed flashing at the transition. Transition flashing will be a minimum of 3 inches wide. This maintains the 3-inch headlap necessary to keep the roof watertight. The layouts for a shallow pitch to steep pitch and a steep pitch to shallow pitch roof are the same.

Snap a reference line where the roof changes pitch. This will become the layout line for the last full course of slate. Adjust the prior course layout to tie in with this transition reference line.

A finishing course will be installed over the last full course. This finishing course will be the same size as the starter course, which is exposure plus headlap, and will be laid to the heads of the last full course. The transition flashing is then installed and will maintain proper headlap.

The roof area above the transition is treated as if you are starting a new roof surface at an eave. It will be necessary to lay out and install a cant strip and starter slate. If you are laying out a shallow to steep roof, the layout line for the starter slate and first full course of the upper roof surface will be measured from a point two slate thicknesses above the transition reference line.

A steep to shallow or gambrel roof transition will have the upper roof slate overhang the transition reference line two slate thicknesses plus 1 1/2 inches, typical of a new roof layout.

Vented Ridges

If a slate roof project requires a vented ridge, there are two options. The roofer can fabricate a vented

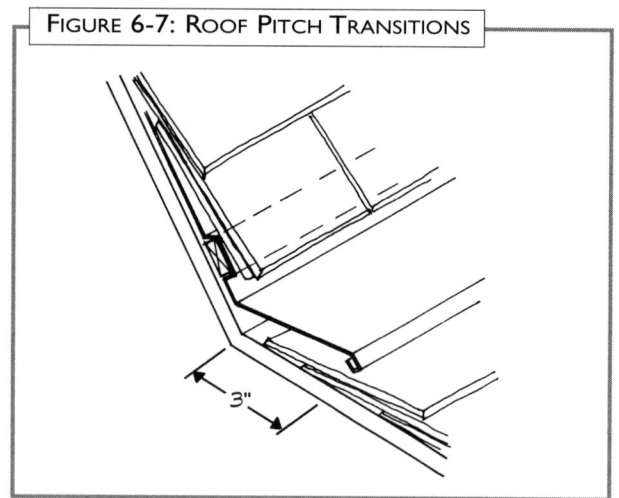

FIGURE 6-7: ROOF PITCH TRANSITIONS

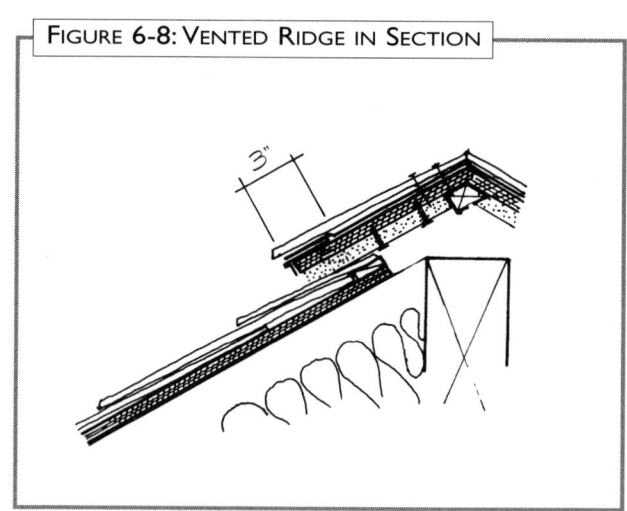

FIGURE 6-8: VENTED RIDGE IN SECTION

Roof Layout

metal ridge cap or fabricate a wooden venting structure to which the slate can be attached. (See Chapter 8 for instructions on installing vented ridge caps.)

On most surfaces designed to receive a slate roof, the slate will be laid with a 3-inch headlap to prevent water infiltration. This means that the ridge cap being used must cover the head of the last full course of slate by a minimum of 3 inches. On most projects, the roof sheathing will be set back at the ridge 1 1/2 inches on either side of the ridge pole. In order to cover the head of the last full course of slate by a minimum of 3 inches, cover the 1 1/2-inch opening left for the vent area, cover the area taken up by the ridge pole, and meet at the actual peak, the vent will need to provide for a minimum 8-inch wide piece of slate on each side of the peak.

Some things to keep in mind when completing the layout for a vented ridge:

1) For slate from 12 inches long to 18 inches long, snap the line for the last full course of slate at a point 3 1/2 inches down from the actual peak.

2) For slate 20 inches long or longer, the standard roof layout will work as long as the cap size chosen is equal to the field slate exposure plus 1/2 inch.

3) If the cap size desired is 8 inches but the field exposure is less than 8 inches, follow #1 above.

4) The finishing course will always be the length of the concealed portion of slate that matches the size of the starter slate. The finishing course will have to be nailed using a notching technique (see Chapter 8), keeping the nails within the top 1 inch of the finishing slate.

Dormers

In terms of roof layout, dormer roofs need to be treated as individual roof surfaces. The butt end of the starter and first course of slate should line up with one of the courses in the adjacent roof surface. To accomplish this, it is easiest to lay out the dormer roof starter and first full field slate course first (Figure 6-9). Where the first full field slate course on the dormer roof intersects the valley reference line, measure up to the nearest reference line on the main roof surface. Use this measurement to lay out a line on the main roof surface parallel to the reference line. This new line will become a layout line for the field slate on the main roof surface which intersects with the first course on the dormer side. Finally, lay out the rest of the dormer roof surface the same way you would any other roof.

If an open valley is being used, proceed with the dormer roof layout as though it were a standard slate roof layout. If a closed valley is being used and the slate courses are

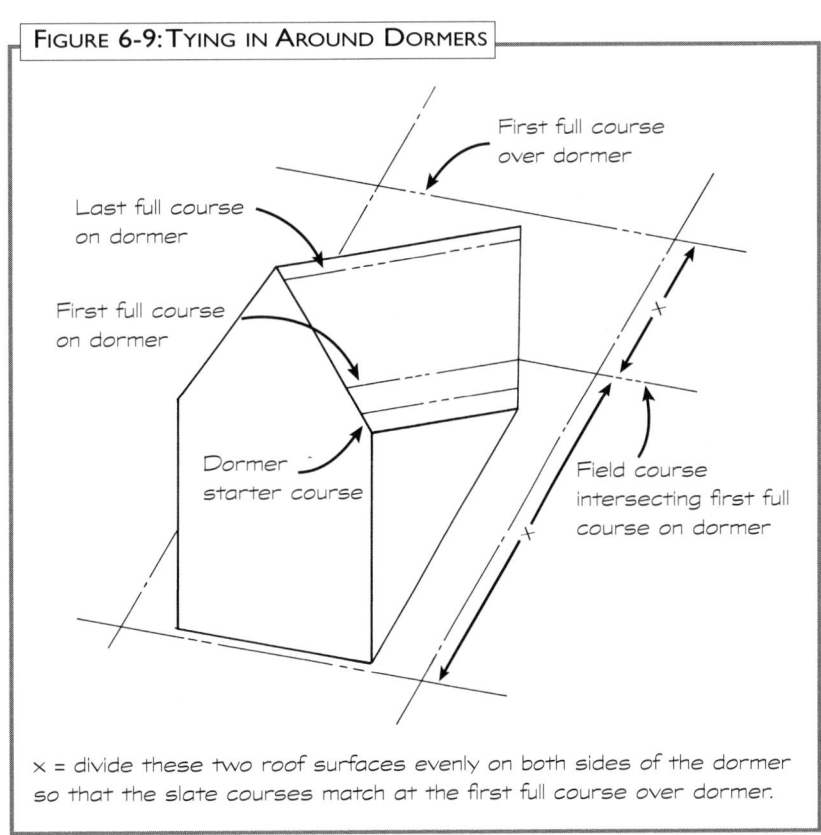

FIGURE 6-9: TYING IN AROUND DORMERS

x = divide these two roof surfaces evenly on both sides of the dormer so that the slate courses match at the first full course over dormer.

Chapter 6

meant to line up at the valley, the dormer ridge size may need to vary.

The importance of following this layout is not just for aesthetics. Proper dormer layout will insure that the valley flashing covers adjacent roof surfaces evenly and properly.

Dormers create a situation where the main roof layout will be divided into sections. The first section is from the eave up to the dormer first full course reference line. The second section is from the dormer first full course reference line to the last full course reference line at the ridge. The layout and adjustment of layout lines for each section is the same as detailed above. These reference lines are the lines for laying the heads of the slate shingles. Intersecting roofs like dormers will govern the location of the main roof layout lines, so these intersecting roofs will need to be laid out far enough in advance to permit proper adjustment of the main roof layout.

If the pitch of the dormer roof is different from the pitch of the main roof, it is even more important to follow the procedure outlined above. This will allow the installer to set the valley flashing properly in relation to the dormer roof and the main roof.

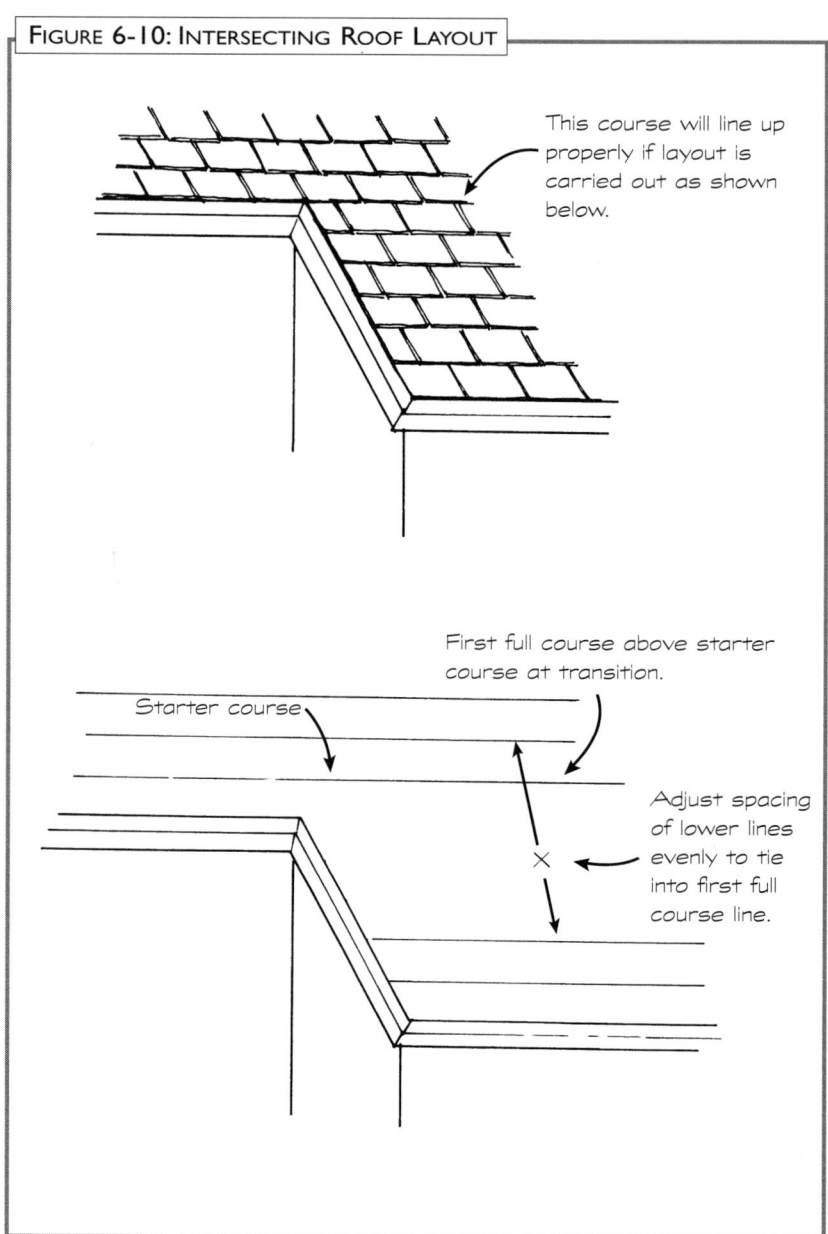

Intersecting Roofs

Anytime one roof surface intersects another roof surface at a point above the eaves, treat the layout as though it were an intersecting dormer roof. Establish the location of the butt of the shingle from the intersecting roof. Then measure up from the butt to establish the reference line for the head of that course. Tie the lower roof into that upper reference line so that the slate courses line up at the intersection.

CHAPTER 7
Slate Tools

Slate roofing projects require a number of tools that are standard in the building trades, such as chalk lines, tape measures, utility knives, and tin snips. Leather gloves also come in handy. Slate roofing also requires some special tools, particularly hammers, punches, cutting equipment, and repair tools. This chapter describes the tools needed for a slate roof application.

Every slate roofing mechanic should have a set of tools that is kept in a tool box or bag and maintained in good working order. The minimum equipment needed is:

1. Tool apron – should have at least one large nail pouch.
2. Tape measure – 25-foot to 30-foot works best.
3. Chalk line – if possible keep two lines: one for red chalk; one for blue chalk.
4. Utility knife – try to keep extra blades as well.
5. Tin snips – offset snips work best. Remember that red-handle snips are easiest for a right-handed person to use, and green-handle snips are easiest for a left-handed person to use.
6. Framing hammer – 16 oz. or 20 oz. Heavier hammers tend to break slate if you miss the nail head.
7. Slate hammer – they are expensive but very handy.
8. Nail set – 3/32-inch to 1/8-inch works best for most nails.
9. Slate cutters – there are many varieties, but remember to choose one that will be easy to use on the roof scaffolding without it being in the way.
10. Slate ripper
11. Small handsaw – for cutting lath to length.
12. Safety harness

In addition to this minimum list, you should have these tools available on site:

1. Circular saw
2. Drill – cordless drills with extra batteries are handy, but you should also have a more powerful corded drill for those areas where it is needed.
3. Selection of masonry drill bits, regular bits, and screwdriver bits.
4. 100-foot extension cord – the better the quality, the longer it will last.
5. Hand seamers – for bending sheet metal
6. Portable sheet metal break – for making flashing pieces

Chapter 7

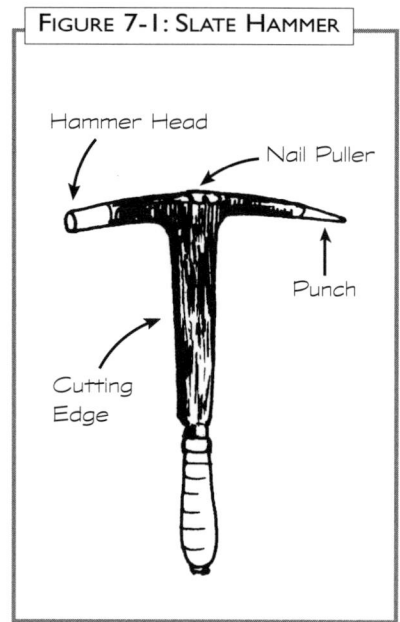

FIGURE 7-1: SLATE HAMMER

Hammers

Slate is most commonly installed with nails, and slate roofing generally requires two different types of hammers: a framing hammer and a slater's hammer. Nail guns cannot be used under any circumstances for installing slate.

Any hammer with a flat head and of a weight that the mechanic can control may be used to install slate. Many slate roofers will carry both a common framing hammer and a slate hammer. Most roofers are comfortable using a framing hammer. They can balance and control a framing hammer much better than they can a slate hammer. It is easier to set nails with a framing hammer because of its weight, and the claws are easier to use for pulling nails. A framing hammer is also useful for those needs related to scaffolding and other roof-related carpentry. Once the roofer becomes familiar with the slate hammer, however, the framing hammer will be used less and less.

Slater's Hammer

The main disadvantage of a common hammer is that it does not have a pointed punch end, as does a slater's hammer. The traditional slate hammer has evolved into a multi-functional tool. It has a relatively small hammer head, a punch point opposite the hammer head, a nail-pulling claw on the side of the hammer shaft, and a beveled cutting edge along the shaft itself. This tool is probably the single most valuable tool to the slate roofer and will definitely make the installer more efficient.

In the past, the best quality slate hammers were a single piece of forged steel. They usually had a pear-shaped leather handle. This pear shape allowed the hammer to roll in the palm of the hand, allowing the installer to gauge the setting of the nails. The punch end allowed the installer to punch a hole in a piece of slate quickly and sharply using the same tool he held in his hand for nailing. The nail claw was used for removing nails, and the cutting edge allowed the installer to cut slate by chopping at the edge with the shaft of the hammer. Today, the most common slater's hammer available is imported. It has all of the same qualities as the old, but the handle is a little different. Slate hammers are also expensive; however, with proper care, they should last a lifetime. Try to avoid punching thick slate with your slate hammer, as the thick slate will dull the point, can break the hammer when punching, and in general cause unnecessary wear.

Punches

Using the punch end of the slate hammer is the easiest and fastest way for the slater to punch a hole in a slate shingle. Using the hammer leaves one hand free to hold the slate. However, the slate hammer does not work well for slate 1/2 inch or greater in thickness. For thicker slate, other kinds of punching tools will be needed.

FIGURE 7-2: PUNCHING MACHINE

Slate Tools

One such tool is the treadle-operated **punching machine**, but these machines are typically used only at the quarry (see Figure 7-2).

A **nail set** can be invaluable as a secondary punch. If the installer forgets to punch a valley or a hip prior to cutting, the strike from a slate hammer will often break the cut piece. However, a fine point nail set will often work with greater success. The nail set is also the preferred tool for punching thick slate. The fine point will focus the punch, and there is much less wear on the slate hammer. Nail sets are available at any local building supply and are inexpensive compared with the slate hammer.

A nail set can also be used to punch a series of relief holes for cutting thick slate (as explained in Chapter 8).

When a slate hammer, nail set, or other common punching tool is not available, a large **framing nail** will often work, but only with standard-thickness material. A **mason's hammer** with a fine point will also work most of the time.

Instead of being punched, slate is often drilled with a **masonry drill bit**. Although not commonly thought of as standard slate tools, drills, with the advent of cordless tools, have become much easier to use on the job. The obvious drawback is that the hole produced by a drill bit is not countersunk.

FIGURE 7-3: TRIMMING MACHINE

Cutting Tools

There are several types of tools that can be used for cutting slate. Most quarries use **trimming machines**. These machines are very heavy and not practical for use on a job site. Also, they are very difficult to find. Most quarry owners will not part with them and they are no longer commercially available.

Guillotine machines, which were used in the past but are rarely used today, work on the same principle as the trimming machine. The difference is that a long heavy blade is brought down on the slate to trim it, much the same way as with a paper cutter. This machine is operated with no safety guards and is very dangerous. There are few, if any, of these machines still in use today.

In some quarries, **pneumatic shears** have been introduced. The end result is the same as that produced by older-style trimming machines. As these new machines prove cost-effective, they will probably be used more widely.

Hand-Operated Cutters

Shown in Figures 7-4 through 7-7 are several of the more common hand-operated cutters available. They all work on the same principle as a paper cutter. The lever action allows the cutting blade to crush through the slate, and the cutting progresses from one end to the other. This cutting or crushing action will leave a chamfered edge on the opposite side of the slate from that which is facing up. Thus, if a piece of slate were cut from the back side, the chamfered edge would be created on the front side.

Having a slate cutter on-site is a must for modern applications. Among its other benefits, this type of cutter will usually leave the piece of slate on either side of the cut intact. When choosing a cutter, look for one with a narrow side support. The cutters with wide side supports tend to collect slate debris, which then acts as a fulcrum, causing the piece of slate being cut to break over this debris.

Most of these hand-operated cutters will have a concave cutting blade that appears to have two cutting edges. A flat blade will be much more difficult to force through the slate.

During the days of asbestos roof shingles, roofers sometimes used a large, heavy cutter called an **asbestos shear** (or "green machine"). Some roofers confuse this machine with a slate cutter. The asbestos shear crushes the shingle by applying weight to the center and then distributing it outwards as the shingle crushes. This technique may have worked well for asbestos, but it does not work well for slate. The slate will break in this type of cutter.

On the top of this tool is a small nibbling jaw. This small jaw will cut slate, but it is very inefficient. In addition, cutting tools must be moved all over the roof as needed, and this heavy tool will prove very difficult to move and balance.

The **slater's tee**, the chosen cutter of yesteryear, is still used today by a few mechanics who have not adapted to the changes in the industry since the original slate book was published in the 1920's. The leg of the tee is pointed so that it can be stuck into a scaffold plank, or in some cases into the roof. The slate is then laid across the tee and is chopped with the cutting edge of the slate hammer. Although this method works well, it has some drawbacks. The piece of slate being cut off is usually broken and unusable, the cut line is more difficult to follow, and the tool itself may be easy to trip over, depending on where it is placed on the job site.

The **slate hammer** has a cutting edge on it that can be used to chop at the slate, as described above with the cutting tee. Some roofers will take a piece of slate, stand it on its edge to mimic the slater's tee, lay the piece to be cut over the vertical piece, and chop at it with the hammer. However, the slate hammer chopping method works much better over a solid piece of wood or steel.

If a standard cutter is not available, a variety of other tool combinations will work for most standard-thickness materials. The principle is to support the piece of slate that you want to keep and chop along the cut line with a hammer shaft, a piece of steel, or even a hatchet. Remember to wear eye protection whenever you cut slate.

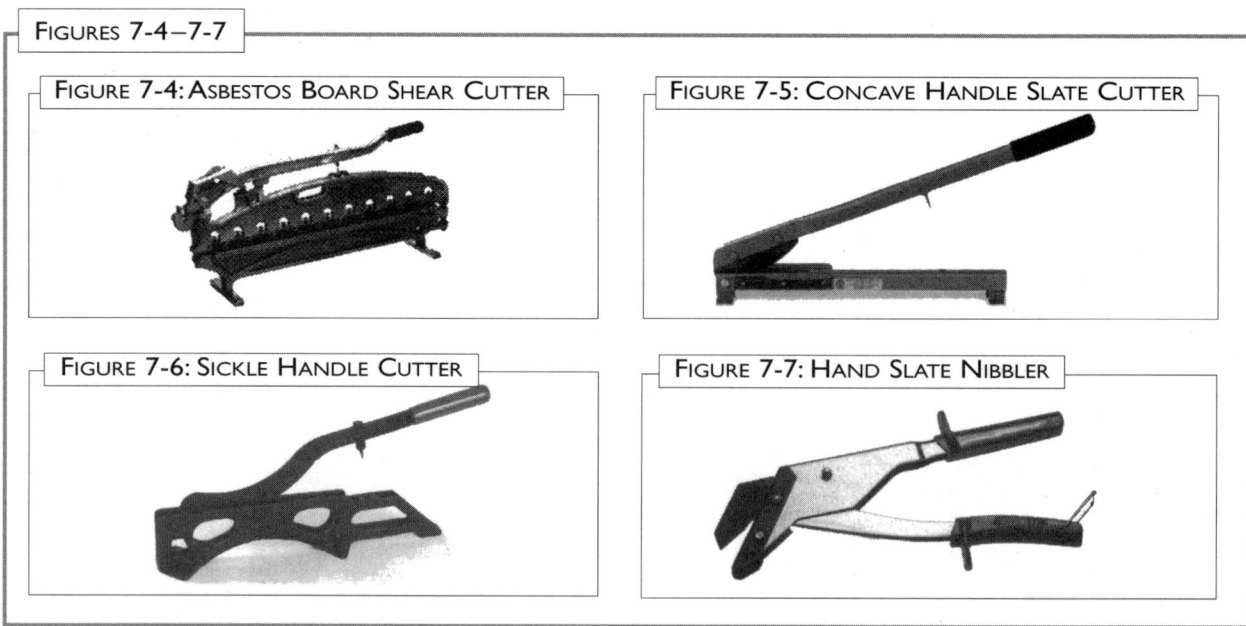

FIGURES 7-4—7-7

FIGURE 7-4: ASBESTOS BOARD SHEAR CUTTER

FIGURE 7-5: CONCAVE HANDLE SLATE CUTTER

FIGURE 7-6: SICKLE HANDLE CUTTER

FIGURE 7-7: HAND SLATE NIBBLER

Wet Saws and Diamond-blade Saws

You should never cut standard-thickness slate with a saw. Using a saw takes longer than cutting with a hand cutter, but more important, it will leave a smooth, saw-cut edge. This smooth edge is not consistent with the rest of the roof's chamfered edge. Never use a saw cut on an exposed edge. However, if a closed valley is desired with a thick slate, a saw can be a very handy tool. Because the cut edge is concealed by the mitered valley, saw-cutting is acceptable for this application. Further, the piece of slate which is cut off will often be usable at the hip. Of course, if it is used at the hip, it will have to be a hip detail that is covered by a cap so the saw edge is concealed. With thick slate, this tool can save a considerable amount of time and material, considering that the alternative is the chopping technique described earlier.

Slate Rippers

The slate ripper is used to remove broken or damaged slate from the roof surface. There is a handle at one end, and the other end is flat with hooks on either side of the shaft that are designed to hook onto a nail shaft. Nails can be removed by either pulling down sharply or by striking the handle portion of the rip with a hammer. For this reason, when selecting a slate ripper, quality and durability are essential.

Look for a slate ripper which is at least 24 inches long and no more than 1/8 inch thick at the hook end. Look for a flat portion perpendicular to the shaft that can be struck with the hammer. Some rippers have a rounded shaft at the handle attachment area. If, when removing nails, you strike the ripper with a hammer, this rounded portion can allow the hammer to glance off the ripper handle and into your hand.

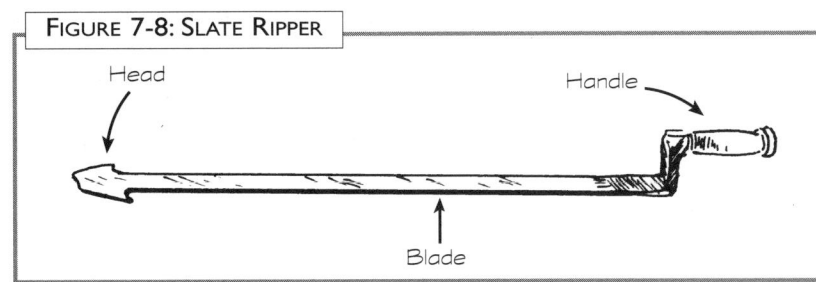

FIGURE 7-8: SLATE RIPPER

NOTES

CHAPTER 8
Installing Slate

After putting considerable effort into planning, ordering, roof layout, and preparation, you are at last ready to begin installing the slate. This chapter explains the procedures and techniques for working on a slate roof, cutting and punching the slate, and installing it.

Working on a Slate Roof

Slate is considered a "steep slope" roof application. For several reasons, the use of roof scaffolding is essential when working on a slate roof. First, unlike some roofing shingles, slate does not stack well on a sloped surface. Because it is smooth and flat, it will slide off a sloped roof. Second, as much as possible, the installer wants to avoid walking over the finished roof. Third, proper scaffolding provides safe access to the roof surface. Whenever using scaffolding, consult OSHA rules and regulations.

With composition shingles, cedar shakes and shingles, and with most tiles, the roof is prepared as specified, coursing is laid out, and flashing details that can be installed are. The roof shingles are then distributed in small amounts over the entire roof. As the installer works, he simply grabs shingles from the nearest pile and installs them. These roofs are often installed with the roofers working from above and/or beside the shingles. Because the shingles can be stacked all over the roof without sliding off, this material loading approach makes perfect sense. Slate, however, needs to be stacked on a platform of some kind to keep it from sliding.

Unlike some roofing materials, such as tile and standing seam metal that can be installed vertically, slate must be installed horizontally. The roof must be scaffolded as the installers proceed up slope. Standard slate roof brackets can be used to install the scaffold. Once the brackets are properly installed, a plank is laid from bracket to bracket to create a "bridge" or roof scaffold. The slate can then be stacked flat on the planks and used as needed.

Chapter 8

FIGURE 8-1: ROOF SCAFFOLD BRACKET

The first set of scaffold brackets is normally installed on top of the third course of slate. This may vary when small slate, such as 14-inch and 12-inch long material, is used. The butt end of the bracket should never overhang the eaves. Also, try to avoid spacing the brackets more than 8 feet apart horizontally. This helps minimize stress on the planks and distributes loads more evenly over the brackets.

Roof scaffolding brackets are installed approximately every 4 to 5 feet up the slope. Attempting to stretch the vertical spacing much more than this will result in excessive traffic over finished roofing. As progressive tiers of roof scaffolding are installed, it is wise to provide a short ladder between tiers in order to minimize traffic on finished areas and thereby minimize repair later.

The bracket should be installed with a minimum 1 1/2-inch gap between the bracket top and the butt end of the next course of slate. This will allow the installer enough room to more easily slide the bracket attachment strap away from the fasteners when it comes time to remove it.

The fasteners used to attach the bracket to the roof are installed through the vertical joints of the slate from the preceding course. The heads of the fasteners must be set so that they do not sit above the surface of the slate on either side. The subsequent slate shingles are then installed over the top of the scaffolding bracket attachment strap. When the roof is completed, the strap is then slid out from under the fasteners which remain in place.

It is common for a roofer who is not familiar with these slate roofing brackets to install the bracket properly but leave off the piece of slate directly over the top of the attachment strap until the roof is completed. As the roofer works back down the roof checking for broken slate, the brackets are removed, the fasteners holding them are removed, and the missing pieces of slate are installed in the voids as repair pieces. This is not necessary. Even though the installer is attempting to be conscientious about the proper setting of fastener heads, the result is a new roof with unnecessary and excessive repair. In addition, during the installation process, the corners of the slate in the course above the void created by the missing slate will often get broken. The result is that even more repair will be needed.

Slate is commonly broken under these brackets and must be repaired as the installer removes them. During the installation, any weight put on the scaffolding planks will be focused on the brackets. Due to the focused weight at these scaffolding attachment points, the slate in these areas is susceptible to damage. This should be anticipated. In general, not more than one piece in ten is broken in relation to the scaffold brackets.

These scaffolding brackets are more costly than typical shingle brackets; however, they are reusable and will last for many projects if they are properly cared for. Because of their expense, some installers will remove brackets as they progress up slope, leaving only what is needed for easy access. Although this is fine in theory, it can create a problem. Once the peak of the roof is reached, it can be very difficult to remove the top scaffolding if the tier below is missing. In addition, leaving the progressive tiers will offer some protection against potentially dangerous falls. Consider your project scaffolding needs carefully. Once an entire roof section has been completed, the brackets can be removed and used on the next section.

A stipulation should be added to every contract to cover the expense of brackets that need to be left in a finished roof for other trades people to use. Ideally, when the roof surface is completed, no other trades people should have to go on it. But this is not always the case. Painters, for example, will often have work to do around dormers and other details well after the slate work is completed. If the roofer is requested to leave brackets for others, the owner or builder should purchase those units and take responsibility for their use and removal. It should be agreed that any slate broken or damaged with relation to these remaining brackets will be repaired or replaced at the owner's expense.

Loading the Roof

Once the scaffolding is in place, you can begin moving slate up on to the roof. As stated previously, you usually won't be able to distribute all the slate on the roof for use as needed. Instead, it will be most practical to place handfuls of slate on the scaffolding. The quantity of each load on the scaffolding will be dictated by the size of the roof area being roofed and the ability of the scaffold to hold a given weight. Consult OSHA and manufacturer's specifications for maximum loading.

When placing piles of slate on the roof scaffolding planks, be sure to leave space between each pile for walking. Never stand on top of a pile of slate. It will shift! Also, this is the only time when the slate is stacked flat. These flat piles should not exceed 8 inches in height. This will allow for easy and safe movement by the mechanics.

One of the differences between slate roofing and many other types of shingle applications is the speed at which it is installed. On average each mechanic can install one square of slate per day. Bring the slate up to the roof as needed, and try to avoid leaving piles of slate on the scaffolding overnight.

It is important to keep in mind the weight of the slate. It may be possible to overload a structure like a church by completely roofing one side of it. On a building with many cuts and valleys, etc., the framing is probably such that it can be loaded from any given side without concern. However, on a large structure with a cathedral ceiling, loading one side without considering the effect on the structure as a whole could cause the structure to shift.

Basic Slate Laying Techniques

The following rules of laying slate are "basic" in the sense that they apply to all slate roofing installations. Calling them basic does not mean that there are more-advanced techniques yet to learn. It means that these basic rules need to be followed in order to install any slate roof properly.

Hanging Slate

Roofing slate is hung, not nailed. The slate rests against the shaft of the nail; it is not held down by the head of the nail. However, the large head of a roofing nail is important to prevent the slate from being lifted up in extremely windy conditions.

When slate is punched for nail holes, it is punched from the back side so that each hole will be countersunk on the front side. This countersink allows the head of the roofing nail to be set flush with the top surface of the slate. If the head of the fastener is set too tight, the slate can break. If the head of the fastener is not drawn down to the surface of the slate, the shingle in the next course will cantilever over the nail head and

Chapter 8

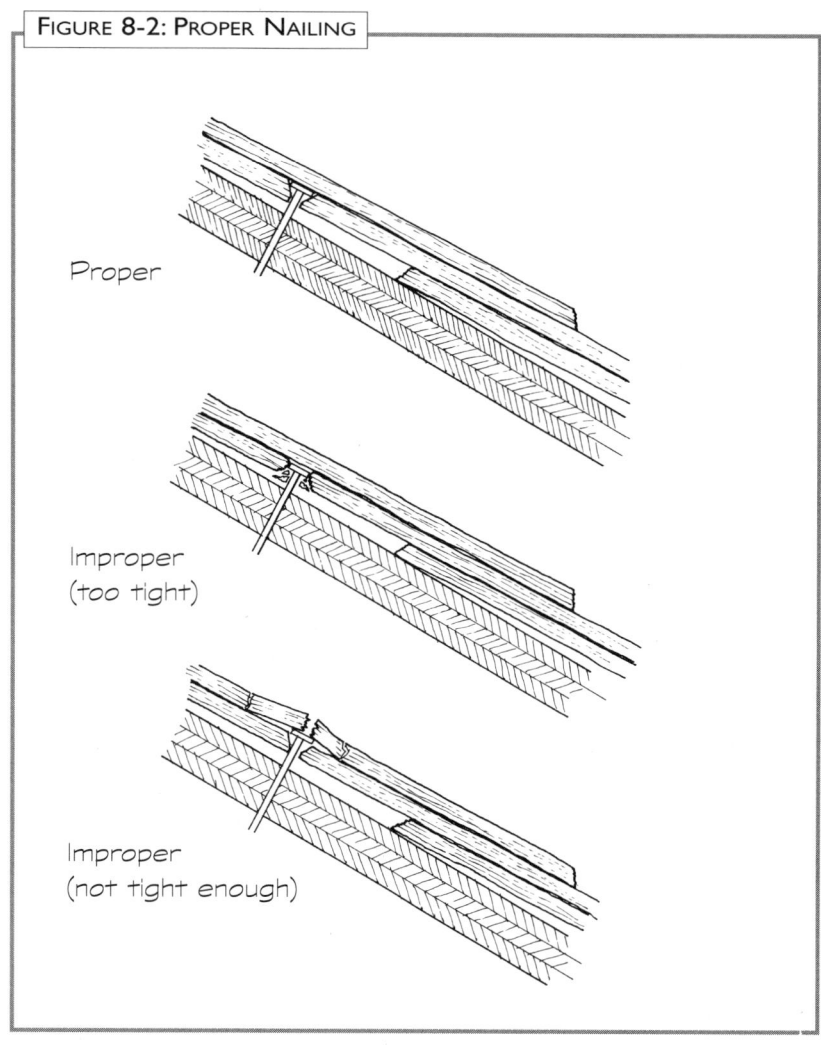

FIGURE 8-2: PROPER NAILING

Proper

Improper (too tight)

Improper (not tight enough)

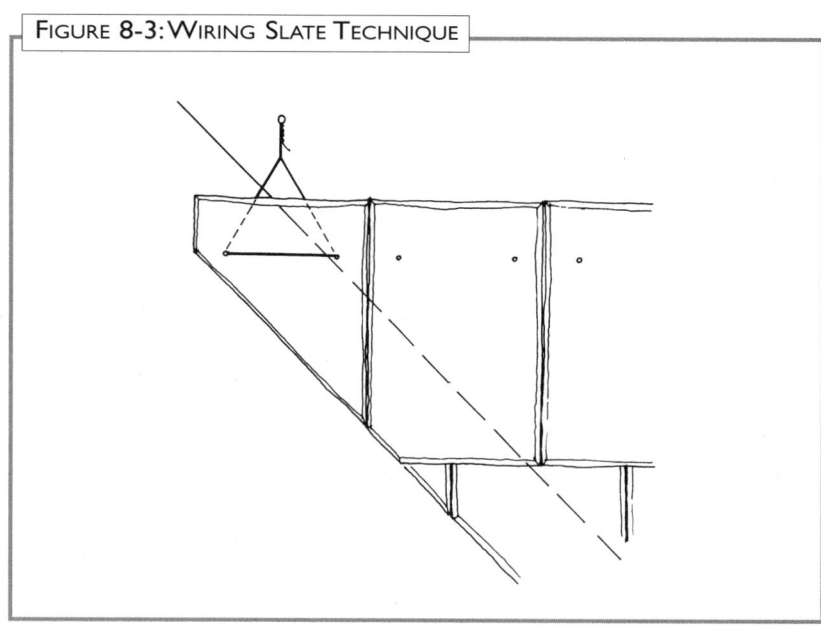

FIGURE 8-3: WIRING SLATE TECHNIQUE

eventually break. When properly nailed, the heads of the nails should be flush with the surface of the slate, yet there should be enough play so that the installer can wiggle the butt of the slate with his fingertips. If the slate is over-nailed, it will not wiggle.

The proper fastening of roofing slate is somewhat of an artform. Because slate thickness varies somewhat within shipments, care must be taken to hang each piece properly. This practice is one of the telltale differences between a mechanic who knows and cares about what he is doing and one who does not. Mechanics who are not familiar with proper slate fastening technique have a tendency to set the nails too tight. The installer must refrain from the final hit of the hammer that is normally associated with the proper nailing of a shingle roof.

In some cases, it may be necessary to wire the slate in place. This may be the case when working with small pieces along open valleys. The installer will feed wire through the two nail holes and secure the end of the wire back to itself. The slate is then hung by wrapping the loose end of the wire to a nail placed outside of the valley flashing. After wrapping the wire around the nail, the nail is driven down so that the nail head will not interfere with the proper setting on the next course of slate.

Installing Slate

Other fastening techniques have come and gone over the years, including the use of wooden pegs, brass machine bolts, and several hook configurations. In Europe, for example, some slate roofs are installed with hooks similar to those used for repair.

Hooks are usually used only with thin, uniform slate. The theory is that the hooks will prevent wind uplift or expedite production at the quarry by eliminating the need for punching nail holes. In addition, the roofer can lay the roof out, install the hooks, then go along and set the slate on the hooks quite rapidly. The down side is that the hooks will usually wear out long before the slate does. The perimeter slate for this application will still need to be punched and nailed. Under certain circumstances, hooks may be used in conjunction with nails to prevent wind uplift, as in the apex of a curve on a bell shape.

Punching Slate

When roofing slate arrives from the quarry, it will have two holes punched in it (see Figure 1-3 of nail hole locations on page 12). However, there are many situations where the slate will need to be punched in the field.

All field-related slate is always punched from the back side. This will create a countersink on the front side. The only material which

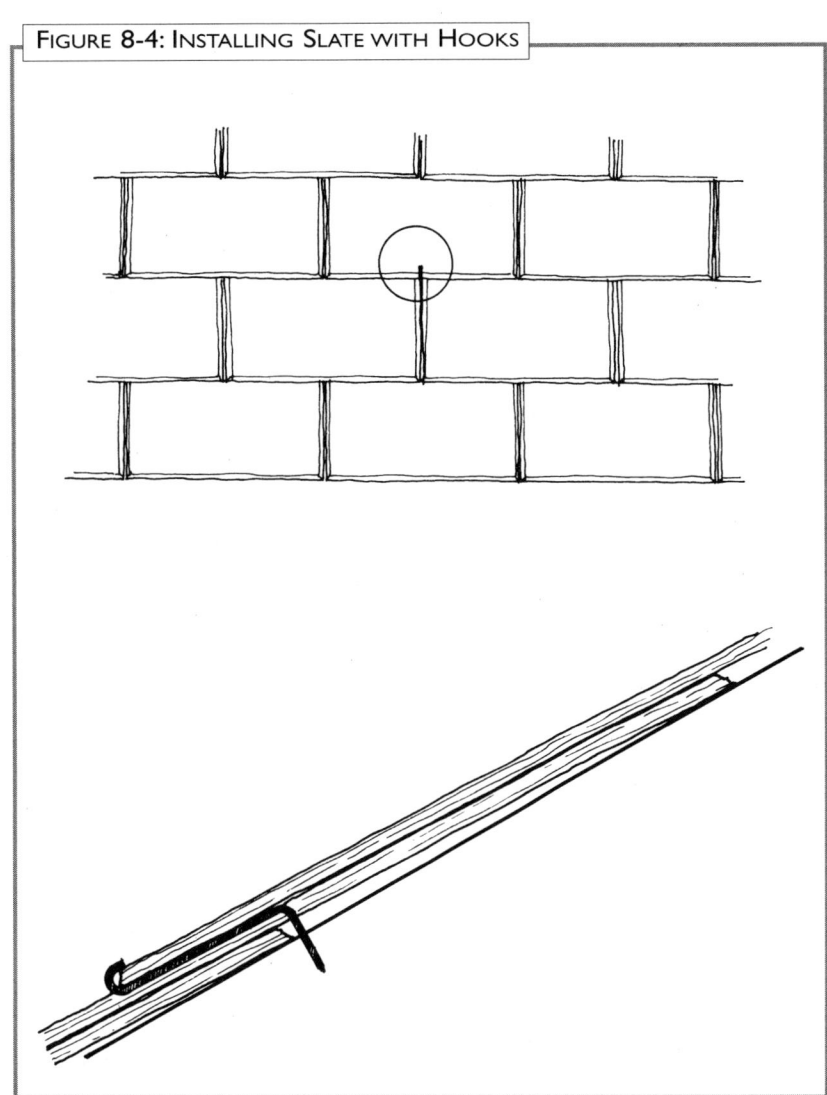

FIGURE 8-4: INSTALLING SLATE WITH HOOKS

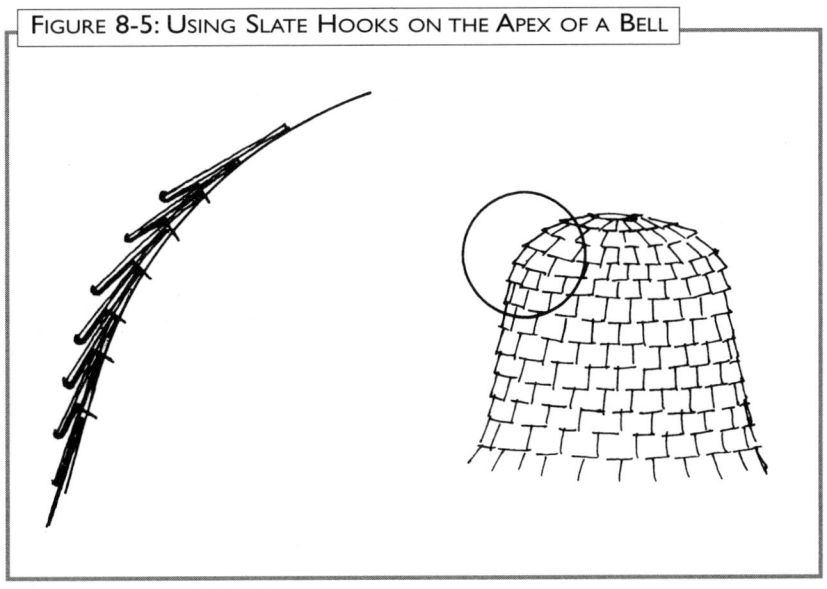

FIGURE 8-5: USING SLATE HOOKS ON THE APEX OF A BELL

Chapter 8

is punched from the front side is the starter slate because it is installed face-side down and requires the countersink on the opposite side.

Always punch slate that you intend to cut before you cut it. It is common for slate to break when punching, but it is much more common for it to break when you try to punch it after you have cut it. This is especially important with relation to a hip cut. Because the nail holes are in a narrow end of the slate, it is very easy to break the nail end off when punching.

The most practical and easiest way to punch holes in slate is to use a slate hammer. Holding the slate away from your body, firmly strike the piece with the pointed end of the hammer. The point will then punch out a hole similar to the quarry-punched hole, although the countersink left on the opposite side of the punch will not be as concise and uniform. You will notice that the punch end of the hammer is not a sharp point. If the point is too sharp, it will punch a very small diameter hole. A hole too small will not allow for the diameter of the nail shaft. In addition, a sharp point will tend to bend and round itself when striking hard slate.

Be sure to hold the slate a safe distance away from your body when punching. Never lay it across the palm of your hand or your arm when punching, and never punch slate while standing above another person. Not only can you be cut by the fragments punched out, but it is also common for the piece to break when punching. Broken edges can be razor-sharp.

It takes a little practice to become proficient in punching. Extremely thin slate will obviously require less of a hit than textural material. The most common mistake made when learning to punch slate is to hit it too softly. Practice a couple of times on broken or scrap material. Standard thickness slate is very easy to punch, but the punching process will cause some breakage. This has to be expected and accounted for when ordering material.

If a slate hammer is not available, you can punch slate with a nail set or even a 16d framing nail. The process is basically the same as that for punching with a slate hammer, except that the slate has to be laid down so the mechanic can hold the nail set with one hand and the hammer with the other hand.

Lay the slate down so that the material punched out will fall away. Try laying it over the side of a rigid surface, such as a scaffolding plank or a piece of angle iron. If the slate is placed flat on a flat surface, striking it for the purpose of punching a hole will often break it.

Punching with a nail set may not be as fast as punching with a slate hammer, but it is handy for several specific applications-and, it is far less expensive to replace a nail set than a broken slate hammer. When making repairs using the bib method (see Chapter 11), you are more likely to punch a hole accurately in an installed piece of slate with a nail set than with a slate hammer. If you are working with a slate that is 1/2 inch thick or thicker, the nail set will work better than the slate hammer and will cause less breakage.

Drilling Slate

Sometimes thick material will be drilled for holes at the quarry instead of being punched. If this is the case, the quarry will usually countersink the drill holes. Should you decide to drill thick material in the field, this can be done with a masonry drill bit. Be sure to countersink the nail holes so that the head of the nail can set even with the surface of piece being fastened.

For slate with a thickness of 3/4 inch or more, it may not be necessary to countersink the holes. Slate of this thickness is not likely to break if the nails are set tight or if it rests on the head of a nail below. Regardless of thickness, set the nail head as flush with the surface of the slate as possible.

Installing Slate

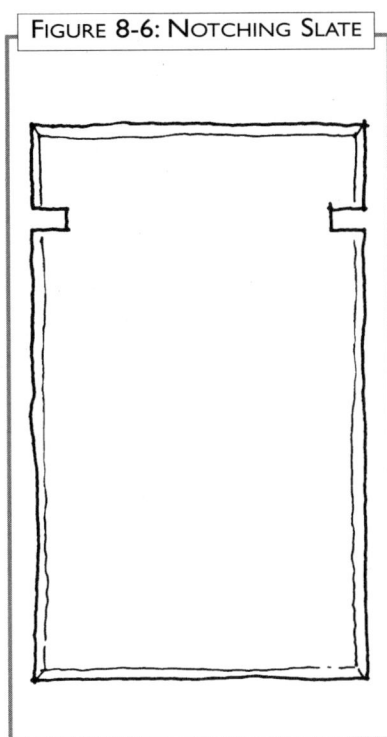

FIGURE 8-6: NOTCHING SLATE

Notching Slate

There are alternative methods of fastening slate to the roof deck which do not require punching. In one of these methods, a notch is cut into the sides of the slate at the quarry. This method is probably most common with slate that is one size. The concept is to cut a notch with a saw blade prior to splitting the stone.

Notching allows the quarry to speed up production and cut down on waste created by the punching operation. If this method is being considered, you need to understand that all of the perimeter slate as well as hips, ridges, and valleys will still need to be punched and nailed. Unfortunately, the saw-cut notches do not provide for countersinking the nail heads.

Cutting Slate

Standard thickness slate is most often cut in the field with a hand-operated cutter, as described in Chapter 7, "Slate Tools." It is wise to practice cutting some slate on the ground. You will find that some slate cuts more easily than others due to the chemical composition of the material. You will also find that the hand cutters work well up to slate of 1/2-inch thickness. Beyond 1/2-inch thickness, you will need to employ an alternative method of cutting.

The following procedures apply to **cutting standard thickness slate.**

Field slate is always cut from the back side so that the chamfered edge created is facing up when installed.

When cutting hips and valleys, it is common for the point of the cut to break off on the end which is last to be cut. This is due to the fact that the weight of the piece being cut off will cause the small pointed end to snap. Try cutting a relief cut at the pointed end first. This cut would normally be 1 1/2 to 2 inches. Now if the slate breaks while cutting, it is more likely to break along the cut line.

When precutting hips and valleys, look for even-textured material. The cuts will be more consistent.

Keep the cutting area around the cutting blade free from debris. The accumulation of debris in this

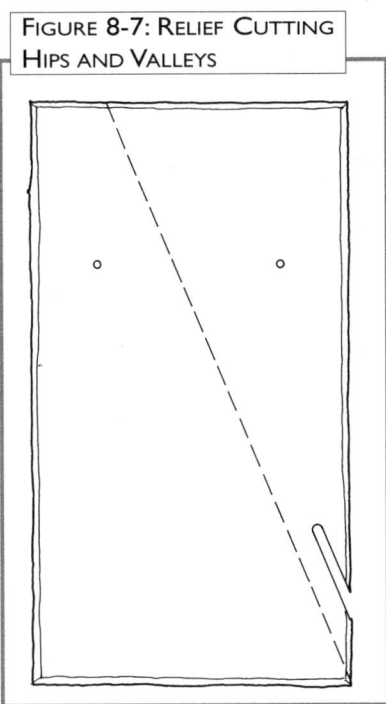

FIGURE 8-7: RELIEF CUTTING HIPS AND VALLEYS

area will often cause breakage.

Mark all slate to be cut with a straightedge. This line will be easier to follow than any guides that may be provided on the cutter.

When cutting a rounded piece for a plumbing vent, punch several relief holes along the round cut line. This can be done with the slate hammer or a nail set. Then cut several relief cuts into the round area. The resulting cut will probably be somewhat jagged. Finish the cut by placing the slate over a solid support and carefully rounding the edges with the head of the slate hammer.

For **cutting thick and rough textured slate:**

Scribe a straight line on the slate where it needs to be cut.

Lay the slate over a solid cutting block along the cut line. A piece of steel such as an I-beam works well for this. The bigger and thicker the slate, the bigger and stronger the cutting base you will need. The cutting base should be elevated enough so that the chopping action of a slate hammer will not be impeded.

Using the cutting edge of the slate hammer, chop at the slate along the cut line. Complete the cut by dressing up the ragged edge along the cut line by chopping with the slate hammer.

In some cases, you will need to punch a series of relief holes along the cut line before chopping the slate. The punching will blow out the slate on the opposite side of the punch, thereby reducing the thickness of the slate at the cut line. These holes will need to be placed every 1 to 2 inches depending upon the thickness of the slate. Sometimes the slate will crack along the cut line after these holes have been punched. Other times it may require a couple of light chops or taps to cause the slate to break along the cut line. Finish the cut by dressing the rough edge with a slate hammer.

Installing Slate

First, you must understand the basic rules of installing slate.

- Slate is hung, not nailed
- Starters are laid chamfered edge down
- Field slate is laid chamfered edge up
- Layout should be done prior to starting the slate application
- Vertical joints must be offset at least 3 inches from preceding course (this is also referred to as side lap)
- Vertical edges must be at least 1 inch inside of nail holes from the previous course
- Try to align the joint as close as possible to the middle of the slate in the previous course

Starter Slate

As its name implies, starter slate is the first course to be installed. Starter slate is installed chamfered edge down over a cant strip of appropriate thickness. (See cant strip in Chapter 3.) When the field slate is then installed over the starter slate chamfered edge up, the opposite chamfers will create somewhat of a mitered edge. This tight miter will help to drain water away from the roof edge. If the starter slate were installed front side up, water would have a tendency to collect at the edge where the chamfers of the stacked slate create a pocket.

Starter slate will arrive from the quarry with the nail holes punched from the front side so that the countersink is facing up when the slate is installed. Starter slates are the only slates on the job installed in this way and the only slates that will be shipped with front-side punches.

The size of the starter slate used will vary depending on the length and exposure of the slate being used on the first course of the roof (see Chapter 3). Many experienced slate roofers will simply use some field slate of the proper width as their starter slate.

Starter slate supplied by the quarry has one function because it is punched up-side down. If properly sized, field slate and any scrap slate of sufficient size when punched properly, can be used as starter slate. Slate for use as starters can also be ordered unpunched so that any material left over when the starters are completed is still usable as field slate.

Installing Field Slate

Field slate is always installed "front side" or chamfered edge up. There is no exception to this rule. The nail holes at the top of the slate are provided high enough above the top of the previous course so that the nails do not penetrate the top of the previous course when installed with a 3-inch headlap, which is standard for field slate (see Chapter 3).

There are two exceptions to this 3-inch headlap rule. If the roof is steeper than a 20:12 pitch, the headlap can be reduced to 2 inches, thereby increasing the exposed surface. This may be a factor when estimating percentages of additional material needed for cutting loss and breakage. If the roof pitch is less than 6:12, the headlap must be increased to 4 inches. This will be a factor when estimating percentages of additional material needed for concealed material, cutting loss, and breakage. Any roof pitch less than 6:12 should have additional waterproof underlayment installed. This is a good area to use modified bitumen, especially in regions where freezing and thawing conditions prevail.

When installing the field slate, try to insure that vertical joints of progressive courses are centered on the slate below. This practice is a must when using slate of one width. However, it is not always possible or practical when laying random-width slate. In all cases, the vertical joint alignment must be a minimum of 1 inch inside of the nail hole from the previous course and at least 3 inches from the vertical joint of the previous course. Under no circumstances should vertical joints in successive courses line up.

Field Slate - Gable Slate

Gable slate is the material exposed at the outside edge along the gable ends of the structure. This slate is installed in much the same way as the field slate. The one primary difference involves the nail placement of the outermost nail.

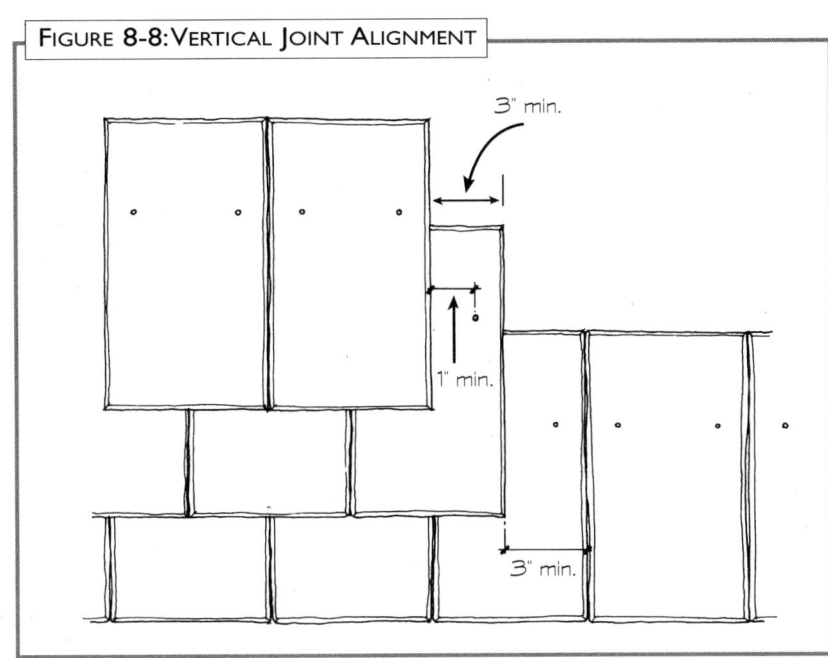

FIGURE 8-8: VERTICAL JOINT ALIGNMENT

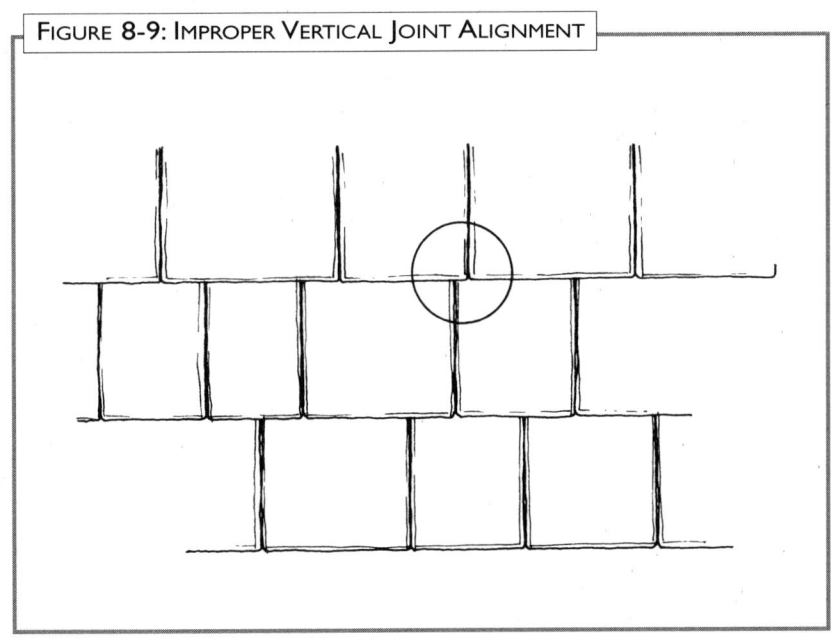

FIGURE 8-9: IMPROPER VERTICAL JOINT ALIGNMENT

The standard overhang for slate along the gable end is 1 1/2 inches, and the standard nail placement of the quarry-punched hole is 1 1/2 inches in from the edge. As a result, the outermost nail hole will not be usable without splitting the gable rake board or completely missing it. The installer must punch a new hole. With the full field slate, the nail hole should be moved in approximately 1 1/2 inches from the gable edge. This will usually allow for good nailing in the roof deck, but the distance may vary and the installer will have to make adjustments accordingly.

With alternating half pieces or

Chapter 8

narrower pieces, the nail holes will have to be placed as needed. If the narrow piece is less than 6 inches wide, punch the new hole directly above the remaining quarry-punched hole. If the narrow piece is 6 inches wide or wider, the new hole will usually work if placed 1 1/2 inches in from the gable edge or 3 inches in from the edge of the slate.

These narrow slates tend to break easily when punched. If the installer avoids punching the new hole on the same horizontal level as the quarry-punched hole, he will have a greater success rate.

Gable end slate greater than 6 inches wide should never be nailed with both nails placed one above the other because the weight of these pieces will cause them to sag. If poor nailing conditions create a need for unusual nailing, then the carpenters must rectify the situation.

On buildings with ornamental gable trim, there will often be a gap between the edge of the finished roof deck and the trim. This creates a problem for the roofer. The slate needs to be nailed as closely as possible to the gable edge. If there is a gap or void at this detail, the slate cannot be properly fastened. Due to the gap left by the trim detail, it becomes very difficult to nail the alternating narrow pieces. Therefore, it is very important that this gap be properly filled while the trim is being installed. (See Chapter 5: Roof Construction and Preparation.)

Although it isn't necessary, it is a good practice to set gable slate in a mastic or adhesive because the slate overhanging the gable is susceptible to wind uplift. Storms that only occur on a 100-year cycle could in fact affect this roof.

Field Slate - Single Width

1) Slate of one width will usually install faster than slate of random-width. However, the slate of one width will have a greater material cost.

2) When using slate of one width, it is best to use properly sized starter slate of the same width. This will avoid any side lap complications.

3) When working on a one-width slate roof, you can start at a gable and work across to the other gable, valley, or hip. The opposite side of the roof will have to be cut to fit whatever the roof measurements demand. This means that on a gable-to-gable roof, the starting side will lay up the gable end perfectly staggered. The gable that you work toward will probably have to be custom-cut to meet the given circumstance. Never use less than a 3-inch wide piece of slate. Slate narrower than this will be very difficult to cut and punch properly. In addition, small pieces are very susceptible to breakage. Experienced slate roofers will plan ahead for these roof width situations and begin to stretch the vertical spacing well before they reach the opposing gable. This is acceptable, but slate should never be spaced side to side more than 1/4 inch. Ideally, if this situation occurs, you should prepare vertical reference lines that guide you from a far enough distance away so that a maximum 1/8" gap is all that is required.

4) Installing a hip roof with slate of one size is probably easier than installing a gable roof. All of the above principles of vertical line layout apply so that the vertical joints alternate evenly. However, the horizontal courses simply get cut into the hip at whatever point they intersect. It is nearly impossible to install a mitered hip on a roof of one-width slate, so you need not worry about proper approach for hip layout. Simply follow the saddle hip installation details.

Field Slate - Random Width

The following technique applies to all random-width slate roof applications. Keep in mind that textured and graduated slate roofs are usually random-width.

Installing a random-width slate roof does not require the use of vertical reference lines because the idea is to install a roof with vertical joints or key ways that do not line up.

A random-width slate roof is probably the easiest slate roof to

Installing Slate

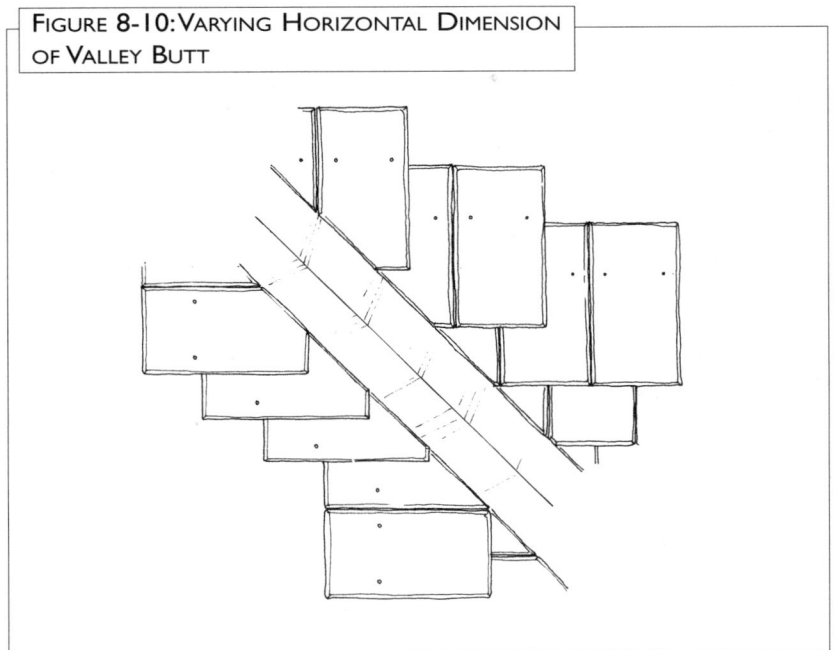

FIGURE 8-10: VARYING HORIZONTAL DIMENSION OF VALLEY BUTT

install because the random vertical joints provide a tremendous amount of flexibility with regard to where the roofer chooses to start the application.

Typically, the starter course will be laid, followed in progression by the first full course. It will help to install a single starter followed by the first full course piece. This will assure that the progressive side lap and nail location are acceptable. This initial layout of starter and first full course will set the stage for most of the rest of the roof.

If the location of the starters dictates that a piece of field slate in the first full course needs to be cut to fit properly, it will generally take several courses to work back out of the cut piece. This means that the cut piece in the first field course will need to be cut in several progressive field courses.

Although it takes a little more time in the beginning to "chase" the starter course with the first full course, it will ultimately speed up the overall installation.

Installing Random-Width Slate on a Gable Roof

For a gable roof with random-width slate, install the first two starter slates. Follow the starter slate installation simultaneously with the first full course of field slate. Begin at the gable and continue this procedure across the roof.

Installing the starter and first full course will often take more time than the progressive courses because the laying of the field slate over the starter needs to be done carefully to follow the basic slate roofing techniques. Go back to the gable end and begin progressive courses. It will now be quite easy to run several courses at a time. However, be sure to use different size pieces up the gable rake to produce a random appearance.

It is a common mistake to install random-width slate on a gable roof yet use the same size gable slate in alternating courses. This looks odd and is very noticeable.

As you progress up and across slope, continually step up the courses at the gable ends of the field surface and work towards the middle. As you lay the field slate between gables, you will have to size the field slate properly to connect the rows. This is done by roofing up to within approximately the last three pieces in a horizontal course. You must then stop and measure the distance between the two ends to see what size slate will fit properly. Doing this will virtually eliminate the need to cut any slate in the field area of the roof. This is why working with random-width slate is often easier than working with one width.

Once you feel confident enough, you will realize that you can precut and punch some random-width gable half pieces (usually on the ground) and virtually eliminate the need for cutting on the roof. Obviously, this will expedite the roof application.

Installing Random-Width Slate Between Valleys

Any time that a random pattern slate roof has a valley inter-

Chapter 8

section, start at the valley and work out. Establish a valley pattern. Because the roof has already been laid out, there are a known number of valley pieces needed.

Using the valley pattern, have a qualified laborer precut and punch the proper number of valleys. Try to vary the horizontal dimension of the valley butt to maintain a random appearance. However, the variation in valley butts is not as critical as it is on the gables. Allow for a large enough valley piece that the nail holes will not penetrate the flashing. The closer to full size the piece is, the easier it will be to punch and less likely to break. If the butt width is varied too much, progressive valley pieces may not have the proper vertical alignment.

When working in an area between valleys, always start at both valleys and work toward the middle. Fit the last three pieces in each course by measuring between connecting slate and installing the necessary random-width slate as needed. Efficiency and minimizing debris on the roof will save time and money.

Installing Random-Width Slate on a Hip Roof

The hip roof is probably the most dreaded of all slate roof applications. However, when you use random-width slate, this installation is quite easy because the you can precut and punch most, if not all, of the hip cut pieces on the ground

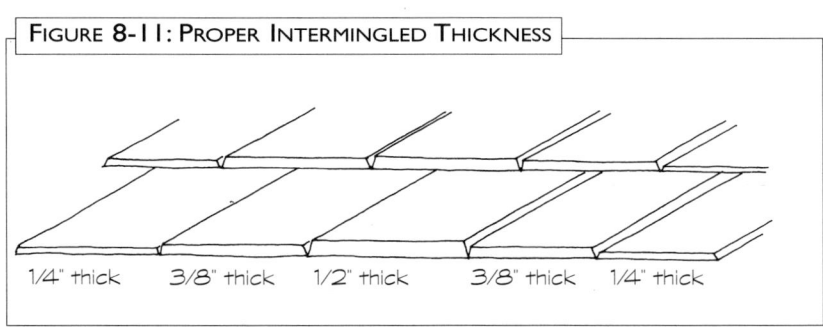

FIGURE 8-11: PROPER INTERMINGLED THICKNESS

1/4" thick   3/8" thick   1/2" thick   3/8" thick   1/4" thick

and fill in the field slate between them. This provides the slate roofer with advantages that virtually no other shingle roof affords. The random-width material, in this situation, makes the installation easier.

Field Slate - Textural

A textural slate roof is designed to show thickness variations. It is important to mix the thicknesses. Avoid installing the slate in any noticeable pattern of thickness. This pattern of thickness will be very visable in the finished roof. For example, stacking the thickest or thinnest piece of slate on top of each other in several consecutive courses will create a band that is noticeable.

When using textural slate, try to cut all hip and valley pieces out of material of similar thickness. This will help to create a clean hip and valley line.

All other rules of application described previously apply to the installation of a Textural Slate roof.

Field Slate - Intermingled Thickness

Whenever slate of varying thickness is being installed. the mechanics need to exercise caution with relation to proper placement of the shingles. If thin slate is installed beside thick slate, a void will occur under the slate installed in successive courses. To avoid this, always place material with a maximum of one thickness variation side by side. Slate that is not properly supported will eventually break. This problem is amplified on an intermingled thickness roof because the courses below support the courses farther up the roof. Improper installation of slate will affect the courses above.

All other rules of application described previously apply to the installation of a Textural Slate roof.

Field Slate - Graduated

The most important aspect of a proper graduated roof application is the layout. All installation techniques described previously apply to installing a graduated slate roof. Some helpful hints for a graduated slate roof application include the following:

If you are using graduated thickness material, try to use the thinest material of each length in

Installing Slate

the last course of that size. This will allow for a smoother graduation to the next size and thickness.

Make sure you are using the proper nail length for each thickness.

When graduating lengths, it may be necessary to punch higher nail holes in the first course of each new size.

Graduated roofs usually diminish in thickness, length, and width. It is therefore important to use progressively narrower pieces as you work up slope. There are two different techniques that can be used to accomplish this. One is to start from the hip, gable, or valley and work towards the middle of the roof. As you begin each progressive course, start within 1-inch of the nail in the previous course instead of aiming for the center of each piece. This technique will take several courses to completely graduate the widths to the narrowest sizes needed. (See Figure 8-12.)

The second technique is to place a narrower piece of slate on top of a wider piece of slate. (See Figure 8-13.) As long as proper nail hole coverage is followed, this is a very effective technique for graduating widths. For example, a 7-inch wide piece will lay on top of a 14-inch wide piece, centered between the nails and still allow for proper nail side lap.

The first course of a smaller size slate at a graduation will usually require using the nail lengths from the slate size of the course below to properly penetrate the roof deck.

Avoid placing roof scaffolding brackets on top of the graduation of sizes. This will break slate. This is especially true for the thinner material near the ridge.

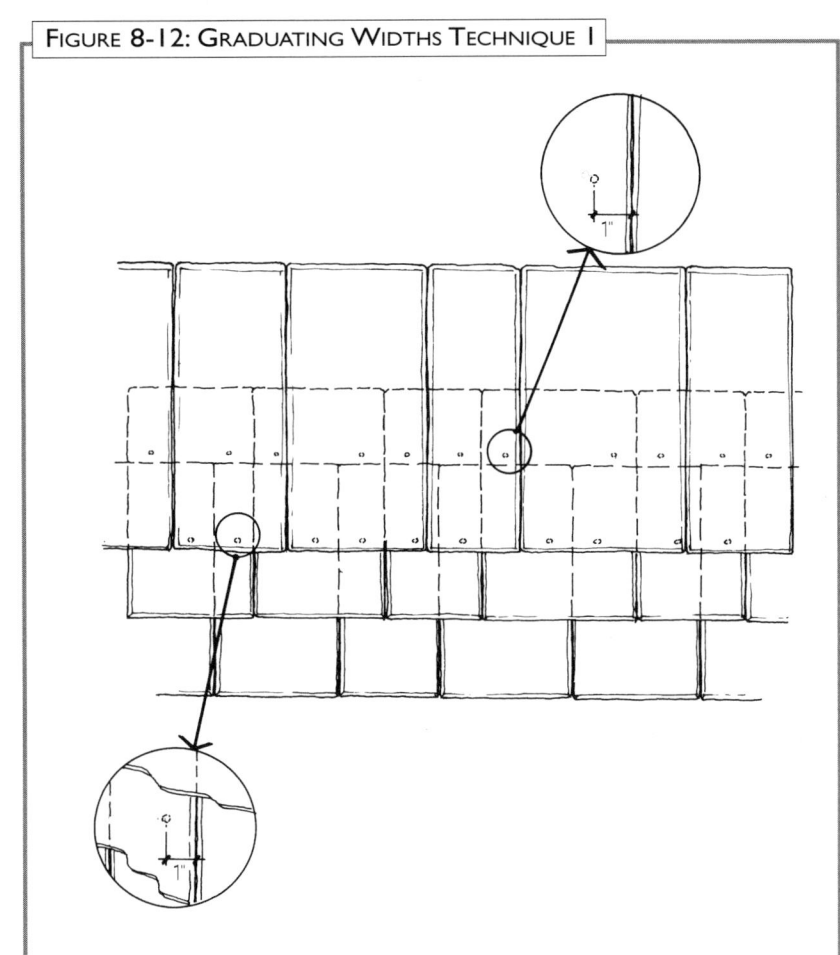

FIGURE 8-12: GRADUATING WIDTHS TECHNIQUE 1

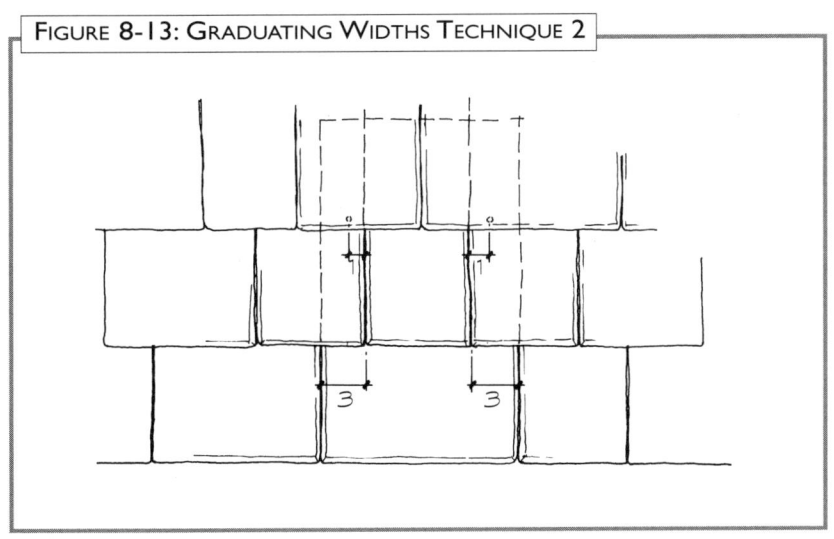

FIGURE 8-13: GRADUATING WIDTHS TECHNIQUE 2

113

Chapter 8

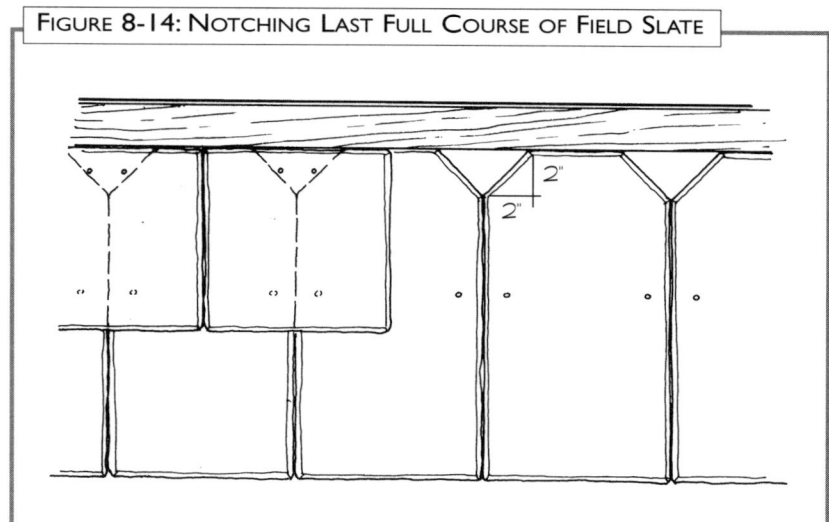

FIGURE 8-14: NOTCHING LAST FULL COURSE OF FIELD SLATE

Field Slate - Finishing Course

The finishing course of slate is the last course of field slate that is installed prior to installing the ridge cap. (To determine the length of the finishing course, see Chapter 3.)

Unlike the starter course, the finishing course slates will be punched from the back side. This will allow the nails to be properly countersunk for a piece of field slate. Some installers prefer to have this material delivered unpunched so they can place the nail holes as needed in the field. Unless directed otherwise, the quarry will provide the holes in these pieces in the same location as those provided for the starter slate.

The finishing course is attached to a nailer just above the head of the last full course, and the thickness of the nailer should be equal to the thickness of the slate being used. The exception to this would be if an unusually small cap were being used. If this is the case, an alternative method of attaching the finishing course may need to be incorporated.

When a vented ridge is specified (or an unusually small cap is used), the finishing course size and layout will have to be approached differently. When the open vent area is taken into account at the ridge and the last full course of field slate is laid at or near the ridge opening, there is no room to attach the nailing strip. Nor is there any room to attach the finishing course.

To provide for a nailing area for the finishing course, you should notch the top corners of the last full course. This will allow a small triangular area of roof deck to nail the finishing course into. (See Figure 8-14.)

Because the finishing course is supported over its entire length, it will not require a nailing strip to support it at the head. Keep the nail holes as far apart as possible when attaching the finishing course. These pieces should also be set in a bed of mastic. This notching technique may not be necessary for the 22-inch and 24-inch lengths, since these pieces may be long enough to be nailed to a nailing strip if the ridge vent opening allows for this.

Installing Valley Slate

When installing slate in any kind of valley, try to follow the procedures explained here. Start slating from the valley out whenever possible. Try to establish a pattern which allows for at least 2 inches of width to remain on the butt edge of valley slate when cut. This will provide a strong valley slate that is not likely to break.

Whenever possible, establish a valley pattern and cut as many of the valleys as possible on the ground. This will minimize the amount of debris falling from the roof.

If the piece of slate being cut off to create the valley slate can be used as a hip piece, be sure to save it. When working on a roof that has all the same roof pitches, the valley slate and the hip slate will be cut at the same angle. If this is the case, try to establish a valley pattern that will allow you to cut each piece of valley slate and always have a hip slate left over. However, be sure to punch both the valley piece and the hip piece prior to cutting. Slate that is 12

Installing Slate

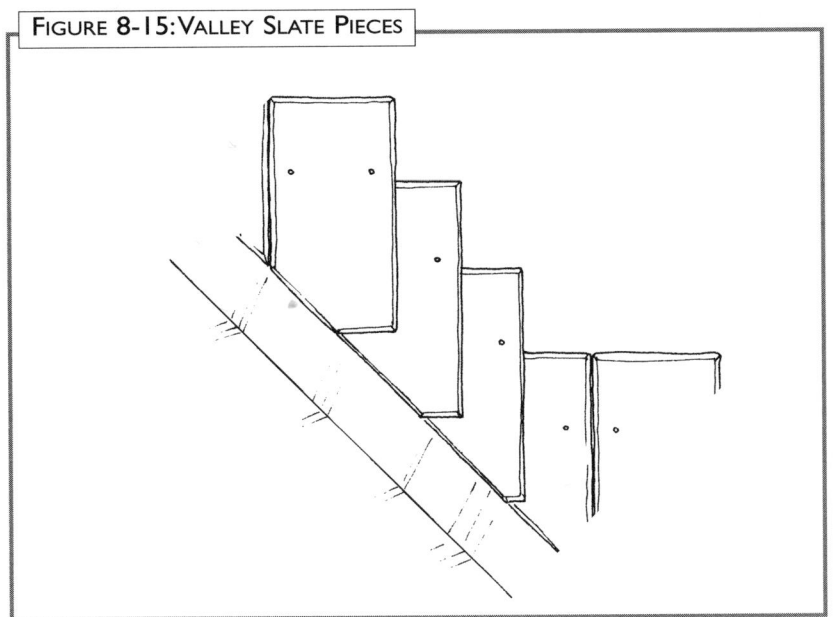

FIGURE 8-15: VALLEY SLATE PIECES

Valley Slate - Open Valley

In an "open" valley, the valley flashing is exposed. This is probably the most common valley used with slate roofing. As described in the estimating section, open valleys are always installed with a taper which increases from top to bottom. This taper amounts to 1/2 inch per 8 feet. The reason for this taper is to allow the valley to clear itself as water, snow, and ice flow down it. In order for the open valley to function properly, the flashing cannot be penetrated by any fasteners.

When the valley slate is cut, the quarry-punched nail hole above the valley cut should not be used. This is due to the fact that this portion of the slate will be laying over the valley flashing. Using this hole will result in penetrating the valley flashing. Instead, a hole will have to be punched directly above the one usable quarry-punched hole. This will allow the installer to use two fasteners per valley slate without

inches wide or wider will be required to get both a hip and a valley cut out of one piece. Set aside the widest material shipped to the project for this purpose.

To look right aesthetically, the valley must be straight. Always lay the slate to a straight reference line. This is true for both open and closed valley details.

Never install the field slate on a slate roof with the intent of going back later and cutting in the valleys. This is a common practice in tile roofing, but it will not work on a new slate installation. The only time that a slate valley should be seen incomplete with relation to the field of the roof is in a repair situation.

Once a valley has been installed, try to avoid walking on it. Unlike field slate that is easy to replace, valley slate is more difficult. Because valleys provide the easiest access to a roof, they are often damaged by other trades people walking up them.

Some contractors choose to set their valleys in a bed of plastic roof cement. This is not necessary with a slate roof. If you choose to use a mastic with the valley slate, use it only to adhere one piece of slate to the other. Keep the adhesive off the valley flashing. You do not want to restrict the thermal movement of the valley metal.

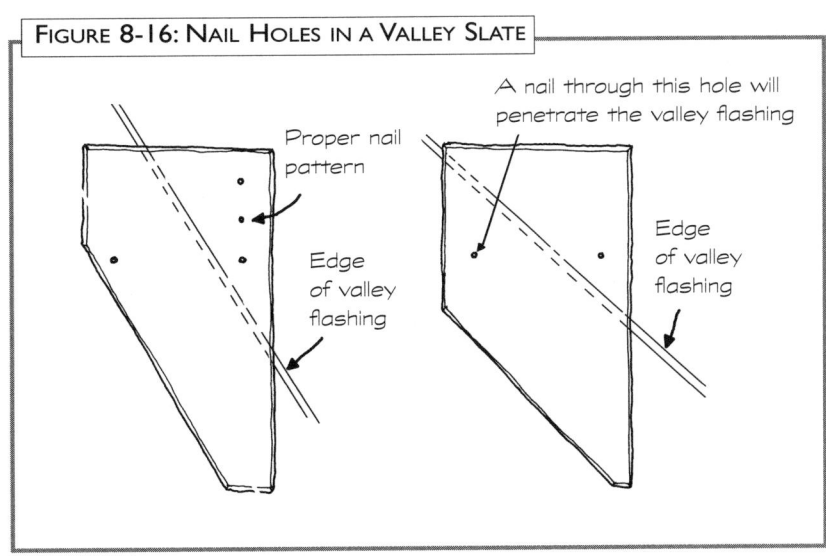

FIGURE 8-16: NAIL HOLES IN A VALLEY SLATE

Chapter 8

penetrating the valley flashing. A common mistake when punching additional nail holes for valley cuts is to punch the hole in the center of the slate. This will cause a problem with vertical joint alignment in the following course. When using large or heavy slate, it may be necessary to punch two new holes to allow for three nails to hold the valley slate securely. These new holes should be stacked above the appropriate quarry-punched hole. As always, these new holes should be punched from the back side. Try to make a habit of punching these holes prior to cutting the slate. The field slate is less likely to break when punched than the smaller valley cut piece. Further, if the slate is going to break during the punching process, it will save time if you determine this prior to cutting.

Always cut the open valley slate from the back side so the chamfered edge will face up like the rest of the field slate. If the open valley slate is cut from the front, water will tend to follow the direction of the chamfer back under the back side of the slate. If you have set aside pieces with broken corners that are unusable as field slate but are usable as valley pieces, you may be able to lay those pieces flat in the valley and mark the valley angle with a pencil or scratch mark. These marks will then need to be transferred to the back side of the slate so that the cut can be made from the back. If you are using full field slate, try the method shown in Figure 8-17.

This method is especially helpful when working with slate of one width. Because the alternating vertical joints on this type of roof have to line up, the valley slates will have to be cut individually for each course. This is not the case

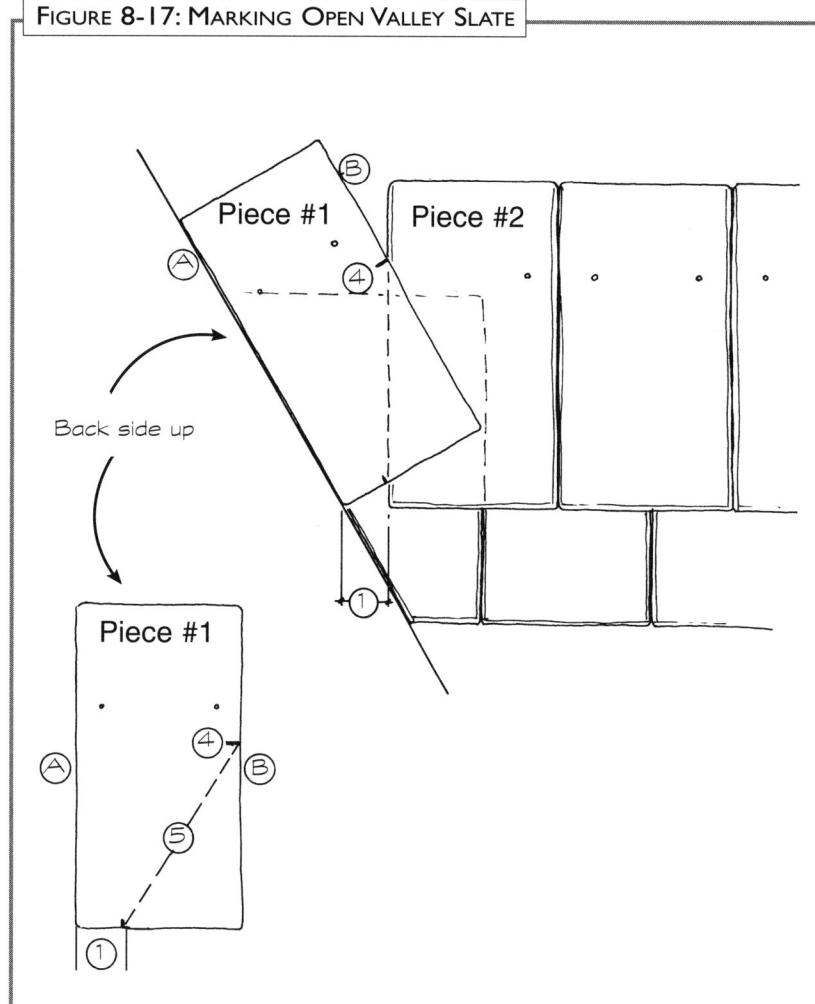

FIGURE 8-17: MARKING OPEN VALLEY SLATE

1. Measure horizontal dimension of missing piece.
2. Transfer dimension from Step 1 to back side of piece #1.
3. Lay piece #1 into valley so that side "A" lines up with the valley reference line.
4. Scratch a mark where the right side "B" of piece #1 intersects with piece #2.
5. Connect point #1 and point #4 with a cut line. Note: these cut marks are all made on the back side of piece #1. Therefore, the cut will be made from the back side as it should be.
6. Turn the piece over and install it.

Installing Slate

when working with random-width slate. On a random-width slate roof, a valley pattern can be established, all of the valleys can be cut on the ground, and the valleys can all be installed to the previously set layout lines.

There are several important details that the installer needs to keep in mind in an open valley detail: (1) Do not nail in the flashing, (2) maintain the proper headlap, and (3) maintain a smooth line.

On a steep complex curve, the valley curve will be accomplished progressively by cutting each slate over its exposed surface relative to the valley line. (See Figure 8-18.)

Valley Slate - Closed Valley- Continuous Flashing

In a "closed" valley with continuous flashing, the flashing is there but is unseen. (See Figure 8-19.) This continuous flashing method is probably the least desirable method of flashing a closed valley. This type of closed valley flashing is very similar to that of the open valley, but the valley flashing is concealed by the valley slate. In addition, the valley flashing is narrower than that of the open valley to avoid penetrating the flashing with the slate fasteners.

In a closed valley with continuous flashing, the "W" flashing configuration is most commonly used. The slate is then cut to fit the val-

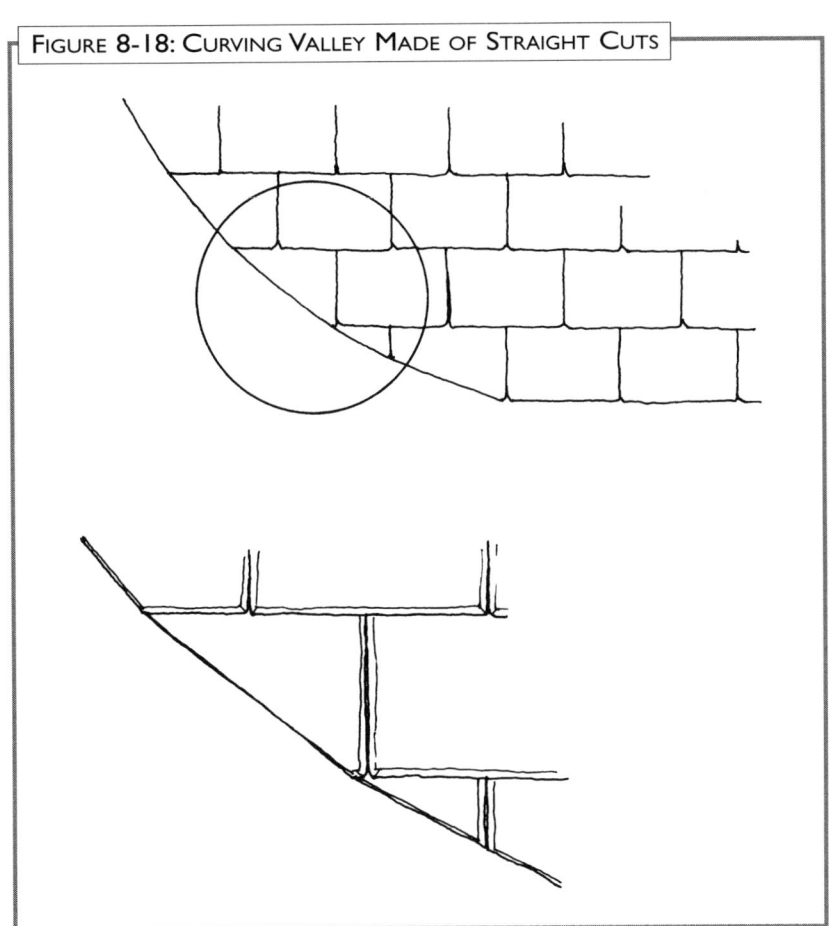

FIGURE 8-18: CURVING VALLEY MADE OF STRAIGHT CUTS

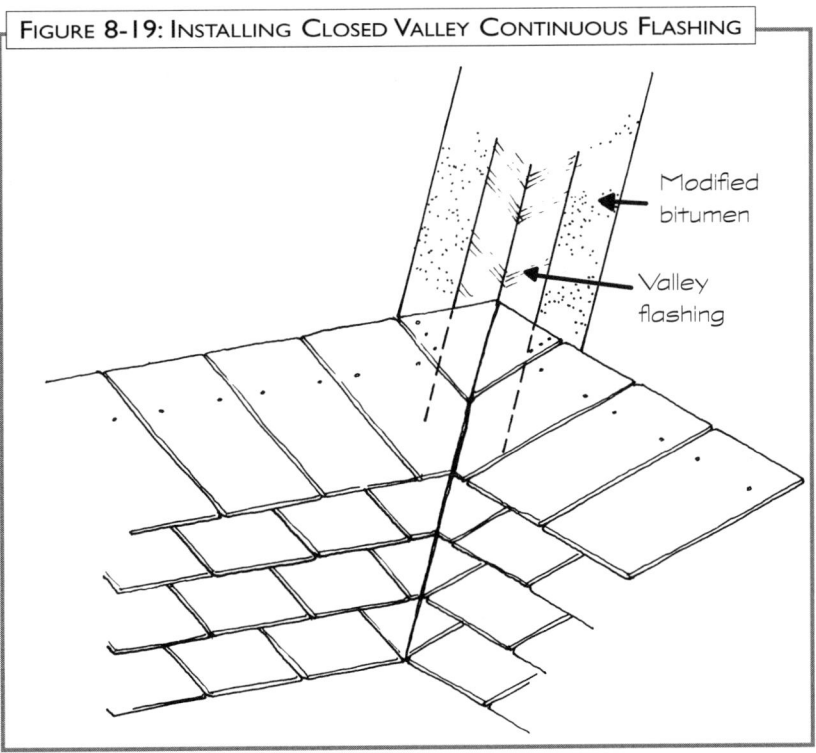

FIGURE 8-19: INSTALLING CLOSED VALLEY CONTINUOUS FLASHING

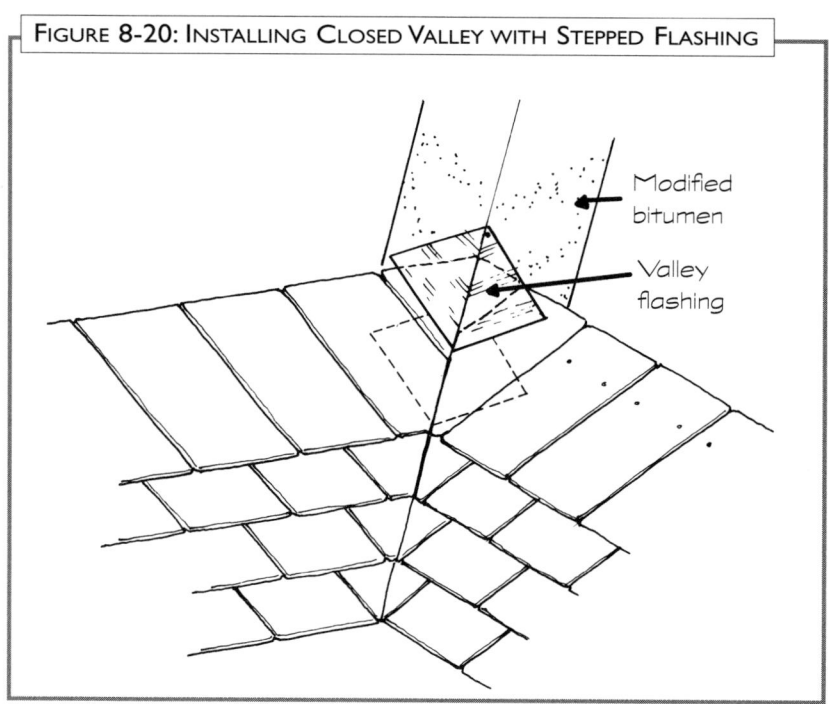

FIGURE 8-20: INSTALLING CLOSED VALLEY WITH STEPPED FLASHING

ley angle and butted up tight to the "W." In this application, the valley slate is cut from the front side so the chamfer created from the cutting process will form a miter with the "W" shape. This helps to conceal the valley flashing in the center of the valley. Otherwise, the slate installation procedure is the same as with an open valley.

Valley Slate - Closed Valley - Stepped Flashing

Stepped flashing is probably the most desirable method of flashing a closed valley. The slate for a closed valley is always cut from the front side, as described above, and the stepped flashing is then laced into each individual course. (See Figure 8-20.)

The procedure for installing a closed valley with stepped flashing with two different roof pitches is the same as the procedure shown above with one exception. If the roof pitch on one side of the valley is different from the roof pitch on the other side, always install the closed valley slate and stepped flashing on the shallow pitch first. The stepped flashing laid on the steeper pitch will then be covered and counter flashed by the valley slate laid on the steeper pitch. Do not attempt to lace the stepped flashing into the slate courses on the steeper pitch. Because the roof pitches are not equal, the butts of adjoining courses will not always line up. By laying the slate on the steeper slope over the slate and stepped flashing on the shallow slope, the water running off from the steep slope will run out on to the valley slate on the shallow slope. If the opposite approach is used, water running off the steep slope will tend to shoot behind the valley slate and flashing on the shallow slope.

If differing pitches are a consideration, it will not work well to precut valley slate for the steeper pitch. It will work well to precut the valley slate for the shallow pitch. However, the steep-to-shallow pitch will cause a situation where the valley cuts on the steep surface will have a long gradual cut that may extend out over more than one piece of slate. (See Figure 8-21.)

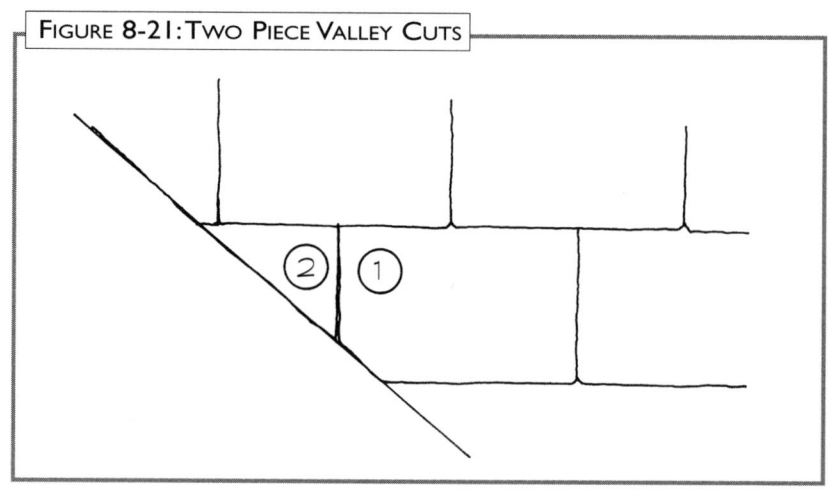

FIGURE 8-21: TWO PIECE VALLEY CUTS

Installing Slate

Valley Slate - Complex Valley

A complex valley is one in which the slate must be laid in a curved fashion. This is not meant to imply that the application itself is complex. Examples of this type of valley include valleys laid in a sweeping eave detail, open valleys that arch around a barrel vault (see Figure 8-22), and large unusual cricket details where the cricket roof changes pitch.

The only time that a complex valley would be encountered in a closed valley would be in the case of concave sweeping detail. The installer would not be able to snap an accurate chalk line through the belly of such a sweep. A straight line must then be transcribed into the sweep so that the finished valley line is straight. This can be done with a short straightedge. The step flashing pieces should be increased by 2 inches in each direction until the valley sweep has been completed. The application technique is then the same for any closed valley.

It will not be necessary to attempt to cut a curve into each valley piece on a steep valley with curves. However, it may become necessary to cut a curve into each valley slate when working on a shallow valley or the top of a barrel dormer valley.

When working with a barrel dormer, for example, there may be courses that require three or four pieces of field slate to be cut per course to fit the valley properly. To determine when to start the next progressive course of valley slates, you need to consider what keeps the valley watertight at that point. Typically, the 3-inch headlap from the preceding course of slate is what provides a watertight roof. Because the nailing of the preceding courses in a shallow valley may penetrate the valley flashing, it is not always possible to install these

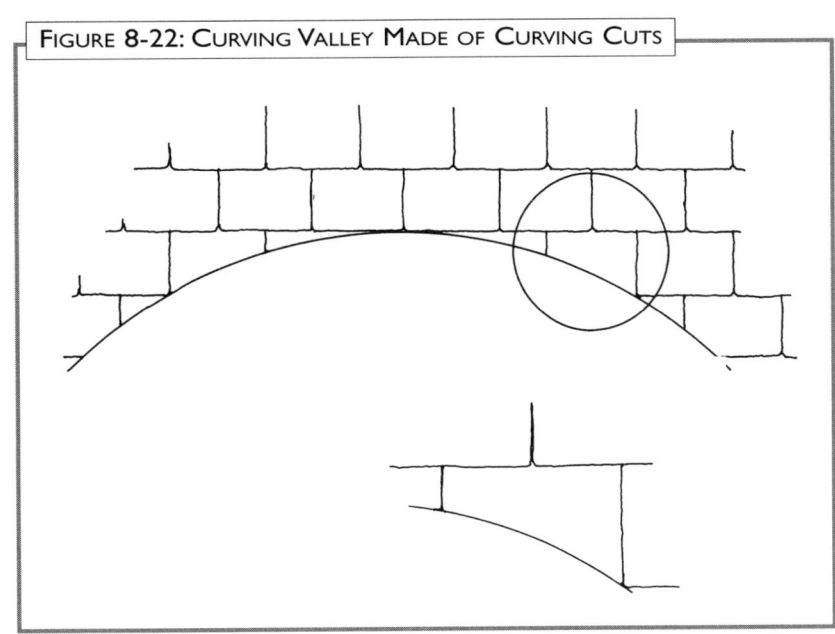

FIGURE 8-22: CURVING VALLEY MADE OF CURVING CUTS

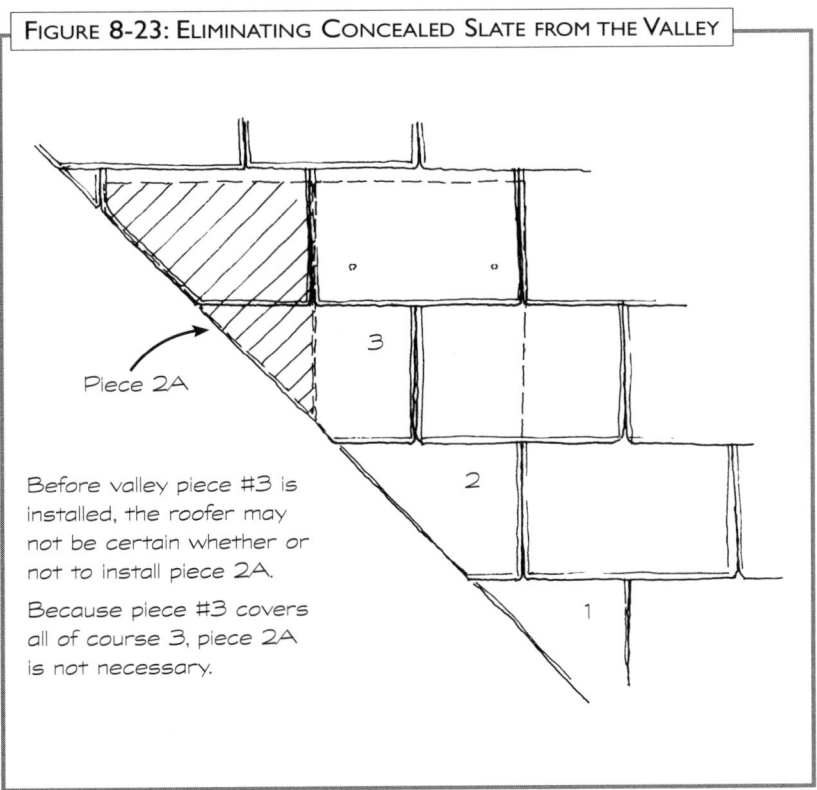

FIGURE 8-23: ELIMINATING CONCEALED SLATE FROM THE VALLEY

Piece 2A

Before valley piece #3 is installed, the roofer may not be certain whether or not to install piece 2A.

Because piece #3 covers all of course 3, piece 2A is not necessary.

preceding pieces in the valley. Therefore, the valley flashing itself is treated as the head of the preceding course of slate. When the valley cut extends through the concealed portion of the slate, it is not necessary to install any more valley pieces in that course because the next course above will completely cover any of the additional pieces in the valley. (See Figure 8-23.)

The width of the valley flashing in a complex curve will be determined using the same method as that for a standard valley (see Chapter 3). The slate, cut to fit the complex curve, will also taper to allow for water flow. The only time that this will change is in an open valley concave curve, such as a sweeping eave detail. In these circumstances, the installer will need to increase the width of the valley slate as well as the width of the exposed valley flashing to account for the complex curves in the sweep.

Installing Hip Slate

Hip slate is material cut or fitted to finish the hip. The term "hip slate" defines two different types of pieces of slate depending upon the hip style chosen, capped hip or mitered hip.

Hip caps are designed to cover the hip cuts and prevent water infiltration, but, as stated earlier, it is the base flashing material that really keeps this joint waterproof. Without proper base and/or step flashing, the water that would infiltrate at the joint created where the two hips meet could eventually enter the structure. In addition, in the case of a saddle hip, water entering at this joint will cause the hip nailer to rot, and the hip caps will sag as the nailer deteriorates. If the hip cap is not properly flashed, it will not last as long as the rest of the slate roof. Evidence of this can be seen on some older buildings.

Capped Hips

A capped hip will be required on any roof where: 1) the roof planes which meet at the hip are of differing pitches, 2) a vented hip is necessary, 3) slate of one width has been specified, 4) the specified slate is more than 3/8 inch thick, 5) the building owner prefers the look of a saddle hip.

Any hip that is capped with a slate, tile, or metal cap has base pieces of field slate cut to fit in place next to the hip nailer. These pieces of slate are referred to as "hip cuts."

Hip cuts must be installed after the related field slate has been installed for each course. Unlike valley slate, hip cuts cannot be precut from patterns and installed one on top of the other ahead of the field slate. Some installers will finish the field slate and go back later to cut in all of the hips, but you should try to avoid this approach. It is easier to finish each course and hip cut while roof scaffolding is in place and material is close at hand. The field of the roof is a large percentage of the roof area, and it is not uncommon for a roofer to install all of the possible field slate and request payment for that percentage of completion. Don't be fooled by this game. The hip, valley, and ridge details are where most of the work is required on a slate roof installation.

When installing a slate roof that requires a hip cap, the hip will have nailers installed up its length for securing the cap slate. These nailers will be started approximately 3 inches up from the eaves. The starter slate and first course of field slate will then be notched around the hip nailer so that they meet at the center of the actual hip at the eave. (See Figure 8-24.)

By using this technique, the hip nailer will not be visible from the ground once the hip caps are installed. Next, the field slate is run over to the hip and marked to be cut such that the hip cut will lie next to but not on the hip nailer. The width of the hip cuts will vary with each course. (See Figure 8-25.)

This will create a problem on some roofs related to nail placement in the hip cut. These hip cuts are the only pieces of slate within the field of the roof that may have exposed nails necessary to secure the slate properly. These nails must be covered by the hip cap. Therefore, to determine an acceptable location of a fastener for a hip cut, the installer can measure out perpendicular from the hip reference line the width of the chosen hip cap. Any exposed nail must be at least one inch less than this width so that the flashing material placed underneath the hip cap will flash these nails without being exposed once the hip cap is installed. This means that very small pieces of hip cuts may require nail placement such that the nails will end up going between vertical joints of preceding courses.

The same principle applies for determining whether a hip cut is needed or not. If the hip cut is so small that it will fall within this one-inch coverage of the hip cap, the hip cut will not be necessary. As always, the nail holes that need to be punched will have to be punched from the back side.

Hip cuts can be cut from the front side of the slate. When the piece to be cut is laid in place to be marked for cutting in relation to the hip nailer, it will be marked from the front side. Because the cut will be covered by the hip cap, the cut does not need to be made from the back side. This is a great area to use scrap material or pieces cut off from valley cuts. Whenever a need to cut slate arises, try to use a large piece of broken or scrap slate before cutting a good piece of field slate. You can get carried away attempting to

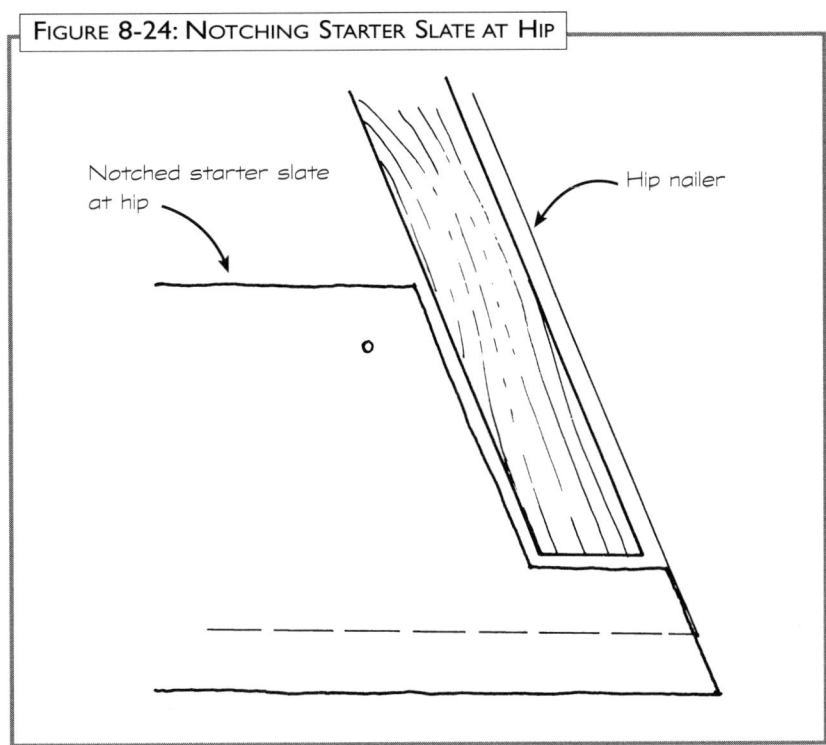

FIGURE 8-24: NOTCHING STARTER SLATE AT HIP

use scrap, though. Good judgment will dictate proper use of scrap.

Hip Slate - Saddle Hip

Slate that is attached to the hip nailer is referred to as the hip slate. The size of a saddle hip slate will usually be the size of the exposed surface of the field slate. For example, an 18-inch slate will normally be laid at 7 1/2 inches of exposed surface. Since the slate is not provided in 1/2-inch increments, plan on using the next size larger, or an 8-inch cap

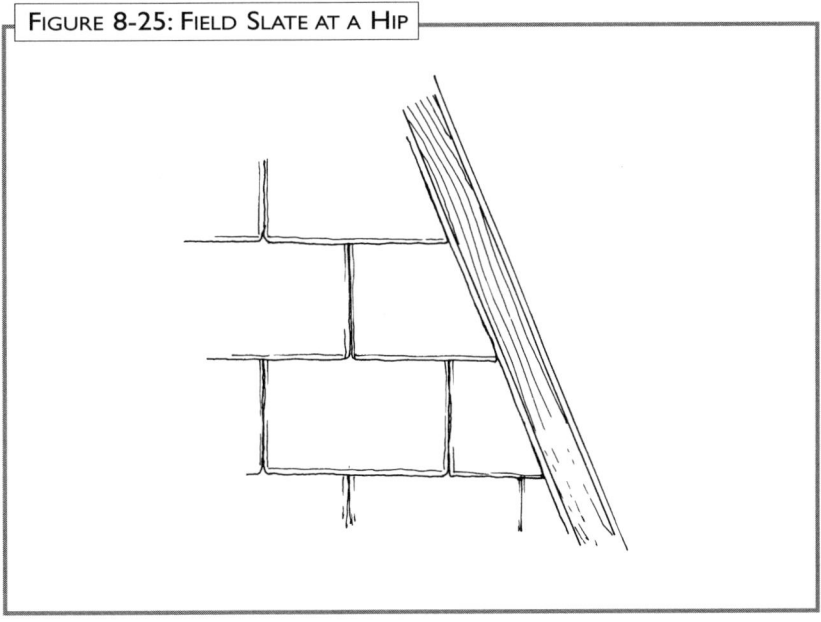

FIGURE 8-25: FIELD SLATE AT A HIP

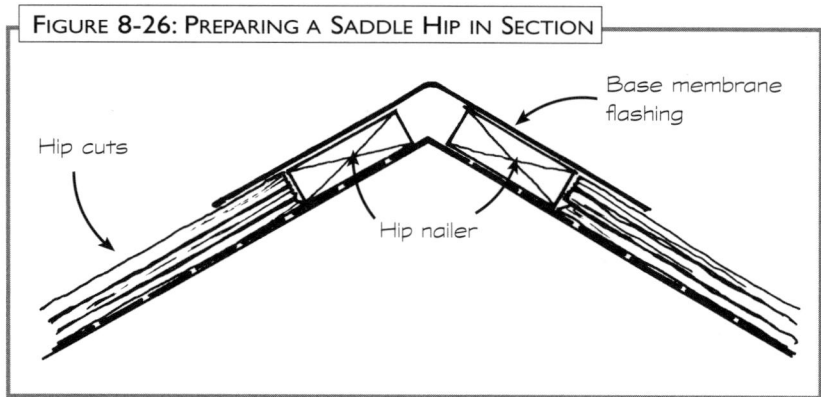

FIGURE 8-26: PREPARING A SADDLE HIP IN SECTION

slate. Keep in mind that this indicates 8 inches on both sides of the hip. The only exception to the rule of using a larger increment for the hip caps would be for 12-inch long slate. Since the minimum standard slate width is 6 inches, you may have to go slightly larger for this size.

To determine the length of the hip caps, multiply the hip cap width by 2. Thus, the 8-inch wide hip cap will be 16 inches long. These hip caps serve as a slate trim detail. The waterproofing membrane and/or flashing underneath is what keeps the hip waterproof. It is not necessary that these caps overlap the way that the field slate does.

Once the field slate and hip cuts have been completed on both sides of the hip nailer, the installer can prepare to install the hip caps. The first step is to install a waterproof membrane up the entire length of the hip. When installed, this membrane will be one inch narrower on either side of the hip than the hip caps. Using the same example of the 18-inch field slate with a 7 1/2-inch exposed surface, yielding an 8-inch wide cap, the membrane should be 7 inches on either side of the hip or a total of 14 inches wide. Some specifiers will ask for a continuous metal base cap at this point. However, the metal will be penetrated by the nails used to fasten the hip caps to the nailers, so the metal at this point will not perform properly. Once the base waterproofing material has been installed, snap a hip reference line from the eaves to the peak. This reference line will help the installer to keep the hip caps straight.

The application of the hip slate begins the same way as the slate roof: with a starter piece. The hip caps will be laid with an exposure equal to their width. Thus, a 16-inch cap will be laid with an 8-inch exposed surface. The hip cap starter slate will need to be 8 inches long to mimic the concealed surface of the rest of the hip cap slate. The hip cap starter will be installed front-side down just like the starters on the rest of the roof.

All hip caps as well as the hip cap starter will be nailed into the hip nailer. However, the hip caps are shipped to the site without nail holes because hip nailer locations will vary from project to project and the quarries have no way of knowing where to place the hip cap holes.

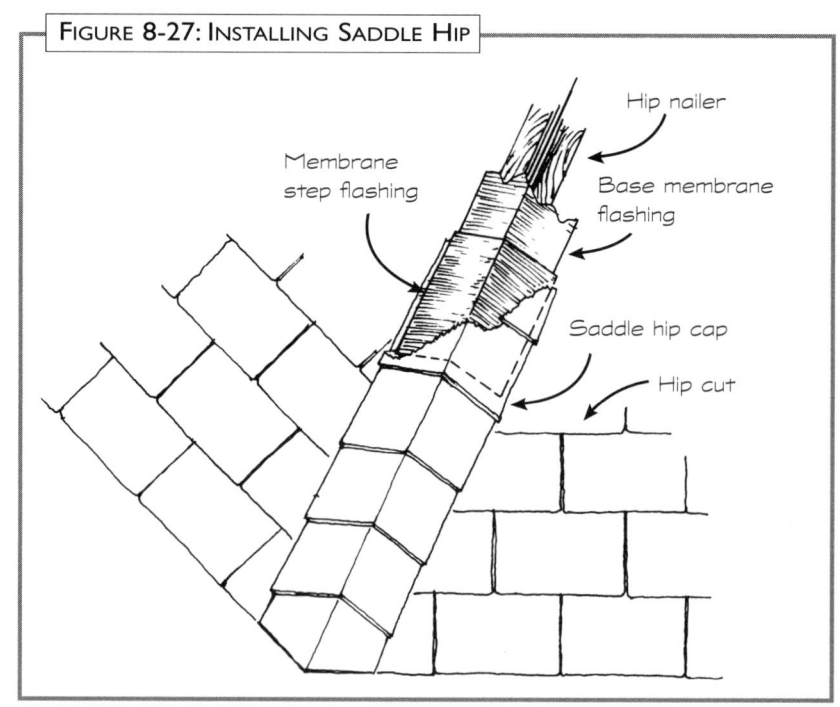

FIGURE 8-27: INSTALLING SADDLE HIP

Installing Slate

To punch nail holes in the hip cap starter, lay the hip cap starter slates on the hip front-side down. Mark the proper location for two nail holes in each starter with relation to the hip nailer. Then mark the eave line on the hip starter. Since the back side of the slate is facing up, this cut will be made from what is usually the back of the slate. The nail holes will be punched from the opposite side to provide the proper countersink.

After punching the hip starters, cut the eave angle. Lay the hip starters back in place on the hip to see that they fit properly. If they fit properly, use them as a pattern to mark the angle of the cut on the first full length hip caps prior to nailing them in place. Be sure to mark and cut the first full pieces from the back side. Then nail the hip starters in place. Location of hip cap nails may vary due to variations in the hip nailer placement. Punch 5 or 6 pairs of hip caps and install them before punching more. Make adjustments to nail hole locations as needed every 5 or 6 pairs.

The hip caps should be stepped flashed to prevent water infiltration. Not all slate roofers bother with this additional flashing step because the base waterproofing membrane will probably prevent any damage. However, most of these membranes are not UV-resistant, so the ultraviolet light coming through the joint created where

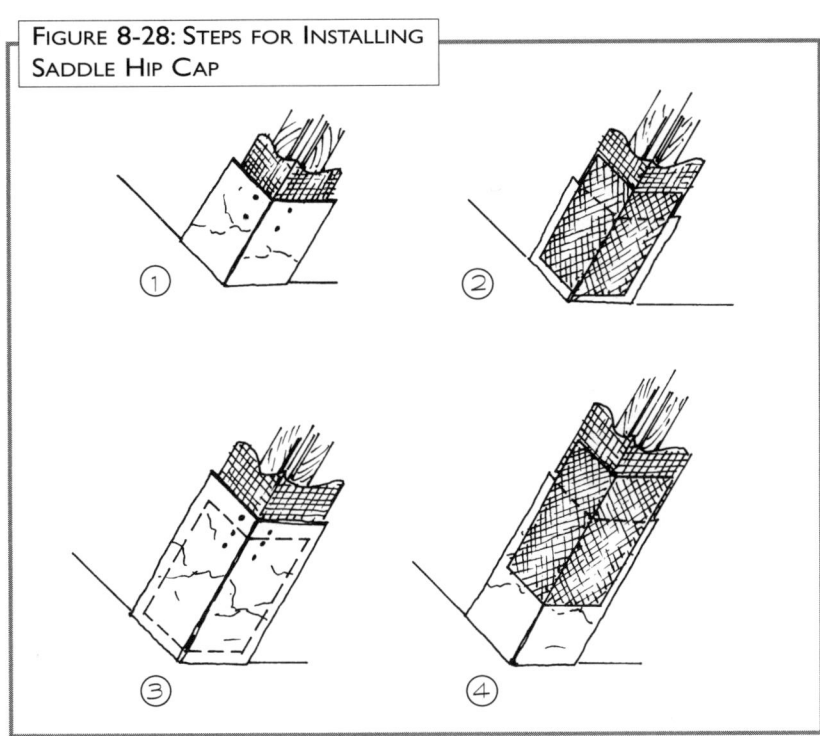

FIGURE 8-28: STEPS FOR INSTALLING SADDLE HIP CAP

the caps meet at the true hip will probably break down the membrane. By incorporating stepped flashing between these caps, the base membrane will be protected.

There are two approaches to this task that work well. The first is to step flash with the same metal flashing that is being used on the rest of the job. However, this metal

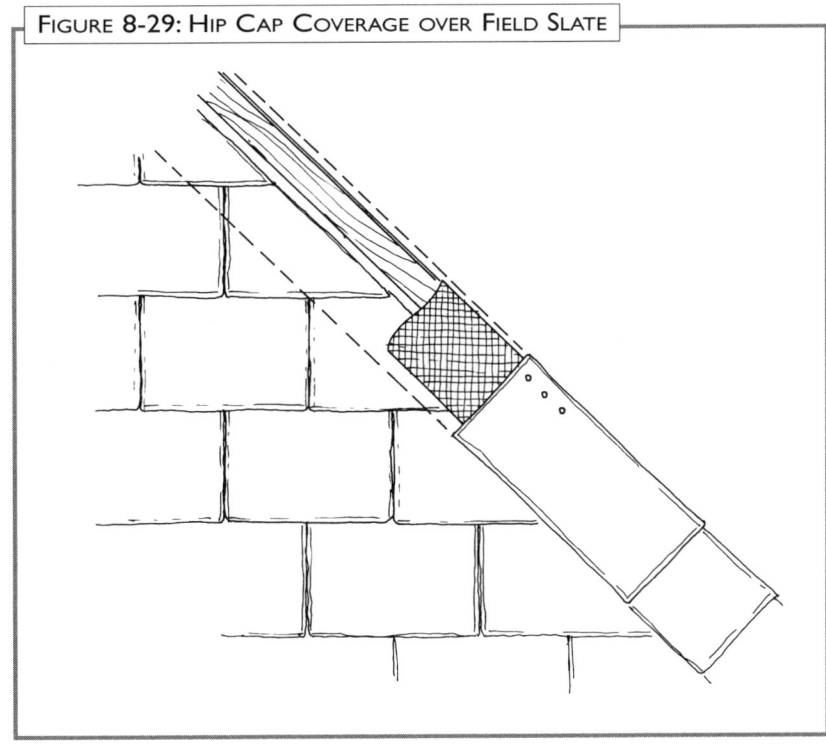

FIGURE 8-29: HIP CAP COVERAGE OVER FIELD SLATE

flashing detail allows the hip caps to slide apart as the progressive hip caps are installed. The second is to step flash with the same waterproofing membrane or a similar waterproofing membrane as that used for the base flashing. If a self-adhering membrane is used, it will help to hold each progressive set of hips in place. To prevent UV breakdown, the installer can caulk the hip joint upon completion.

Hip caps should have two to three nail holes per piece. Spread the nail holes out as far as possible over the concealed portion of the hip slate. This will help prevent the hip cap slate from sagging.

As the hip slate approaches the ridge, fit the last hip with relation to the ridge so that it is fastened in line with the top of the finishing course. (See Figure 8-30.) This will require that the last hip slate be cut to end at the same point as the finishing slate course. The ridge cap will carry across the top of the hip cap at this point. The nails fastening the last hip cap will have to be in line horizontally with the nails fastening the finishing course. The ridge cap will then cover these nails and counter flash the hip.

Where two hips intersect at the peak, the hip caps should be fitted with a miter cut that is perpendicular to the eaves. This will leave a vertical joint to fasten the finishing hip detail. However, the ridge cap must be installed prior to fitting the final hip piece. This final hip piece becomes both a hip and a ridge cap. There will be two exposed nails fastening this final hip piece in place. These nails should be caulked to prevent water infiltration. This is covered in more detail on pages 132–133. See also Figures 43–45.)

If you are working on a new wood-framed structure, it is common for the hip caps to sag and separate as work progresses. Anything more than a 1/4 inch of separation is excessive. Step flashing with an adhesive type of waterproof membrane, using plastic roof cement between progressive courses, and solid nailing will help to eliminate this problem. In any event, this gap should be resolved prior to removing the roof scaffolding from that surface. Such gaps are not acceptable in the finished roof.

The Saddle Hip Cap slate does not need to be, nor is it intended to be, miter-cut. As long as the hip is straight, it will look proper.

The hip caps will typically match the thickness of the slate roof. The

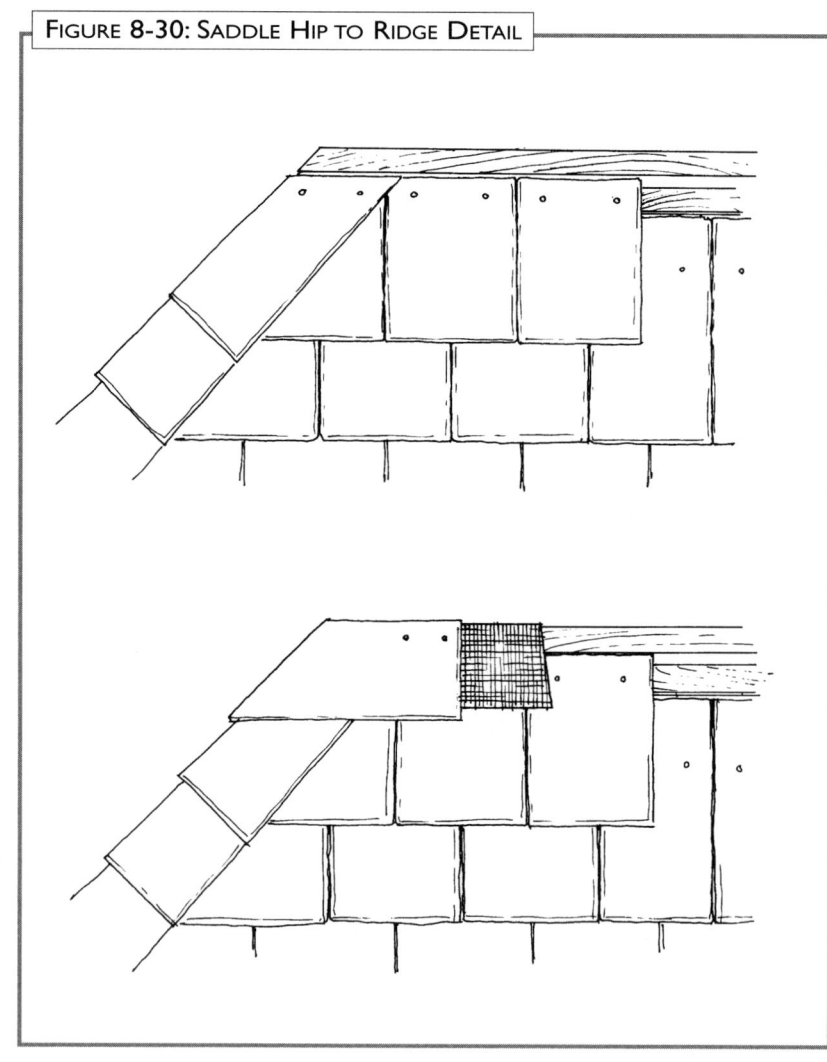

FIGURE 8-30: SADDLE HIP TO RIDGE DETAIL

Installing Slate

exception to this would be on a graduated thickness slate roof. Although you may choose to graduate these hip caps somewhat in thickness, their weight can become a problem. Hip cap slate is usually not thicker than 1/2 inch. Standard or textural thickness is typical.

Some installers like to apply the hip caps as they proceed up the hip area. This is fine, but because the field slate and hip cuts on both sides of the hip need to be in place, it is probably faster to install all of the field slate and hip cuts related to a hip area and then go back and install the entire hip flashing and cap detail all at once. Be sure to follow a hip line when installing the hip cap detail because this is a detail that draws attention.

Hip Slate - Mitered Hip

In a mitered hip, the hip cut as described above actually becomes the finish hip slate. The intent is to provide a smooth hip line that does not have the elevated appearance of other hip caps. In addition, the field course lines from adjacent roof surfaces will carry around the hip at the same level.

A mitered hip will not work on a roof where two different roof pitches meet at the hip. This is due to the fact that the length of the hypotenuse of the triangles created when the hips are cut will vary with the variation in roof pitch. Mitered hips also do not

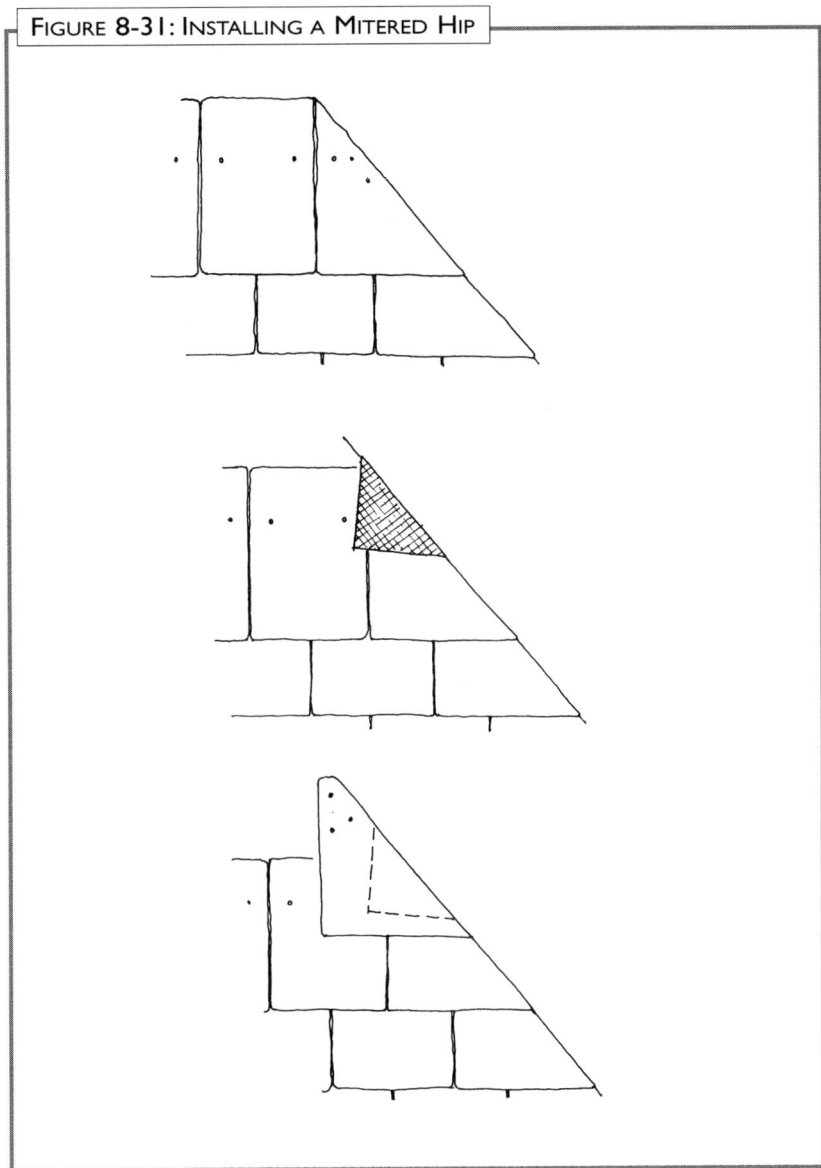

FIGURE 8-31: INSTALLING A MITERED HIP

work well when thick slate is used because thick slate is difficult to cut down to a crisp point.

If a mitered hip is desired on a roof with greater than a 3/8-inch thickness, you will probably need to consider a variation of the miter, such as the Fantail hip (see page 28). In a Fantail hip variation, the exposed points are intentionally clipped to provide a combination mitered hip/saddle hip appearance.

It is very difficult to install a mitered hip on a roof that has been specified with one-width slate. Because these roofs are usually started at a gable or between a hip and a valley, the hip size will vary. It can be done using approach slate, but this is a very difficult technique to master and should only be done by experienced craftspeople. In addition, this is a very expensive method. A

Chapter 8

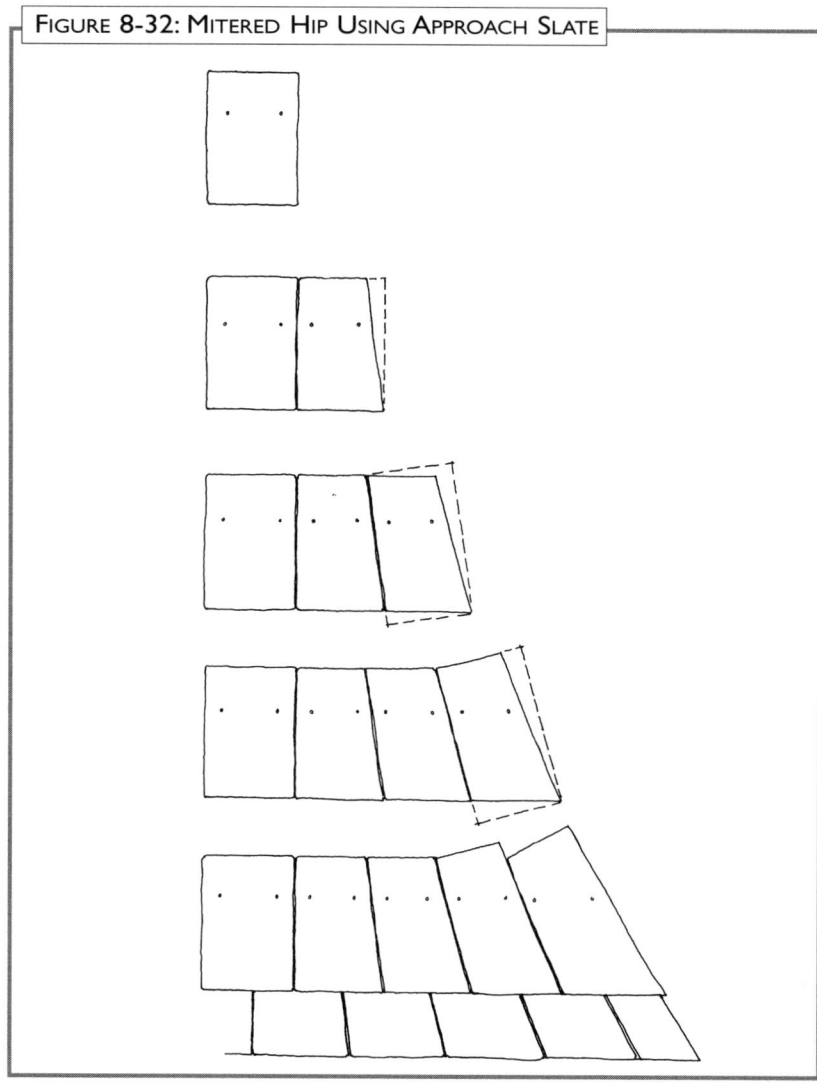

FIGURE 8-32: MITERED HIP USING APPROACH SLATE

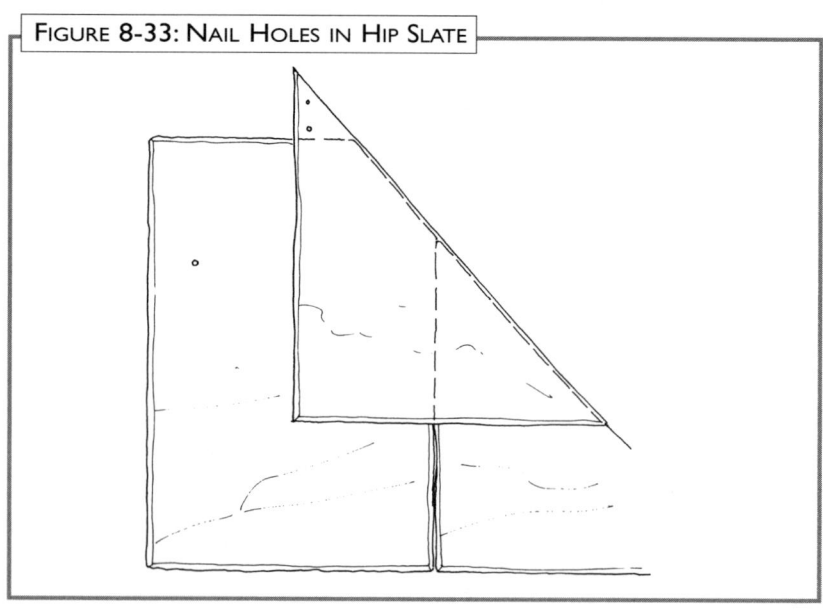

FIGURE 8-33: NAIL HOLES IN HIP SLATE

great deal of slate must be custom-cut to approach the hip prior to the actual hip slate itself, nail holes will have to be relocated in most of these cut pieces, and a fair amount of unusable scrap will be generated. (See Figure 8-32.)

As with the valley slate cuts, it will work best to establish a hip pattern from which to mark all of the mitered hip cuts. These pieces can be marked, punched, and cut on the ground. On projects with closed valleys and mitered hips, it is often possible to get a valley and a hip cut out of one piece of slate. If this seems to work on your project, try to use a slate large enough that when the hip pattern is laid on the front side of the field slate, the remaining valley cut will have at least 2 inches on the horizontal plane.

You will notice that the pattern for the hip slate will probably leave very little room for a second nail hole to be punched. In this particular case, you will need to punch your second hole below the quarry-punched hole. If you punch it above the quarry-punched hole, the piece will probably break there when it is cut.

Due to the weight of the large pieces of slate that are hung from two holes placed closely together, these mitered pieces will have a tendency to sag. Try to keep the holes as far apart as possible yet still covered by the hip slate in the progressive course.

Installing Slate

Mitered hip slate will always be cut from the front of the slate. This will allow the chamfered edge created in the cutting process to create a tight mitered joint with the adjacent hip slate. The initial roof layout is critical when installing a mitered hip roof. If the horizontal reference lines do not line up at the hip, the butts of the hip will not line up either.

Wider slate chosen for the hip cuts will provide a bigger nailing area. For example, on roofs with a 12:12 pitch or less, an 18-inch slate with a 7 1/2-inch exposure will require a minimum of 11-inch wide slate in order to have sufficient slate within the concealed area to punch holes. To get a hip and a valley cut out of one piece of slate, it will also work best to use the widest slate provided. This means that for the 18-inch long slate example, using a 14-inch wide slate as your cutting stock will provide a 12-inch wide hip cut with a valley cut that still has 2 inches remaining on the horizontal butt dimension.

Consider where the quarry-punched nail hole is on the slate to be cut. When cutting a hip slate for a 12:12 pitch that is 12 inches wide at the bottom, both nail holes will be almost if not completely eliminated. If you punch a new nail hole too close to the hip cut line, the slate will often break when cutting. You should attempt to keep the new nail holes as far away from the cut line as possible without leaving the nail heads exposed. The nail holes should be at least 1 1/2 inches above the exposure line. The hip flashing material will then cover the nail heads.

Because of the nature of the new nail placement for a mitered hip, the fasteners will often be below the head of the previous course, so the fasteners to be placed through these new holes will hit the top of the approach slate below. This approach slate is so named because it has to be cut as it approaches the hip. This is not always the case for roofs over 12:12 pitch. When cutting the approach slate, allow room for nailing of the next hip.

Proper flashing of the mitered hip is critical. Years ago most hips were not flashed. The finished joints

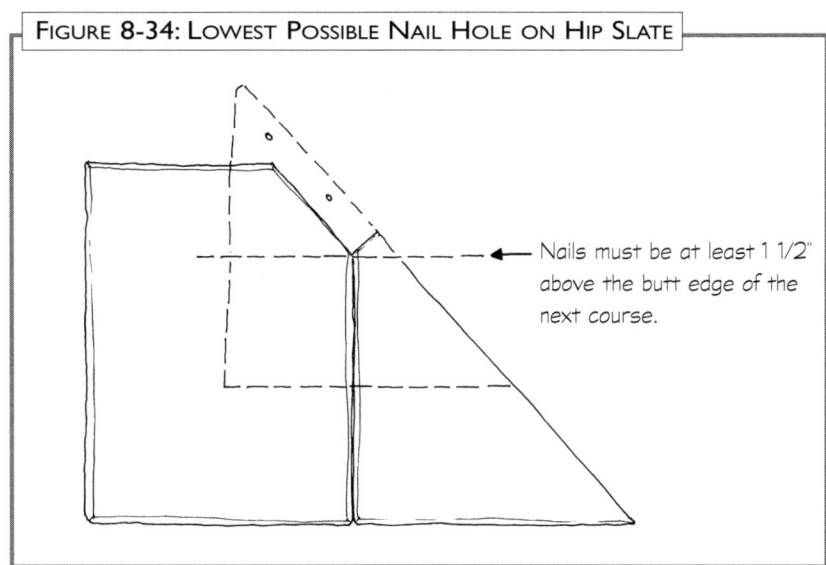

FIGURE 8-34: LOWEST POSSIBLE NAIL HOLE ON HIP SLATE

Nails must be at least 1 1/2" above the butt edge of the next course.

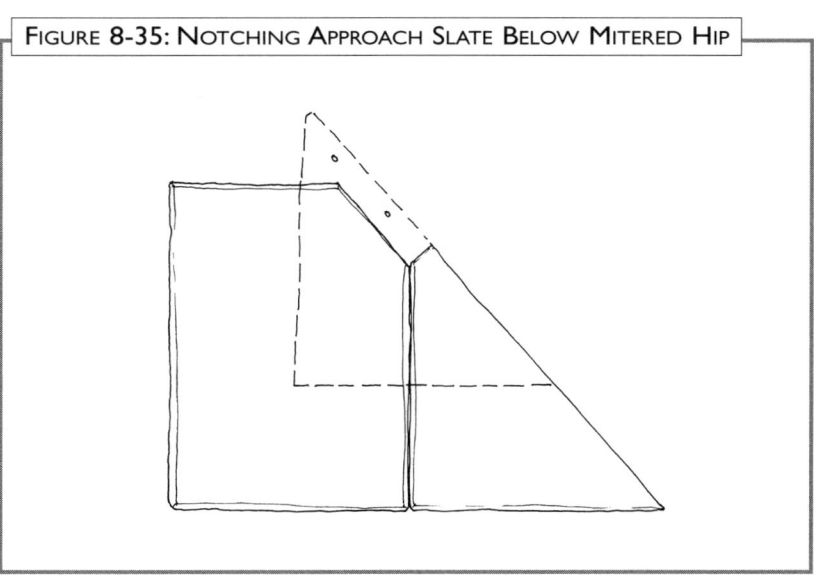

FIGURE 8-35: NOTCHING APPROACH SLATE BELOW MITERED HIP

Chapter 8

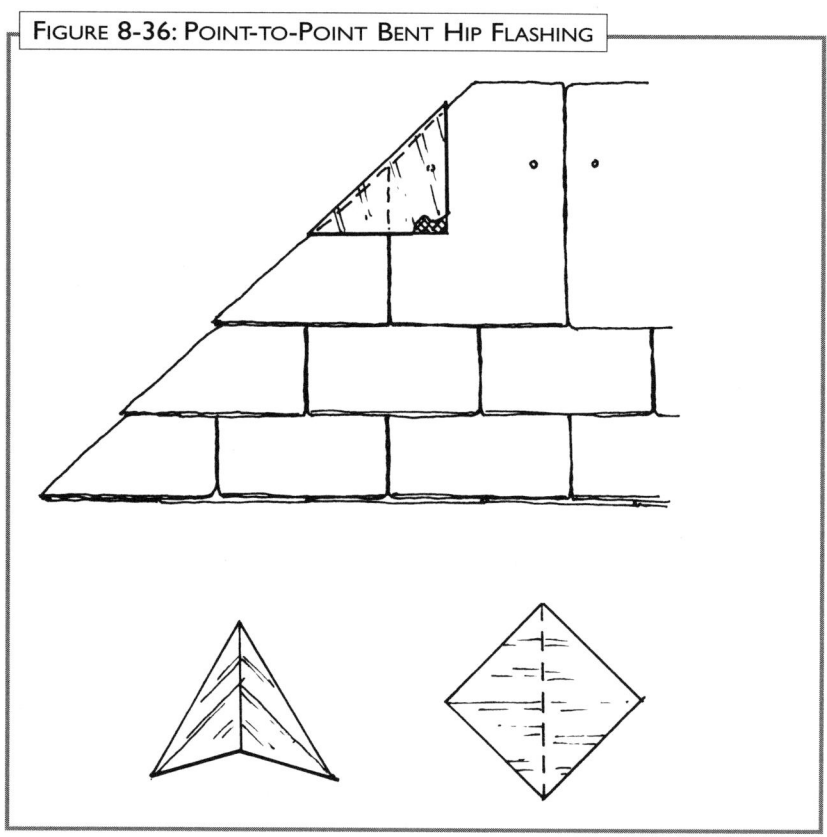

FIGURE 8-36: POINT-TO-POINT BENT HIP FLASHING

were simply pointed with roof cement. This pointing technique is not very attractive and will not last for more than 5 to 10 years. It is much better to step flash these details with concealed flashing. Modified bitumen works well for this application. The membrane will help to glue the hip pairs in place and prevent water infiltration. Some specifiers prefer to see a metal flashing detail incorporated at this joint. Unless the metal is bent and fit perfectly, it will tend to lift these miter cuts off the roof somewhat. In addition, these miter hip pieces have a tendency to sag and slide when metal is used as the base flashing. If metal is chosen, the hip cuts should be set in a bed of mastic.

After choosing the hip flashing material, you will need to determine the size needed (see Chapter 3). These flashings will be formed and installed point to point.

Installed in this way, the flashing will extend over the vertical slate joints of the adjacent approach slate. This flashing will prevent water infiltration where the preceding approach slate had to be notched for nailing the hip pieces.

Mitered hips are a difficult detail for a novice installer. In order for this detail to appear correct, the hip framing must be straight, the slate must be laid to a straight line, the nailing must be secure to prevent separation of adjacent pairs, and base flashing cannot show.

Try to use the smoothest and most even slate available to cut the hip pieces from. It is important that adjacent hips be cut from slate of similar thickness; otherwise, successive hips will begin to wander. Also, because broken corners are extremely noticeable on a mitered hip, do not use any pieces with broken corners over 1 inch.

Installing a mitered hip with rough textured slate can be very difficult. Some dense slate, such as red slate, can be extremely brittle and difficult to cut in this pointed configuration. Be sure to determine whether or not the slate desired will work well for this application.

Mitered hips on roofs steeper than 12:12 are usually easier to install. These pieces are usually easier to cut as well because you do not have to cut both ends to a point. However, it is difficult to precut these pieces from a pattern. These hips will vary in width due to the steep nature of the hip. The installer will often need to mark and cut these hip pieces on the roof as the hip progresses.

The mitered hip does not require a hip nailer. Each hip piece lies in the same plane as the field slate and fastens directly to the roof deck. The finishing course will also require a mitered hip cut to finish the hip. The ridge cap and its base flashing will then cover the nail heads of the hip pieces.

A mitered hip requires careful

planning. When installing a mitered hip on a roof with a 12:12 pitch or less, it is best to precut and punch these pieces on the ground.

Try to work from the mitered hip area out into the field of the roof. This will allow the installer to use the precut hip slate. Always try to put a wide piece of slate beside the mitered hip cut, and try to use the same size piece beside the hip all the way up the slope. This will allow the progressive precut hip pieces to lie with the proper vertical joint alignment with respect to the hip approach slate.

Keep in mind that a mitered hip roof is generally requested on a home where symmetry is important. A mitered hip will almost always be seen in conjunction with random-width slate. By precutting the hips and working from the hip out, this detail can be done easily with minimal cutting required on the roof.

In order to install a mitered hip properly, you must be prepared to install the field slate on both sides of the hip. This means that additional scaffolding as well as manpower may be necessary.

Hip Slate - Fantail Hip

A Fantail Hip is a variation of a Mitered Hip. The only difference is the shape of the miter cut piece. All of the installation, flashing and layout is the same as a Mitered Hip. The critical point is that the clipped corners will all need to be the same which means that a pattern should be used. Also, the flashing will need to be cut or modified to match the clipped corner of the slate.

Hip Slate - Boston Hip

A Boston Hip is a combination of the Saddle Hip and the Mitered Hip. This hip will require the most work and planning of any of the hip details. You will need to precut, from patterns, the hip slate and the approach slate. The roof must be started at the hip for this detail to work properly. This detail will not work well with slate of one width, nor will it work well if the roof pitches on adjoining roof surfaces vary. This installation should only be attempted by experienced installers.

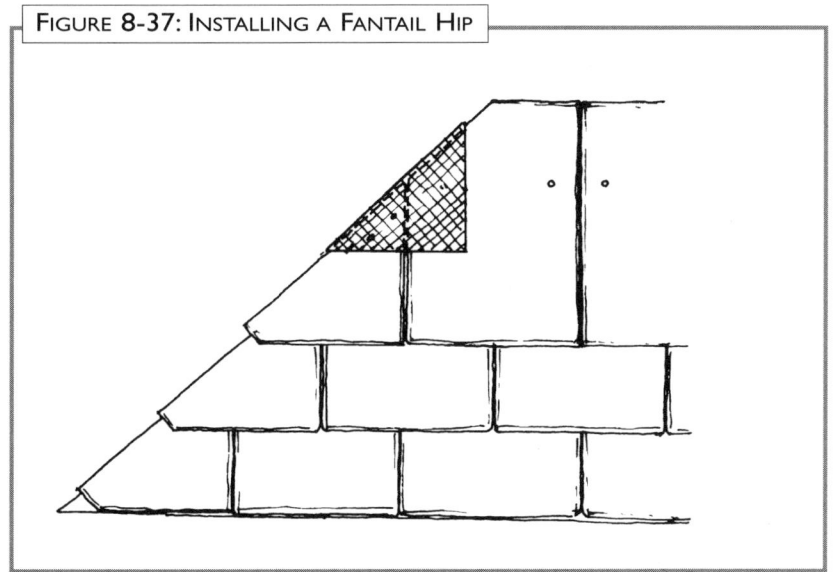

FIGURE 8-37: INSTALLING A FANTAIL HIP

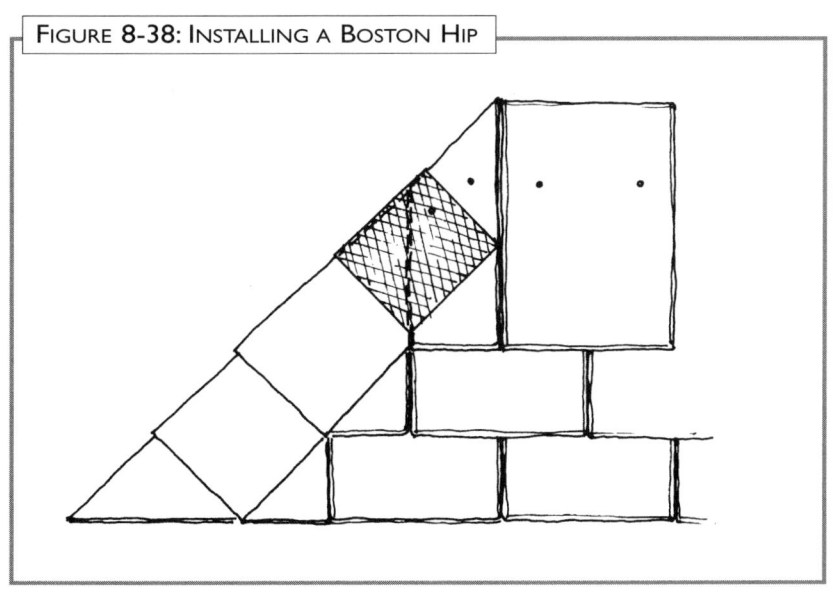

FIGURE 8-38: INSTALLING A BOSTON HIP

Chapter 8

Installing Ridge Slate

Ridge slate is the slate attached to the ridge nailer or vented ridge cap detail. Unless otherwise specified, a slate roof will be finished with a slate ridge cap. Ridge slate must be installed after the related field, valley and hip slate has been installed.

The ridge cap will usually be close in size to that of the field slate exposure, so if you are using 18-inch slate with a 7 1/2-inch reveal, the ridge cap should be at least 7 1/2 inches, rounded up to 8 inches. The ridge slate needs to cover the head of the last full course of field slate by 3 inches to achieve proper headlap.

Ridge Slate - Strip Ridge

The basic installation principle for the strip ridge is to make it appear as though it is a course of field slate. The slates are installed butted side to side, not overlapped the way a saddle ridge would be. This detail will work fine, but if this is the chosen cap, a metal base flashing should be used over the membrane base flashing to prevent UV breakdown of the membrane at all of the exposed joints.

To be well secured, these cap pieces should be set in a bed of mastic and the nails placed as far apart horizontally as possible.

The major drawback to this type of cap is that all of the fasteners attaching the cap slate will be exposed and will need to be covered with caulking. The ridge should be pointed with a similar material. This cap was widely used in years past. However, 100 years of experience has shown that this cap will only last as long as the flashings.

Ridge Slate - Comb Ridge

The comb ridge is almost identical to the Strip Ridge. The only difference being that the ridge slate on one side is installed so that it extends approximately the thickness of the slate beyond the ridge of the other side.

Installation and layout for a comb ridge is the same as that for any other ridge. To finish this detail, point the ridge with caulking or roof cement. In years past, this was a preferred installation technique. The joint created allowed the installer to properly waterproof the ridge. However, with modern flashing materials, this approach is not necessary and rarely used.

Ridge Slate - Saddle Ridge

The saddle ridge is installed in the exact same fashion as the saddle hip. The exposure and width are determined the same way, the base flashings are installed in the same manner, and the punching/ nailing is done the same way. The only major difference between the saddle hip and saddle ridge is their orientation to the field slate. Because the ridge is parallel to the field of the roof, the ridge slate can be a little more difficult to fasten. To properly join the slate at the ridge, the cap pieces will begin to slope or cant backwards.

During the layout process, a reference line was snapped at the top of the roof. This line is the line to which the last full course of slate will be laid. If you are working with slate that is 16 inches long or longer, this reference line is far enough down from the ridge that you will be able to install a finishing course

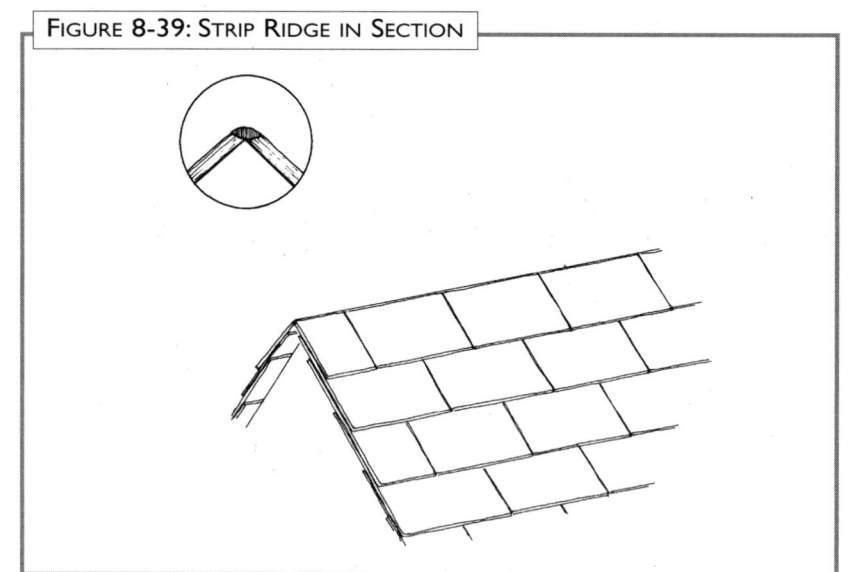

FIGURE 8-39: STRIP RIDGE IN SECTION

Installing Slate

nailing strip as well as a double-thickness nailing strip for the ridge slate to be fastened to. The double-thickness nailing strip will be covered with the waterproofing membrane. This membrane will lap over the top of the finishing course, covering the nail heads and counter flashing the entire peak of the roof. When the ridge slate is later nailed through the membrane into the double nailer below, the membrane will seal around the nails and keep this area watertight.

Using a metal stepped flashing detail on the ridge will not help to shed water. The water will follow the slope of these caps and find its way back to the ridge base flashing. If metal is used as a flashing at the ridge, the slate caps will have a greater tendency to sag as they slide on the metal. Be sure to use an adhesive if metal base flashings are specified. Using a self-adhering membrane flashing at this point will aid in holding the caps together. When the cap is completed, the installer should point the ridge with a UV resistant adhesive to prevent UV light from direct contact with the base membrane.

Installing Ridge Slate at a Valley

Closed Valley. For a closed valley, the ridge detail is relatively easy since the valley base flashing has been installed between the adjacent valley slates. As the valley

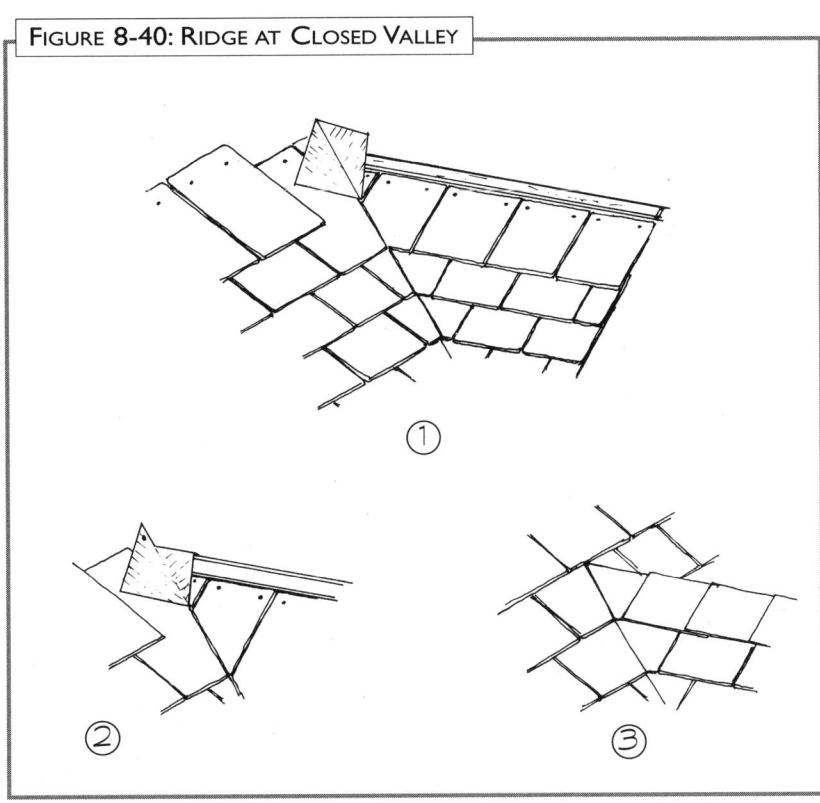

FIGURE 8-40: RIDGE AT CLOSED VALLEY

flashing approaches the ridge it is fitted to form a saddle over the ridge. This flashing saddle is created by installing an additional piece of valley step flashing that lays in the valley on top of the finishing course slate, extends beyond the ridge and is then cut and folded over the ridge. The ridge slate can then go on. This flashing is then covered by the ridge membrane base flashing, followed by the ridge slate. The ridge slate is started at the valley, cut to fit the valley angle, and worked back toward the opposite end.

Try to start all valley ridge details at the valley and work out. This will allow you to keep the nails from the first pair of ridge slates as far away from the center of the valley flashing as possible. A ridge starter piece will usually not be needed at the valley. The build-up of base flashings will elevate the first set of ridge slates adequately. Finish this detail by caulking any joint created where the ridge slate intersects the field slate of any adjacent roof surfaces.

Open valley. For an open valley, the same installation procedures apply as those for a closed valley. The difference is that the valley flashing will be in place and the flashing saddle will be installed and soldered at the peak prior to starting the slate. Extend the ridge cap over the open valley flashing. This will cap both the valley flashing and the adjacent slate. If the ridge cap slate from the intersecting roof is

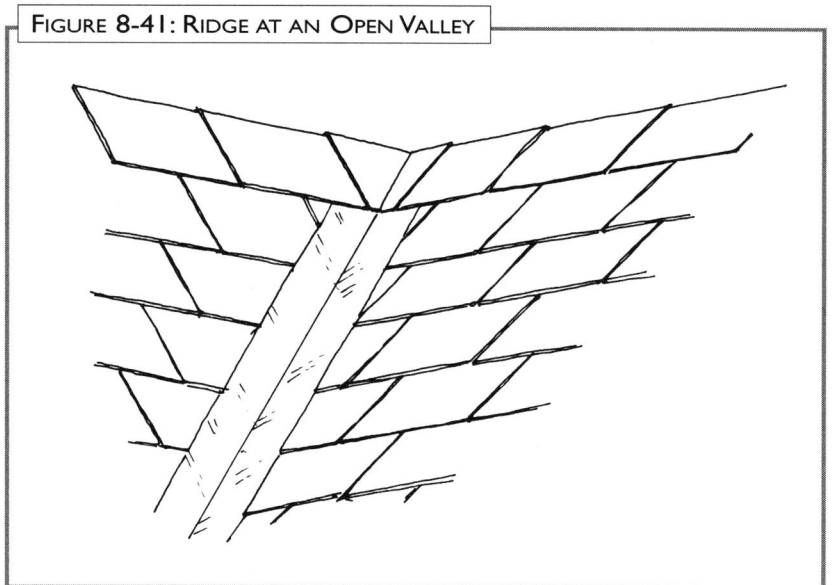

FIGURE 8-41: RIDGE AT AN OPEN VALLEY

left short or cut in line with the open valley line, water will have a tendency to shoot off the field slate of the adjacent roof and under the ridge slate at the valley. In regions with freezing and thawing, snow and ice will have a tendency to break or remove the ridge slate at the end of an open valley if the slate is not installed in this manner.

Finishing the Ridge End

On a gable-end roof, (Figure 8-42) the ridge caps start and finish relatively easily. The ridge is started with a starter slate and ended with the last piece of cap slate that can be attached to the exposed nailer. To insure that all the cap pieces have the same exposure, it may be necessary to stretch or reduce exposure of the last 4 to 5 feet of ridge. In some cases, it is desirable to install one final piece of cap slate that matches the exposure of the rest of the cap; this finishing ridge slate should be the same size as the ridge starter slate. To attach these finishing cap pieces, punch them, lay them over the last full piece of cap slate, drill through the holes in the finishing pieces and through the full piece below with a masonry drill bit, and install the fasteners. These fasteners will need to be caulked to prevent water infiltration.

On a hip roof, the hips will all be completed to the ridge prior to starting the ridge slate. If a saddle hip is being used, it will not be necessary to use a ridge starter slate: the saddle hip slate will be elevated enough to act as the starter slate for the saddle ridge. For a mitered hip, it will be necessary to start the ridge with a starter slate.

First, cover the ridge with the appropriate sized membrane. If the ridge end is a hip detail, be sure to overlap the hip with the membrane to seal that area.

To start the ridge-to-hip connection, take the first two full ridge pieces and lay them in place on the ridge.

Extend these pieces far enough beyond the hip so that the lower points of the ridge slate line up with the hip reference line. Mark the ridge slate along the hip reference line. Cut these two pieces of slate along this hip line. These cuts should be done from the front side of the slate. Keep in mind that only

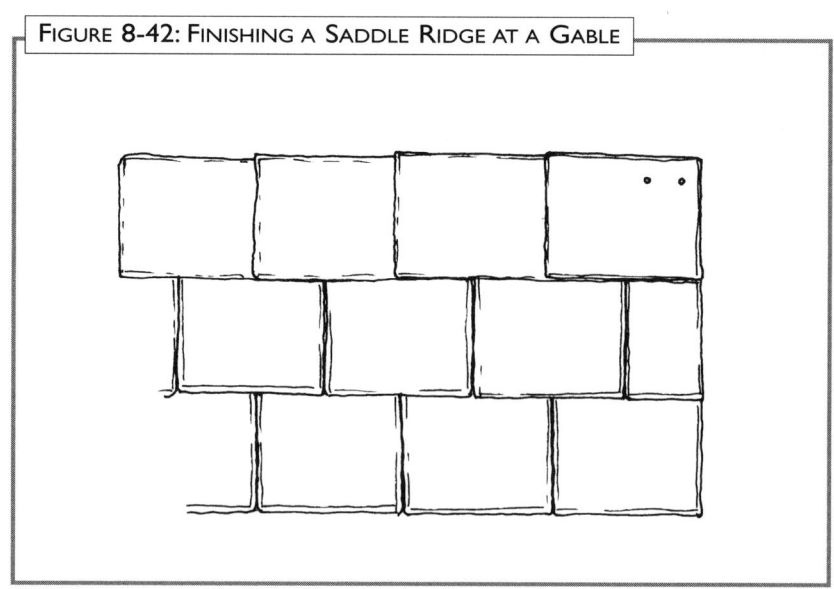

FIGURE 8-42: FINISHING A SADDLE RIDGE AT A GABLE

Installing Slate

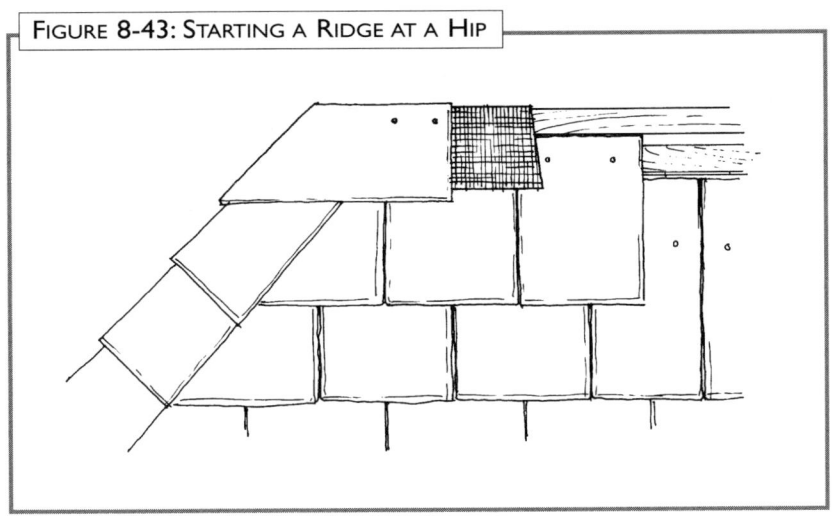

FIGURE 8-43: STARTING A RIDGE AT A HIP

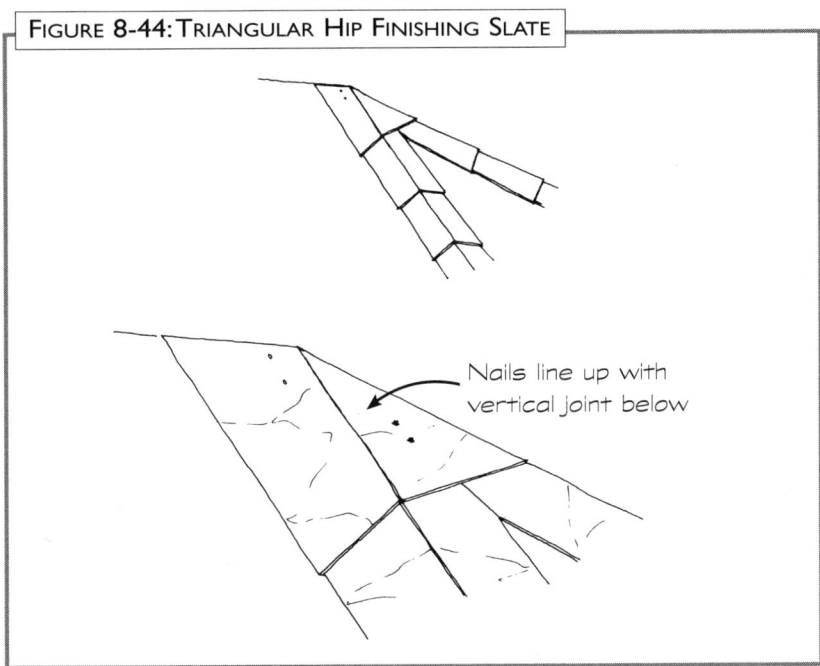

FIGURE 8-44: TRIANGULAR HIP FINISHING SLATE

Nails line up with vertical joint below

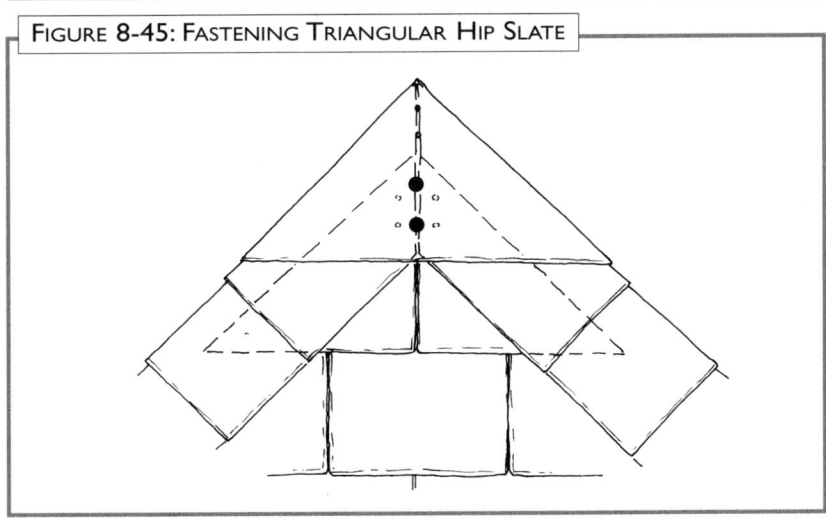

FIGURE 8-45: FASTENING TRIANGULAR HIP SLATE

one half of this piece of slate will be exposed. Place your nail holes accordingly. The ridge can now be installed as usual. (See Figure 8-43.)

On the hip end, there is now a missing piece in the shape of a triangle. Some installers will cut a triangle this shape to finish the hip end. The bottom of this piece should have the chamfered edge up as normal. The two sides will be cut from the front side to form a miter with the ridge cap pieces. Attaching this hip finishing triangle is relatively easy if the hip slate has been finished properly. When the hip is finished properly, there will be a vertical joint perpendicular to the eaves centered with the ridge line. The nails which fasten the hip finishing slate will need to be placed in a vertical line so that they will penetrate the roof through the vertical joint below. (See Figure 8-44.) This is a delicate piece of slate. Be sure to locate and punch these nail holes prior to cutting. It is difficult to punch a triangle-shaped piece after it has been cut. If all else fails, drill it with a masonry drill bit.

On the opposite end of the ridge, the ridge slate will finish the same as it started, but this piece of slate can be difficult to nail because the last full ridge slate will be close to the end of the nailers. There may not be room to nail the finishing ridge slate. You may have to fit, cut, and punch this slate, drill the nail

Chapter 8

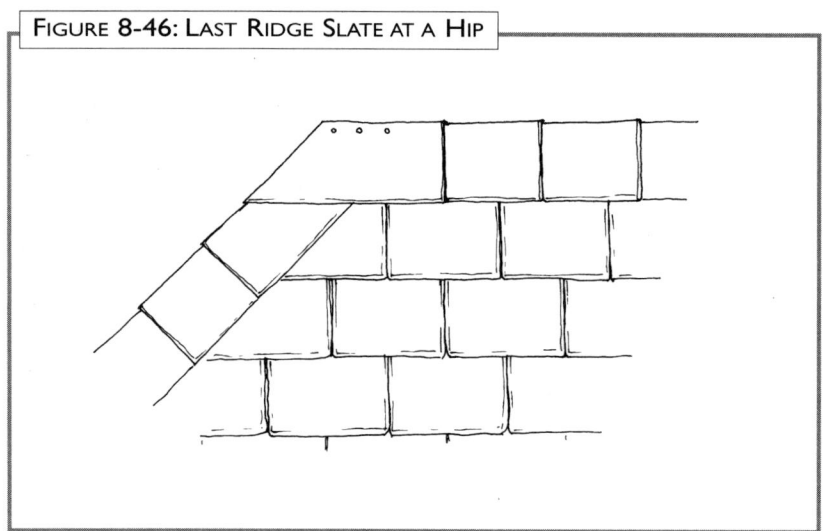

FIGURE 8-46: LAST RIDGE SLATE AT A HIP

holes through the last full ridge slate pieces, and then install the fasteners. Some installers will do all of the above except that they will punch a hole through the last full course with their slate hammer or a nail set. Although this method will work, it does not have a high success rate. If a drill is not available, punch additional holes in the last full ridge slate prior to installation to allow for nailing of the ridge finishing slate.

The ridge finishing slate should always be set in a bed of mastic, and it will also have exposed fasteners that need to be caulked.

When installing a saddle ridge over a mitered hip, you may need a ridge starter slate, depending on how high the last course of hip slate sticks up. If the hip slate elevates the first pair of ridge slate, a ridge starter slate may not be necessary. When possible, avoid using a ridge cap starter when working with a hip roof.

Ridge Vents

To vent a roof, it is possible to install dormer vents, modular vents, and a variety of power vent fans, but most owners and planners agree that they prefer to avoid the appearance of vents because they look non-traditional.

There are a number of ridge vent systems available for composition shingles and wood shingles which allow the installer to fasten the ridge cap shingles through the vent material. In this type of ridge vent, the ridge cap shingles are elevated above the field shingles by 1 inch to 1 1/4 inches. Although it is obvious that the ridge is elevated to allow for ventilation, it is not obtrusive.

Many designers want to achieve a similar appearance for a slate ridge cap, but this technique does not work well with slate. The nails cannot easily pass through the slate below, and this application may cause leaks. The field slate needs to be brought up to the point at which the sheathing ends, so the nails holding the ridge slate will have to penetrate the top of the finishing course to work properly. In addition, slate shingles are much heavier than composition shingles, so the weight of the individual slate caps when fastened by long nails will allow the cap slate to sag. The result is a vented ridge detail that is not likely to last for long.

To install a vented slate ridge, use the following procedure:

Install slate on both sides of the ridge as shown in Figure 8-47.

Determine the ridge vent material or fabric you wish to use.

Build a plywood frame as shown. The width of the plywood should be 1 inch narrower than the ridge slate to be used.

Install drip edge as shown.

Attach the chosen vent material to the underside of the plywood frame. Attachment methods will vary, but when the entire unit is set over the ridge, the pressure from the ridge fasteners will help to hold the vent material in place.

Set the entire unit in place over the finished slate roof.

Using screws of sufficient length to go through the thickness of the plywood, vent area, two thicknesses of slate, and the roof sheathing, fasten the plywood box to the ridge. Screws should be installed at a maximum of 18 inches apart and 2 inches up from the

Installing Slate

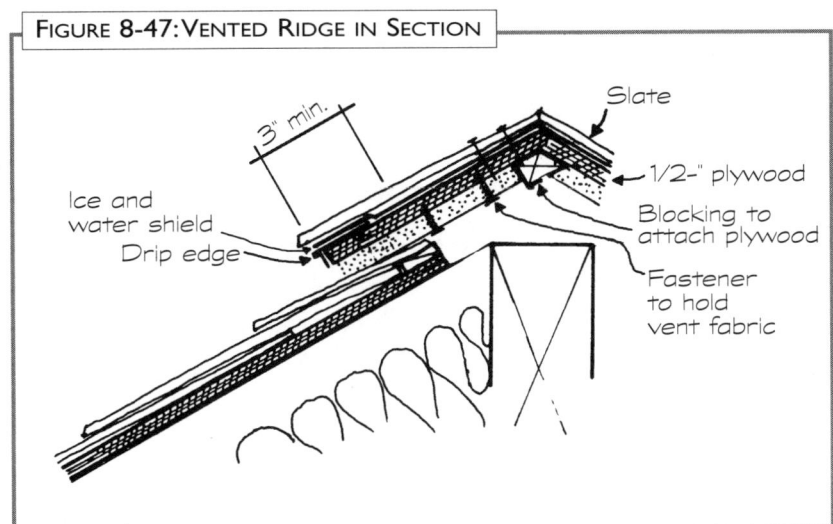

FIGURE 8-47: VENTED RIDGE IN SECTION

outside edge of the drip edge. These screws will need to be installed so that they pass through the vertical joints in the finishing course. They will penetrate the head of the last full course. A pilot hole may need to be punched or drilled through the head of the last full course to allow the screw to pass through to the deck below.

Install membranous flashing over the top of the entire cap as shown.

Install the cap slate as described in the "Saddle Cap" section of this chapter.

Installing Slate around Plumbing Vents and Other Obstructions

Most roofs will have obstructions penetrating the roof sheathing which need to be slated around. Two common obstructions are plumbing vents and attic ventilation units. In most cases, the roofer will have no say in determining where these obstructions

FIGURE 8-48: PLUMBING VENT IN PLAN - STEPS

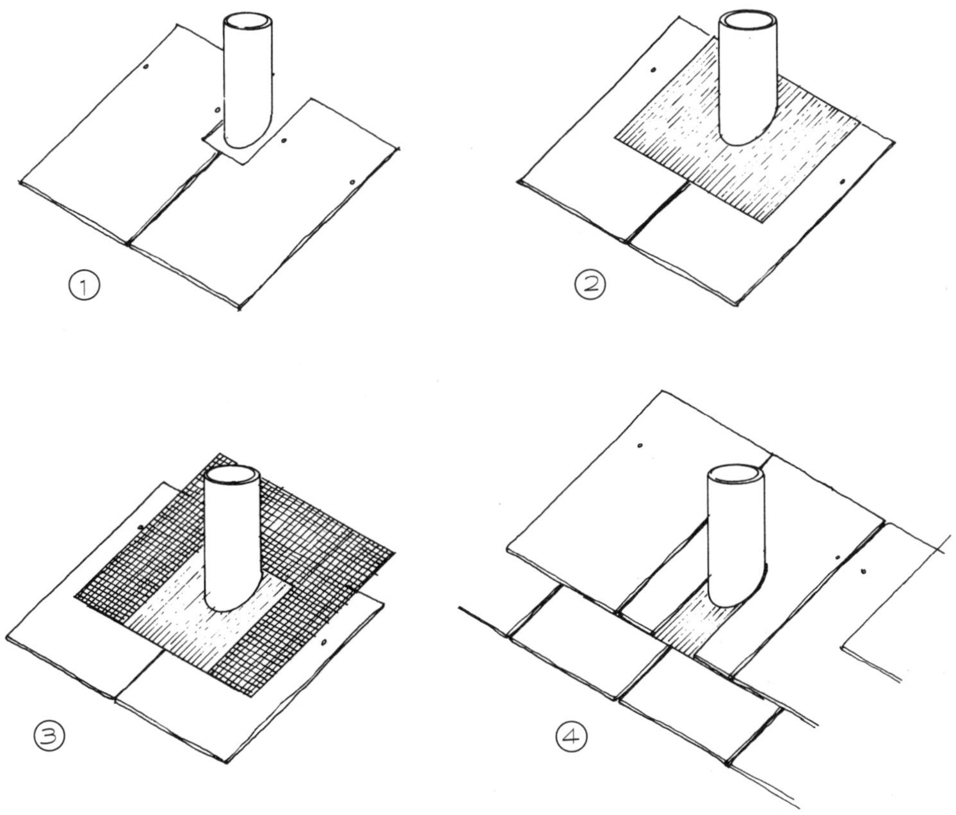

135

Chapter 8

will appear in the roof. Because the horizontal line layout is based on the ridge reference line and the eave reference line, the obstructions generally need to be cut in wherever they happen to be.

The rule for these obstructions is that the butt end of the flashing of these units needs to cover the head of the preceding full course of slate by at least 3 inches. This means that the butt end of the flashing should be fit or cut to line up with the butt end of the nearest course of slate that provides coverage.

To install a plumbing vent, it often helps to set the flashing base in place over the pipe prior to installing the slate. Lay the field slate on either side of the vent flange to the reference line that most closely aligns the butt end of the slate with the butt end of the flange. This will show you which course of slate best lines up with the butt of the flashing flange. The flange is then removed and the slate cut in around the pipe up to this aligning course. The flange is then placed back over the pipe to flash both the pipe and the slate. (See Figure 8-48.) Some installers will strip in these flanges with a strip of modified bitumen. Now the aligning course of slate can be installed over the sides of the flashing, but avoid nailing through this flange. The slate is then cut around the vent flange. If desired, you can notch the slate to fill in the roof area in front of the pipe. This is not necessary and is time-consuming, but some owners and architects prefer this appearance.

This same technique can be employed for almost all obstructions. When in doubt, set the base flashings in place on the roof and determine slate layout accordingly.

It will be helpful with many obstructions such as skylights to do the roof layout prior to installing the obstruction. This will provide the roofer with reference lines on each side of the obstruction which are consistent with the rest of the roof. Otherwise, the roofer will need to tie in the lines around obstructions with the rest of the given reference lines.

Installing Slate at Intersection of Dormer

All of the basic installation techniques described above apply to slating a dormer. The location of the first course of slate on an intersecting roof will determine where the valley flashing will be placed. If prior layout is not carried out properly, the valley flashing may not lay into the slate coursing on the field surface properly. This is an obvious concern with a closed valley. Because a closed valley is best executed with step flashing, the first course from the main roof surface and the first course of field slate on the dormer need to line up; otherwise, the flashing will be exposed. From this point up, standard valley installation technique is used.

Installing Slate on Dormer Sides and Walls

Slate is often specified on dormer cheeks and other vertical walls. The principles of installation are all the same. A cant strip and properly sized starter slate is required, layout prior to installation is a must, nailing is the same, and cutting is the same. On surfaces with a pitch greater than 20:12, the headlap can be reduced to 2 inches. If the vertical surface being slated intersects with another slate roof detail, the flashing should always be laced in with the shallower pitched surface. The slate from the steeper surface then counter flashes the flashing in the shallow pitch.

Because slate is hung and not nailed tightly, it is common for building owners to complain that the slate on the vertical surfaces tends to chatter when it is windy. The solution is not to over-nail but to set each piece with a dab of mastic. The finishing course of a vertical wall is most likely to chatter, but this can be reduced by making sure to install a nailing strip for the finishing course.

Vertical surfaces and dormer cheeks will almost always be finished with some form of metal flashing. Be sure to adjust your layout accordingly to have proper course spacing. Essentially, the flashing cap functions as a slate cap would. However, depending upon the architectural detail, this flashing may intentionally be different in size from the exposure of the slate itself.

Slate used on vertical walls is usually smaller than that used on the field area of a roof. One reason for this is aesthetics. Because of the angle from which a roof is observed, the 18-inch long material applied at 7 1/2 inches on a 12:12 pitch will appear much smaller than the same material observed on a vertical wall. Try to visualize the number of courses of field slate that will intersect the courses on the vertical wall. Using fewer courses than this on the vertical surface may make the wall appear squat.

Installing Snow Guards

Snow guards come in a variety of sizes, shapes, and materials. Installation techniques and specifications vary greatly from one manufacturer to the next. Two sample specifications have been provided in Appendix C.

There are several key things to look for when choosing and installing a snow guard:

Always use a snow guard which attaches to the roof deck independently of the slate itself. There are several designs which hook to the slate and/or the shaft of a nail holding the slate. If and when these snow guards fail they will cause damage to the slate.

Look for a snow guard with a wide base attachment strap. This will help distribute the snow load more evenly over the slate below.

Thick-base straps have a tendency to lift the slate above them. Consequently, when the snow guard is loaded, the slate above it is susceptible to breaking.

Look for a snow guard whose snow pad is perpendicular to the roof surface. There are some designs available that have the snow pad attached at an angle which is less than perpendicular to the roof surface. When the pad is loaded it has a tendency to be pushed over, causing the base strap to lift up, which can cause the slate above it to break.

Some snow pads attach to the base strap on a thin perpendicular post. The snow pad then is wider than the attachment post. These guards have a tendency to twist when loaded. Look for a snow guard whose snow pad is fully supported with a gusset. This support will help distribute the snow load evenly over the entire snow guard.

Each manufacturer should have test results showing how much load each of their units can resist. Load capabilities vary greatly with design. The least expensive cost per unit cost can, in fact, translate into the most expensive snow guard system. Evaluate the price on a "cost per square of roofing" basis.

Installing Lightning Protection

Lightning protection is an important consideration for any tall structure. The attachment of the lightning arresters and grounding cable becomes a concern when working with a slate roof. Slate can be drilled through easily with a masonry drill bit. Any attachment bracket that needs to be installed on a slate roof can be through-fastened by first drilling the necessary holes. Try to do this work in conjunction with the roof application or repair.

Installing Tile Hip and Ridge

The installation of tile hip and ridge on a slate roof will vary depending upon the tile style chosen. You will need to provide for the proper size nailer for these pieces as well as the proper roof layout. When in doubt, construct a mock-up. The mock-up will show you how close to the hip and ridge the slate will need to be laid in order for the tile to cover it and remain watertight. It will also show you how large of a nailer will be needed to properly support and attach the tile. Fasteners will vary with the thickness and style of tile chosen.

NOTES

CHAPTER 9
Flashings

This section provides basic information on flashings and gutters specific to slate roofs. Anyone can install a slate roof. However, the true quality of a slate roof installation is often based on the quality of the flashing installation. There are a number of excellent manuals that cover flashing installation in detail (Sheet Metal and Air Conditioning Contractors National Association (SMACNA), Copper Development Association (CDA), and *Copper & Common Sense* are three outstanding sources for this information). If your flashing detail is more complicated than those shown here, refer to these other manuals.

Stepped Flashings

For any vertical penetration in a roof, such as a wall, chimney, skylight, or dormer side, the point of intersection will need to be flashed. This flashing detail is referred to as "stepped flashing," although some people have different names for it. In some Midwestern states, for example, stepped flashings are referred to as "tins"; in Europe, they are often referred to as "soakers."

All stepped flashing will extend up the vertical surface a minimum of 4 inches and under the slate a minimum of 4 inches, so the raw material width for all stepped flashing will start at a minimum of 8 inches wide and will be bent in the center. (For proper stepped flashing sizes, see Chapter 3.)

Construction

Any wall detail that is less than perpendicular to the roof surface should be treated as a closed valley detail, which requires conventional stepped flashing. In this approach, fasten the stepped valley flashing to the roof deck just above the top of the slate.

Always fasten the stepped flashing to the roof deck. Never attach it to the vertical wall side. Fasten the step so that the bottom edge of the step is concealed by the next course of slate. This can be accomplished by measuring down from the layout line above the length of the slate that you are working with and scratching a reference line on the slate below. The stepped flashing is then fastened not more than 1/2 inch above the scratch mark. The stepped flashing is then completely covered by the next piece of slate.

Apron Flashings

Wherever the slate ends at the face of a penetration, the roof will require an apron flashing with a minimum of 3 inches overlaying the head of the last full course of

Chapter 9

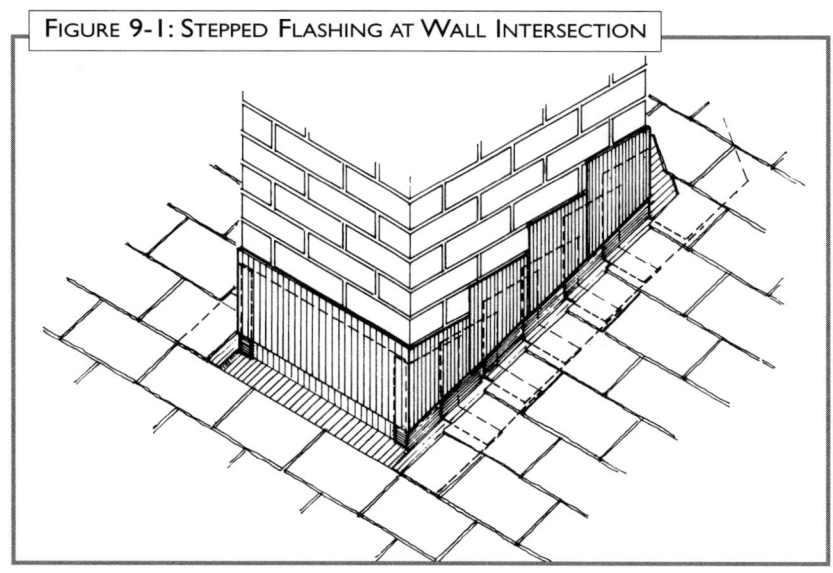

FIGURE 9-1: STEPPED FLASHING AT WALL INTERSECTION

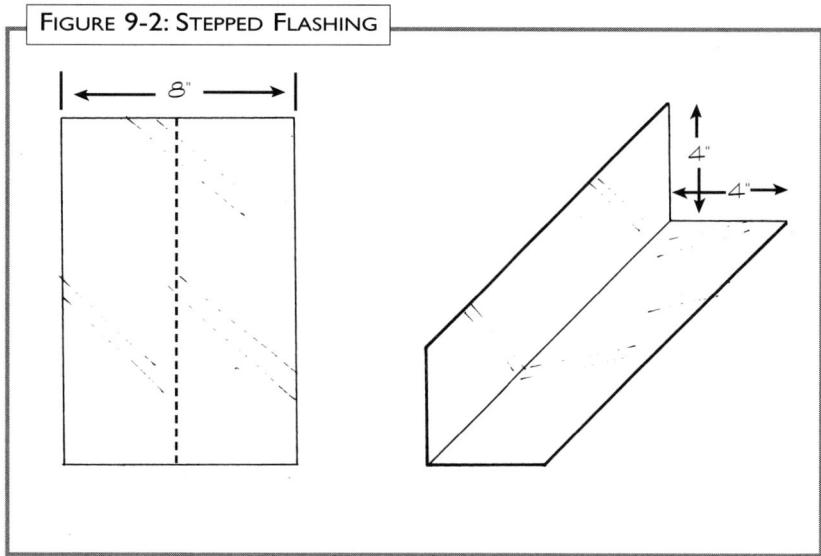

FIGURE 9-2: STEPPED FLASHING

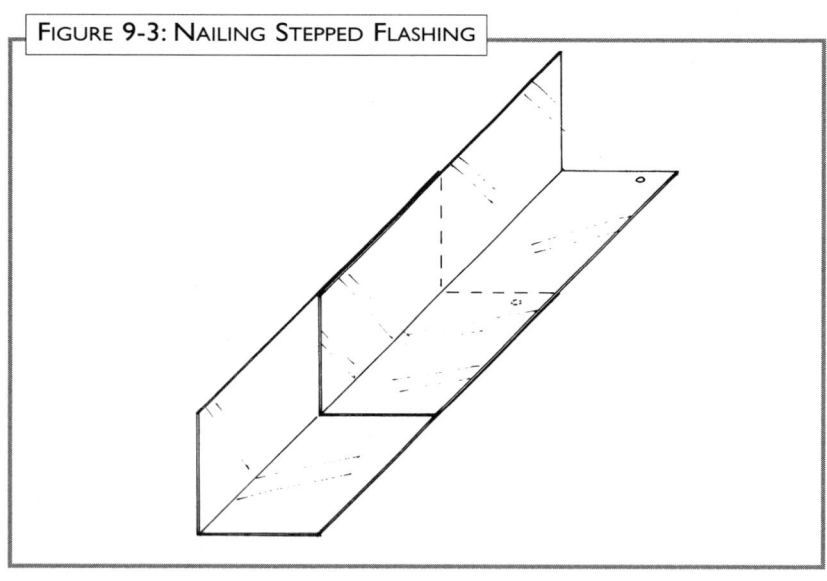

FIGURE 9-3: NAILING STEPPED FLASHING

slate below. The butt edge of the apron flashing will then align with the butt end of the course of slate that it replaces. The apron flashing will need to be a minimum of 4 inches wider on each side of the face of the penetration and will extend up the vertical face a minimum of 6 inches.

Once fabricated, the apron flashing should be set in place. Cut the flashing on the vertical leg so that there will be 4 inches of material remaining on each side. Bend the 4-inch tab down so that it is in the same plane as the roof. Place a piece of step flashing on top of the 4-inch tab on each side in line with the vertical bend. Bend the excess step flashing around the face of the penetration and solder all corners as necessary. The most common leaks that occur on a slate roof are related to the flashings, so proper soldering of these corner details is essential. (See figure 9-4.)

If the apron flashing is more than 3 feet long, it will require hold-down clips, which are 1 1/2 inches wide and 5 inches long. The clips will be installed under the apron flashing such that an area 3/4 inch by 1 1/2 inches is left exposed. These clips should be spaced a maximum of 18 inches apart. To determine the placement of the clips, set the apron in place on the roof. Scratch a line on the slate along the butt edge of the apron

Flashings

flashing and then remove the flashing. Fasten the clips between the vertical joints in the slate course being covered. The fasteners will need to be installed a minimum of 3 inches above the reference line scratched on the slate course below. Place the apron into position on the roof and then fold the exposed portion of the clip up and over the edge of the apron to hold it down. (See Figure 9-5.)

Rosin Paper

Rosin paper is a coated paper product with about the same texture as the construction paper used in schools. It generally comes on rolls 3 feet wide. When the roof underlayment is installed, it is usually installed with ferrous nails, and these nail heads should not come in contact with non-ferrous metals (such as copper or aluminum).

A layer of rosin paper should be installed under all flashings where this contact may occur, such as open valleys, crickets, and standing-seam details. Not only does the rosin paper insulate the non-ferrous flashings from the ferrous nail heads, it also allows the metal to expand and contract without sticking to the underlayment.

Valley Flashings

There are several different flashing methods used with roof valleys. These include open valley flashing and different approaches to

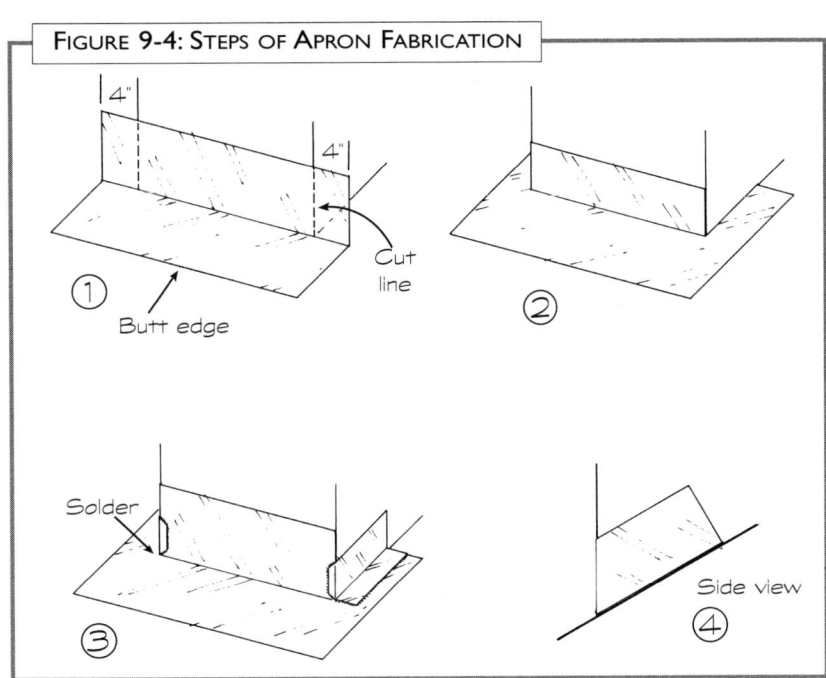

FIGURE 9-4: STEPS OF APRON FABRICATION

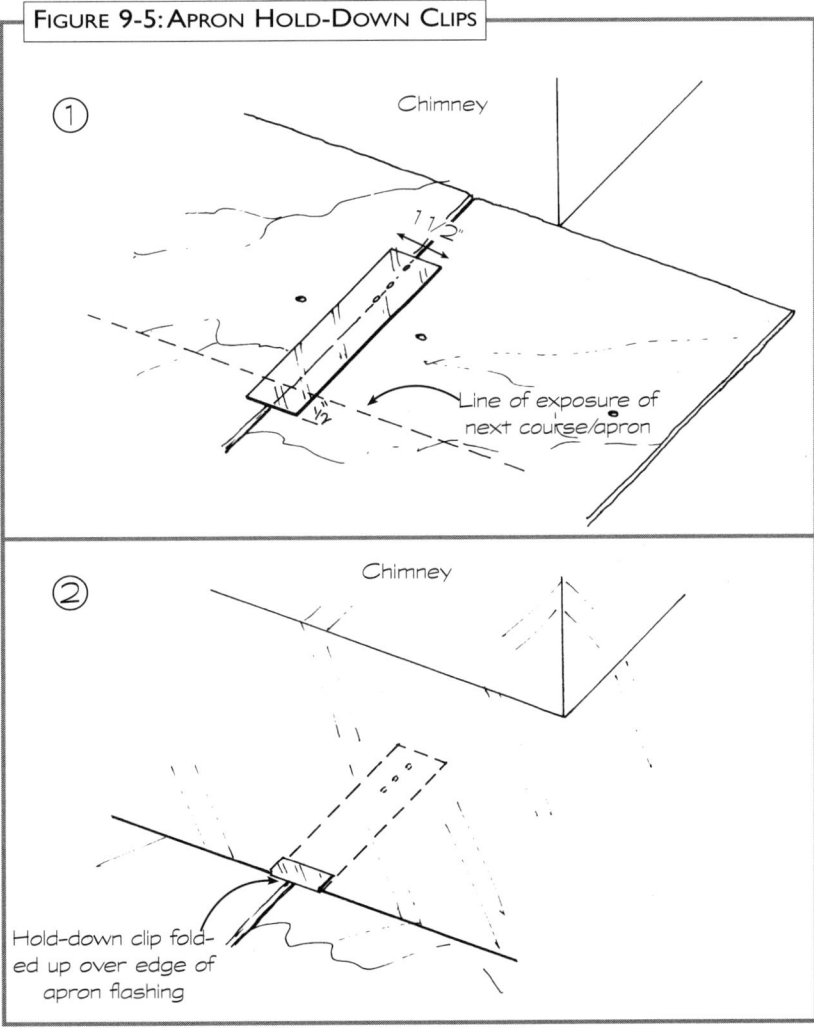

FIGURE 9-5: APRON HOLD-DOWN CLIPS

141

Chapter 9

closed valley flashing, such as continuous flashing and step flashing.

Open Valley Flashing

In an open valley, the flashing is exposed. Open valleys are probably the most common valley details for a slate roof, and there are two styles of open valley flashing: the "V" style and the "W" style. The most desirable is the "W" style in which the "W" bend in the flashing serves to keep water from shooting across the valley and under the slate on the other side.

For open valleys that start at the eaves, open valley flashing materials should be installed prior to the layout and installation of the slate roof. (Note: this will not work for valleys that intersect the field of the roof partway up, as with a dormer roof.) With the valley flashing installed first, all roof layout lines can be snapped into the center of the open valley. Be sure to install rosin paper under all open valley flashing as mentioned previously.

All open valley flashing is fabricated with a 1-inch fold up the length of the side. This fold acts as a water-lock and as a fastening point for the hold down clips. Use the following steps to install open valley flashing: (See Figure 9-9.)

- Snap a line up the length of the center of valley.
- Lay the first section of valley flashing in place at the eave. Center the top of the valley

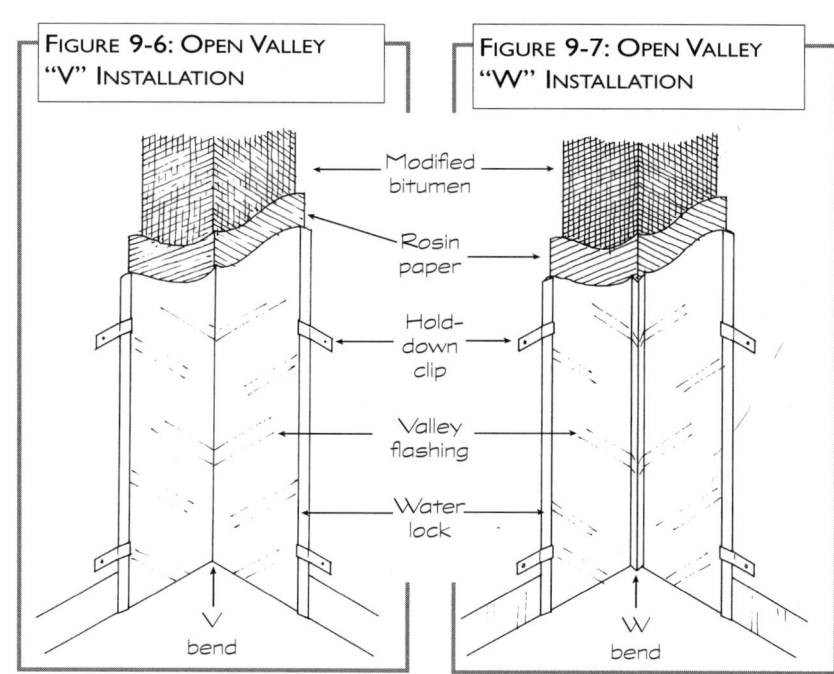

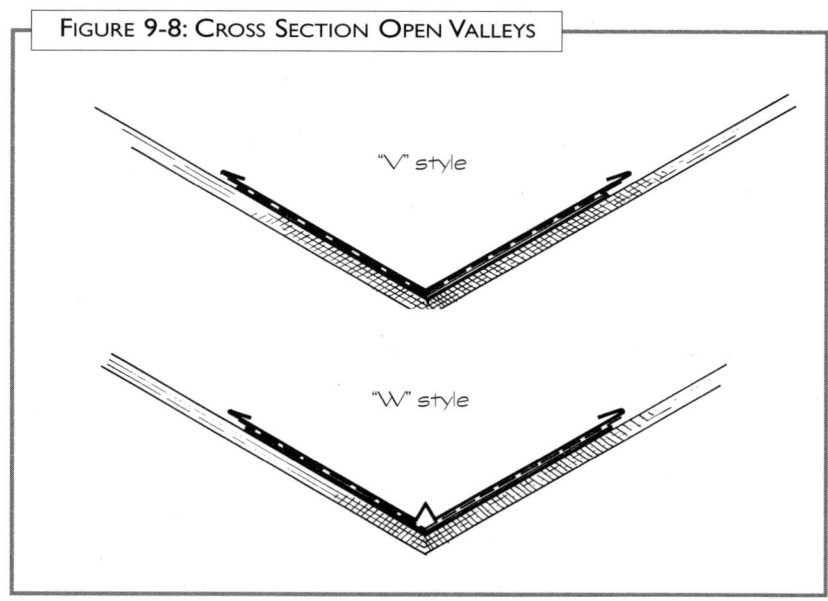

flashing on the valley reference line. Allow the water-lock on each side at the butt end of the valley to extend 1 inch beyond the eave detail.

1) Scribe a line 1 inch beyond the eave from the center of the valley out to the water-lock on each side. This is the cut line for fitting the first piece of valley flashing.

2) Cut away the excess valley flashing and fold the extra 1 inch of valley material back underneath itself 180 degrees.

3) If drip edge has been used at the eaves, this 1 inch of material will lock onto the drip edge. If no drip edge is being used, allow the valley flashing to slide down and overhang the eave the same dis-

Flashings

tance that the slate will be overhanging the eave. This 1-inch hem will stiffen the leading edge of the valley. When a "W" valley is being installed, leave enough material at the "W" bends to close off the end of the "W."

4) After the butt end of the first valley section has been prepared, slide it back into place and nail it on either side 1 inch down from the top edge. These are the only two nails that should ever penetrate an open valley.

5) Use hold-down clips to secure the valley in place every 12 inches up either side. These hold-down clips are 1 1/2 inches by 4 inches. Bend a 3/4-inch hook on the end of each clip. Hook the clip over the water-lock edge of the valley and clamp it in place with pliers or seamers. Fasten each clip to the deck with two nails approximately 1/4 inch beyond the valley edge. Do not penetrate the valley metal. Finish this detail by folding the excess clip material over the exposed nail heads.

Progressive valley sections should overlap previous valley pieces by a minimum of 6 inches. (See Figure 9-10.) On roof pitches less than 6:12, increase the overlap to 8 inches. Progressive valley pieces should slide inside the water-lock of the previous valley sections to counter flash the nail heads at the tops of the valley. It

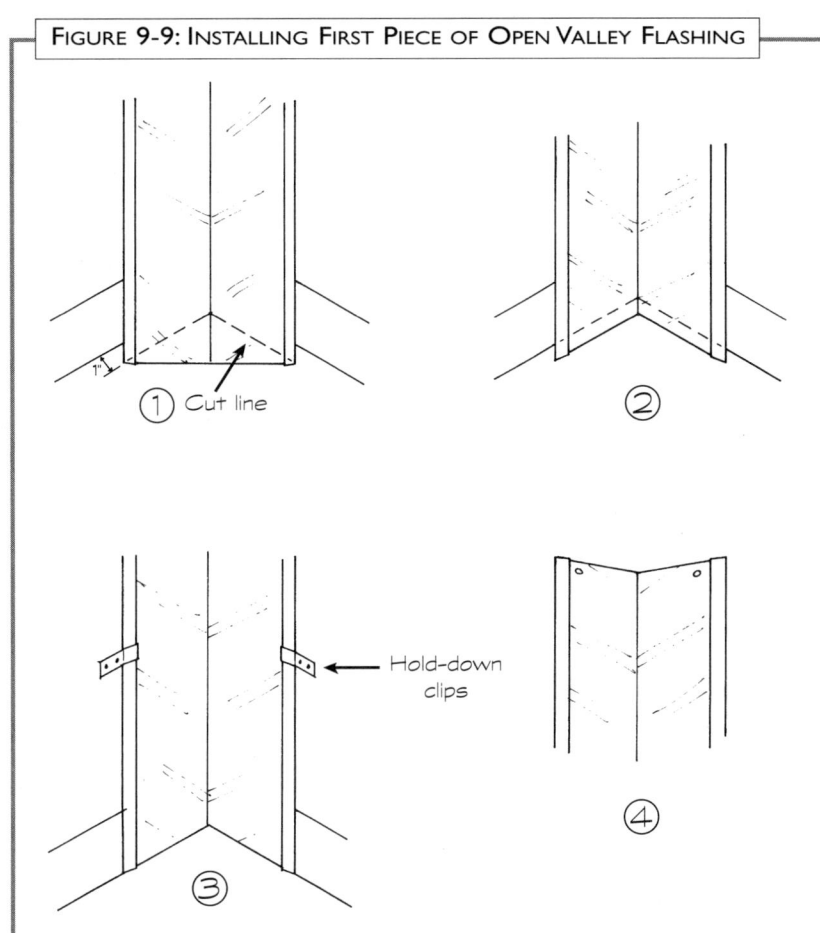

FIGURE 9-9: INSTALLING FIRST PIECE OF OPEN VALLEY FLASHING

may be necessary to open the water-lock on the lower valley section to allow the progressive valley to fit in. Once the progressive valleys are slid into place, the water-locks from the lower valley are then wrapped back into place around the upper valley section.

At the ridge, the valley needs to be cut to form a miter with adjoining valleys or ridge. When adjoining valleys are encountered, this joint will need to be soldered. Long valley sections need to be able to expand and contract. Never solder the overlap of the valley pieces.

It is common for valley flashing to change angles as intersecting roof pitches change. A special flashing detail, which will need to be soldered, will be required here. Valley installation will continue above this flashing as described above.

Once the open valley flashing is in place and the roof is laid out, you can begin laying the slate into the valley. There are several factors to keep in mind as you lay the slate. First, you must plan the width of the valley base flashing so that the nails from the valley slate do not penetrate the valley flashing (see Table 3-12: Open Valley Flashing Width in Chapter 3). There will be times when you will need to install small

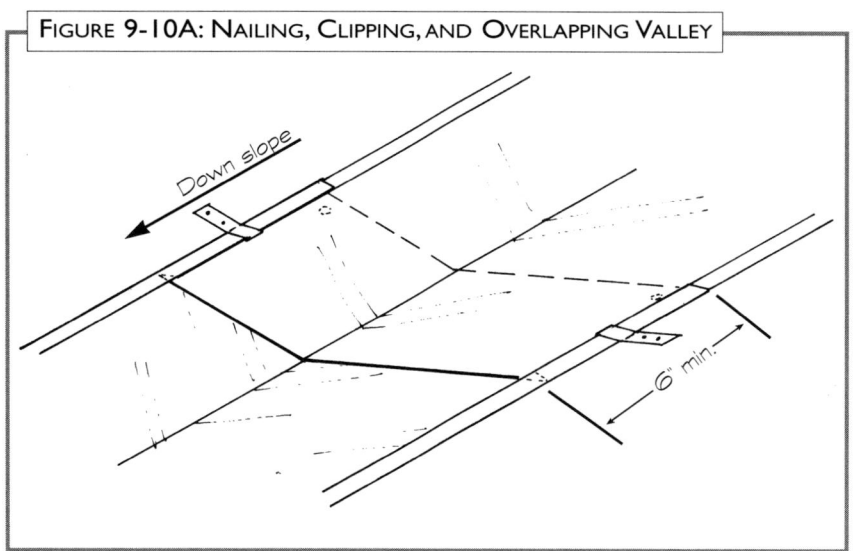

FIGURE 9-10A: NAILING, CLIPPING, AND OVERLAPPING VALLEY

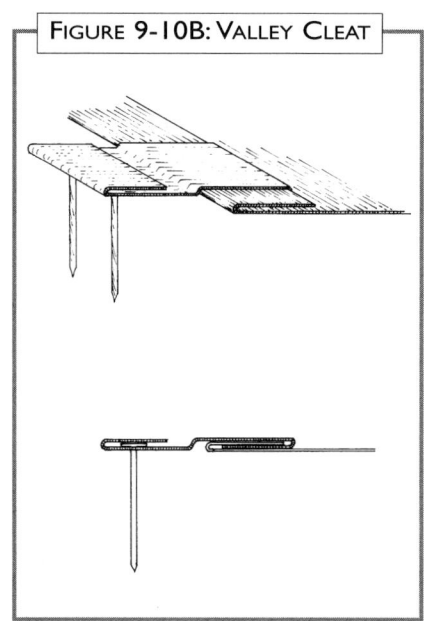

FIGURE 9-10B: VALLEY CLEAT

pieces of valley slate. The only possible place to punch nail holes on these small pieces may result in the nails penetrating the valley flashing. Do not nail these pieces through the valley flashing. Instead, wire these pieces in place using wire of the same material as the valley flashing. Wrap the wire around nails that are outside of the valley flashing, and then drive the nails down so that the nail heads are not too high.

When installing a small piece that needs to be wired in place, you need to set the piece in a bed of mastic to keep it from turning or moving. Be sure that the chosen mastic will not run. Try slater's cement instead of roof cement. This mastic is intended to glue the small wired pieces to larger nailed pieces of slate. Do not apply the mastic to the valley flashing, which is clipped in place to allow for expansion and contraction. Adhering the valley slate to the valley flashing will inhibit this movement.

The water-lock on each outside edge will need to be closed to the point that it will not hold the slate up out of the roof plane at the valley.

With a dormer valley or valley that starts up slope, the slate will have to be installed high enough that the valley flashing will shed water over the intersecting slate. If the roof is properly laid out, the first course of slate on the intersecting surface will line up with the first course of slate on the opposing valley side (see Chapter 6, Roof Layout). This means that when the dormer valley flashing is installed, the edge, which is hemmed to lie on the dormer surface in line with the starter course, will line up with the butt edge of the last course of field slate on the adjacent surface. As a result, the valley flashing will cover the field slate on the lower surface with a proper headlap.

At the tops of valleys, ridge caps will cover the valley intersections. However, the flashings will have to be brought together and soldered before the slate is installed.

Closed Valleys

A closed valley is a valley in which the flashing is covered by the slate so there is no visible flashing. This "closed valley" detail can be installed using continuous flashing or step flashing, either in the conventional style or point-to-point style. (See Chapter 8.)

Closed Valley - Continuous Flashing

Continuous flashing is installed in the same manner as open valley flashing. Always use the "W" configuration for this method of application. Fabricate the "W" flashing from 12-inch wide material. This valley flashing will be the same width over its entire length. Do not penetrate this base flashing with nails. It should be installed with clips in the same way as with open valley flashing. If a piece of slate is too small to nail, wire it in place as described on page 144. There is no need to cement any of these pieces together. Because the slate from each side of the valley touches, is cut in a triangular shape, and is held in place laterally by adjoining slate, this slate will not be easily displaced.

Moisture and debris tend to accumulate in this style of closed valley. Although this type of valley detail is common when working with roof tile, it is not the best closed valley solution for slate. (See Chapter 8.)

Closed Valley - Conventional Stepped Valley Flashing

For conventional stepped valley flashing, each course of adjoining valley slate is step-flashed at the valley. The first piece of valley stepped flashing is placed on top of the starter slate. Align the center of the valley step with the butt end of the starter slate at the center of the valley. After installing the first pair of closed valley slates, continue the stepped-flashing procedure by aligning the center of each piece with the butt of the slate from each progressive course. The length of this stepped flashing is determined by measuring from the point of exposure up the valley to the top of the valley slate and adding 2 inches. (For length and width specifications, see Table 3-13: Conventional Stepped Flashing for a Closed Valley in Chapter 3.)

With this stepped flashing method, the valley pieces do not need to be cemented, and it is not necessary to wire the small pieces of valley slate as when using continuous flashing, because the conventional stepped valley flashing counter flashes the nail heads from each previous course and prevents leaks.

With conventional stepped valley flashing, the water flowing down the valley is continuously brought back out to the surface of the slate on the subsequent courses. This style of closed valley tends

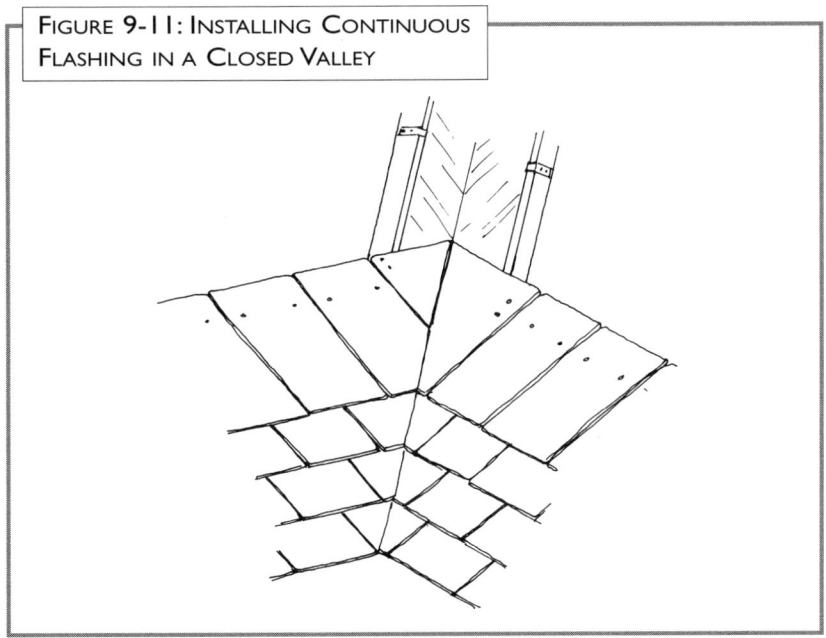

FIGURE 9-11: INSTALLING CONTINUOUS FLASHING IN A CLOSED VALLEY

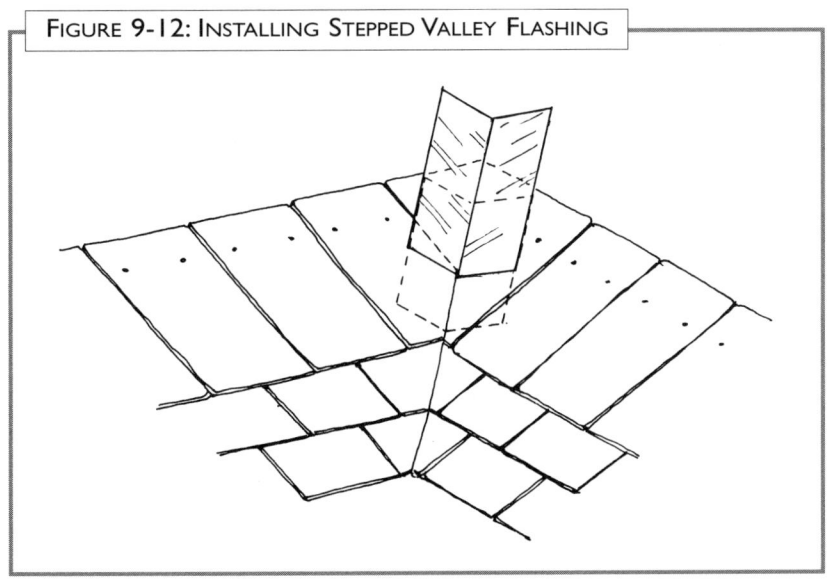

FIGURE 9-12: INSTALLING STEPPED VALLEY FLASHING

to clear itself of debris and moisture better than a closed valley with continuous flashing.

Closed Valley - Point-to-Point Stepped Valley Flashing

An alternative method of flashing a closed valley is to use square pieces of valley flashing bent point to point. (To determine proper flashing size, see Table 3-14: Point-to-Point Flashing for a Closed Valley or Mitered Hip in Chapter 3.) This technique works very well with valleys that join roofs of the same pitch, but it does not always work well on shallow pitches. For adjoining valleys of 6:12 pitch or less, closed valleys are not recommended.

To install point-to-point stepped valley flashing in a closed valley, place the first piece of valley step flashing on top of the starter slate. Align the center of the valley step with the butt end of the starter slate at the center of the valley. After installing the first pair of closed valley slates, continue the step flashing procedure by aligning the center of each piece with the butt of the slate from each progressive course.

When a closed valley is installed on a roof with different roof pitches, the slate course lines will not line up at the valley. Do not use the point-to-point stepped flashing method on a closed valley that has two different adjoining roof pitches. Instead, use the conventional stepped flashing detail for this installation. Lay the stepped flashing with the slate that is on the shallower of the two pitches. Then lay the slate from the steeper pitch over the valley flashing. In a sense, you are actually creating a combination valley flashing. The flashing that is stepped into the shallower pitch is obviously a stepped flashed valley. However, the legs of that same stepped flashing in the shallow pitch lie together on the steeper pitch to form a continuous flashing detail. The slate is then laid over the exposed stepped flashing in the same manner as described in the closed valley continuous flashing detail.

Metal Hip and Ridge Cap

Metal hip and ridge details are commonly installed on slate roofs. Their shape varies from project to project. Some may be manufactured to incorporate ventilation.

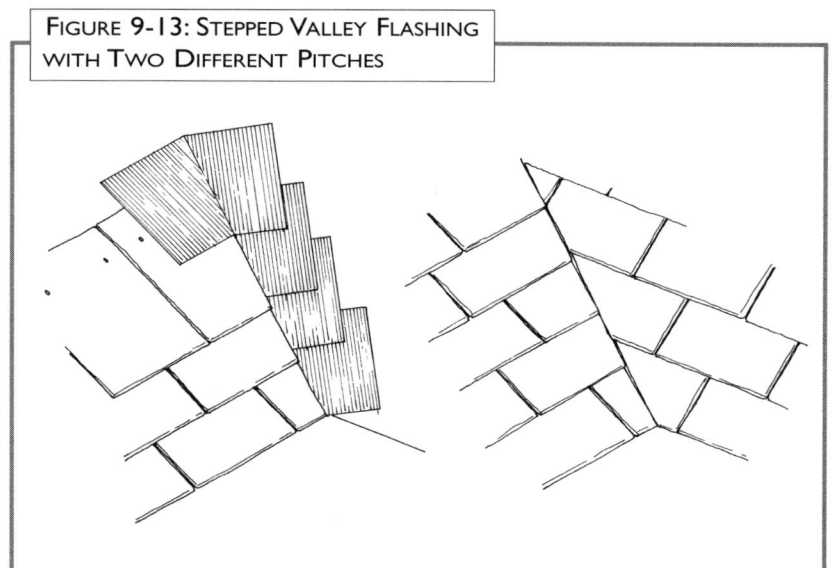

FIGURE 9-13: STEPPED VALLEY FLASHING WITH TWO DIFFERENT PITCHES

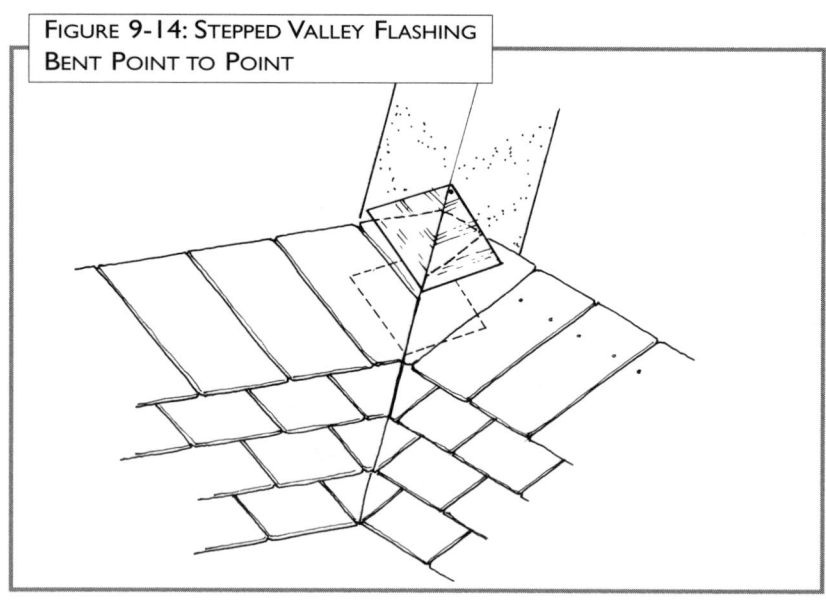

FIGURE 9-14: STEPPED VALLEY FLASHING BENT POINT TO POINT

Flashings

There are several methods that can be used to fasten a hip or ridge cap. The important details to remember are:

- Use concealed fasteners whenever possible.
- Use enough hold-down fasteners to prevent the cap from blowing off.
- If your fasteners must penetrate the slate, drill the holes.
- Properly support any large ornamental caps with wood blocking. This will prevent damage from extreme weather and damage caused by persons performing routine maintenance tasks such as cleaning a chimney.
- Size caps to provide correct headlap of the slate. This needs to be considered during layout.
- Install a modified bitumen base flashing whenever possible.
- Overlap any joints a minimum of 6 inches. It is not necessary to solder the joints.

Crickets

A cricket is a flashing used to shed water out from behind a roof obstruction. Crickets are usually used behind chimneys or in complex valley intersections. Chimney crickets are the most common cricket detail. Figure 9-15 shows a typical chimney cricket. The important points to keep in mind are:

- A cricket must have a substructure to support the metal flashing. This substructure will normally be of the same material and slope as the adjoining roof.
- Install a base flashing of modified bitumen over the entire cricket substructure and extending 6 inches beyond the perimeter including up the back wall of the chimney.
- Install rosin paper prior to installing the cricket flashing.
- Solder all joints in the cricket flashing. This includes soldering the cricket flashing to the top of the last piece of step flashing on either side of the chimney.
- Extend the cricket flashing up the back of the chimney a minimum of 6 inches and bend a 1/2-inch water-lock at the top.
- Extend the valley portion of the cricket flashing a minimum of 12 inches onto the adjacent roof surfaces.
- Fasten the cricket to the roof deck using the hold-down clip method described in the open valley section of this chapter.
- Treat the cricket flashing as though it were an open valley. All slate layout, as described in open valleys, will apply.

In a complex valley cricket, install the flashings so that a minimum of 12 inches is exposed around the entire perimeter after the slate has been installed. As a rule, a minimum of 5 inches of flashing will be covered by the installed slate, which means that your flashing will be a minimum of 17 inches wide around the perimeter of the cricket. This also means that in large complex crickets, the flashings will often have to be lock-seamed and soldered in panels small enough to prevent splitting of the solder joins due to thermal movement (no longer than

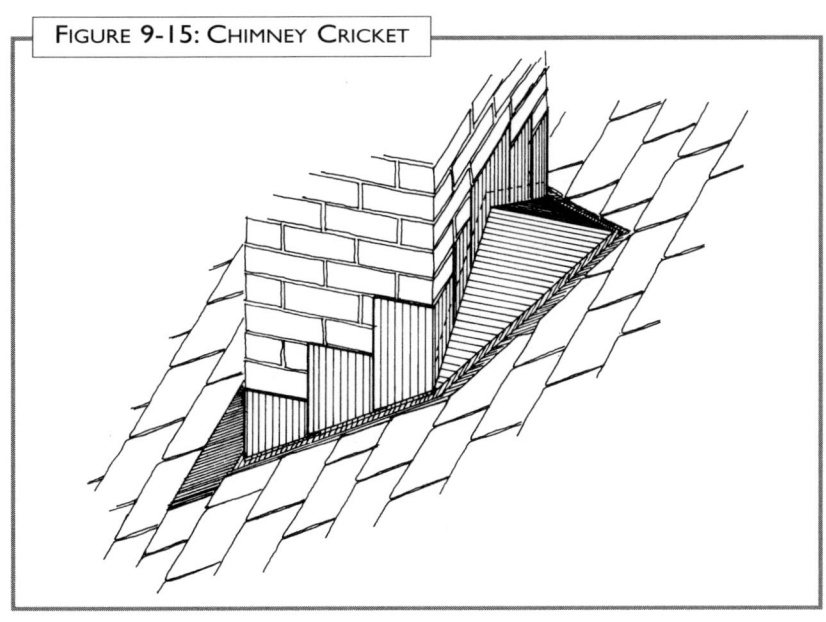

FIGURE 9-15: CHIMNEY CRICKET

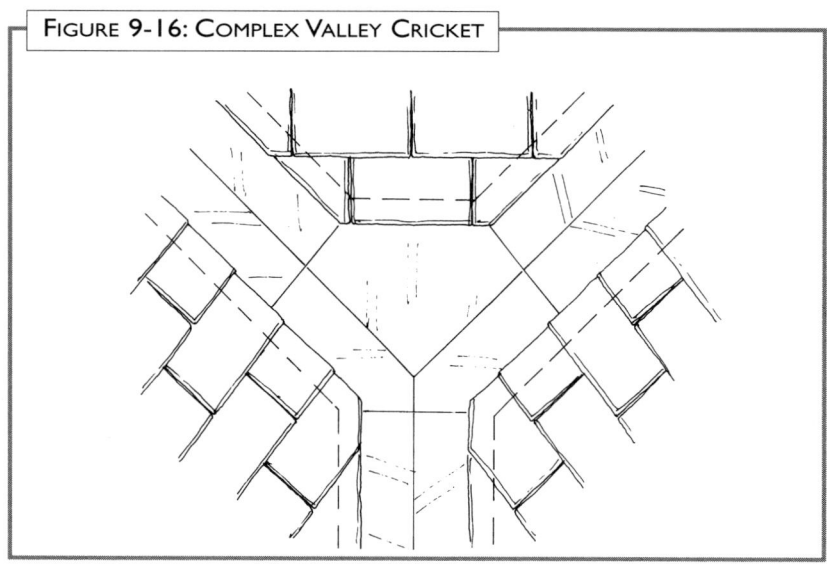

FIGURE 9-16: COMPLEX VALLEY CRICKET

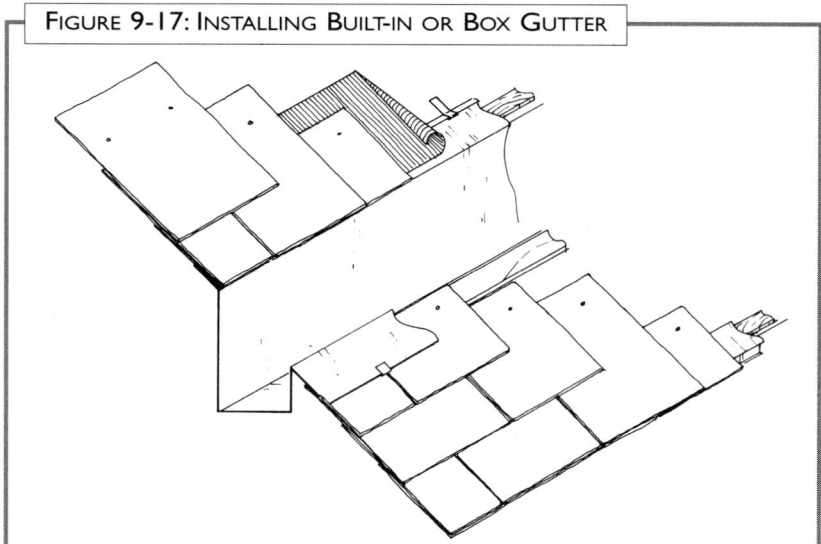

FIGURE 9-17: INSTALLING BUILT-IN OR BOX GUTTER

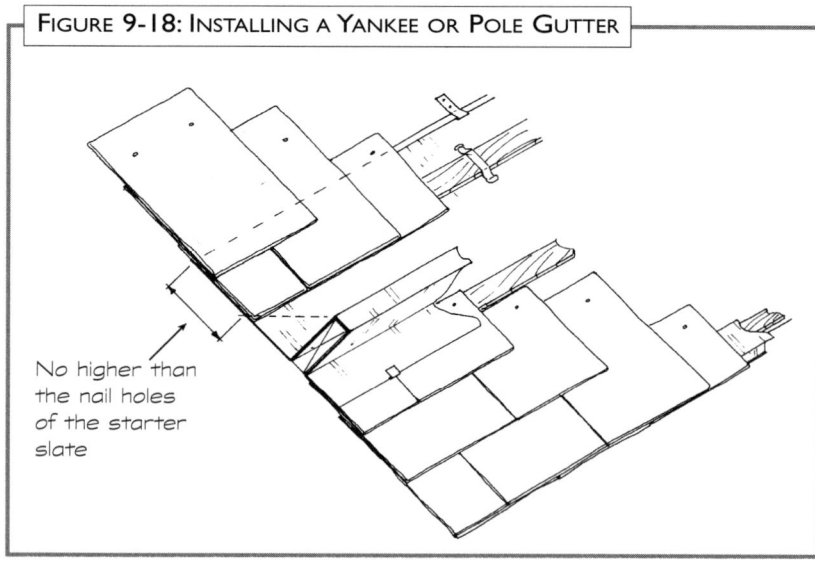

FIGURE 9-18: INSTALLING A YANKEE OR POLE GUTTER

No higher than the nail holes of the starter slate

18 inches x 24 inches). After the cricket has been installed, all rules related to the installation of the slate are the same as those for an open valley.

Gutters

There are several types of gutters, as illustrated in the following details and brief descriptions.

Built-in or Box Gutter

There are a few important points to consider with the installation of this type of gutter.

- Allow for expansion and contraction.
- Use a thick enough gauge flashing material to last as long as the slate.
- Provide a proper cant strip at the start of the slate above the gutter.
- It is not necessary to run the flashing up the roof surface higher than the nail holes in the starter slate.
- Fully support the gutter lining.

Yankee or Pole Gutter

The Yankee or pole gutter is a variation of the built-in or box gutter. The difference is that this style is raised above the finished roof surface. The important points to keep in mind when installing this style gutter are:

- Be sure to support the gutter frame properly. This style gutter is

Flashings

essentially a dam. Debris, snow, and ice will accumulate behind this gutter.

- Do not set this gutter on the finished slate roof. The weight of the debris that accumulates inside it can crush the slate below.
- The exposed gutter flashing on the roof surface should be at least as wide as the height of the supporting gutter framework.
- Make sure that the flashing below the gutter properly covers the head of the slate below.
- Allow for expansion & contraction.
- Use a thick enough gauge flashing material to last as long as the slate.
- Provide a proper cant strip at the start of the slate above the gutter.
- It is not necessary to run the flashing up the roof surface higher than the nail holes in the starter slate.
- Fully support the gutter lining.

If a Yankee gutter is desired in a snow region, you should consider installing snow guards to protect the gutter.

At-the-Eave Gutters

At-the-eave gutters are similar to the other gutter styles mentioned above. The points to keep in mind during installation are the same, with only one important addition. Because these gutters are installed beyond the edge of the roof structure, special care must be taken to ensure that the gutter is well supported.

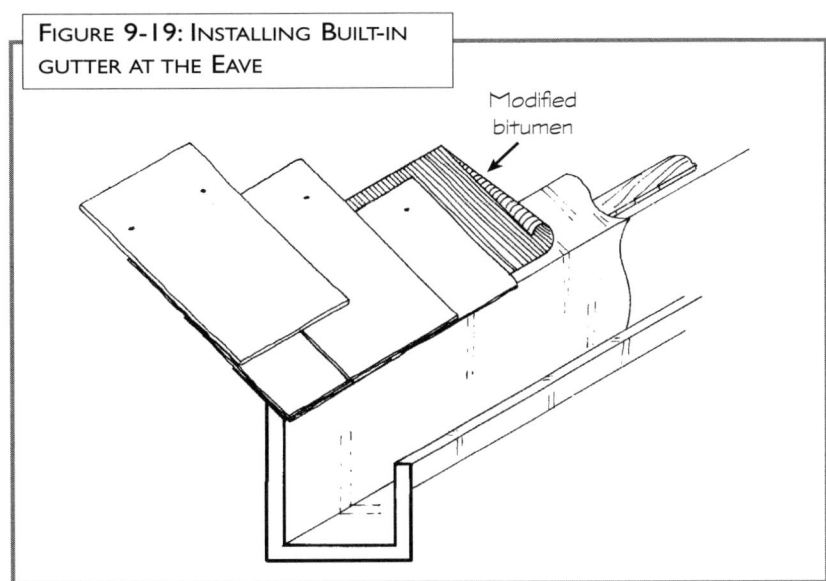

FIGURE 9-19: INSTALLING BUILT-IN GUTTER AT THE EAVE

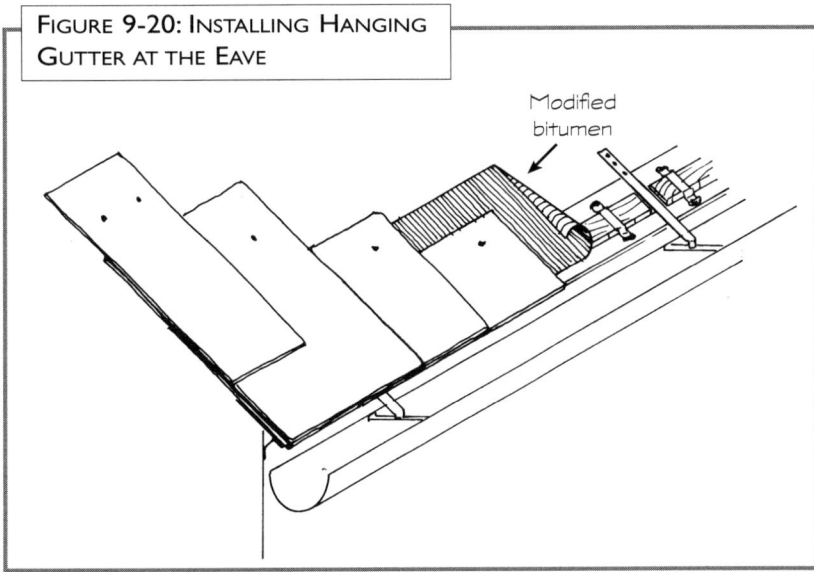

FIGURE 9-20: INSTALLING HANGING GUTTER AT THE EAVE

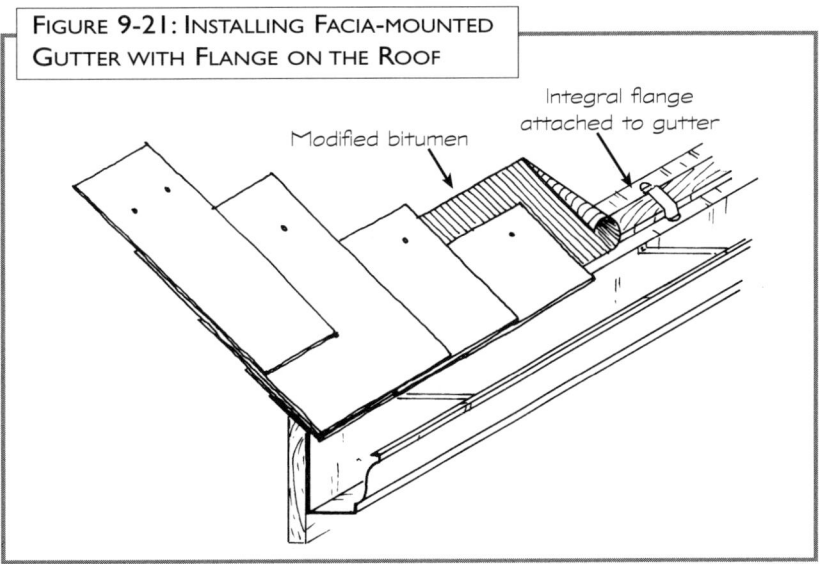

FIGURE 9-21: INSTALLING FACIA-MOUNTED GUTTER WITH FLANGE ON THE ROOF

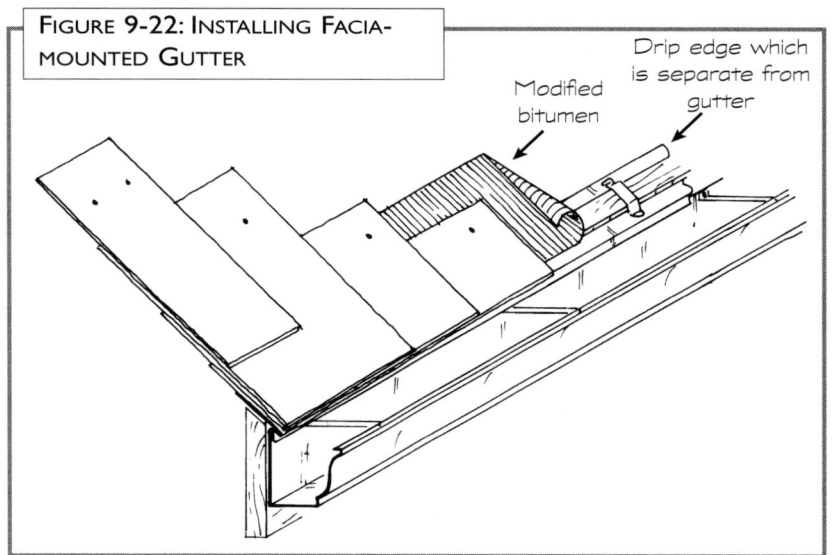

FIGURE 9-22: INSTALLING FACIA-MOUNTED GUTTER

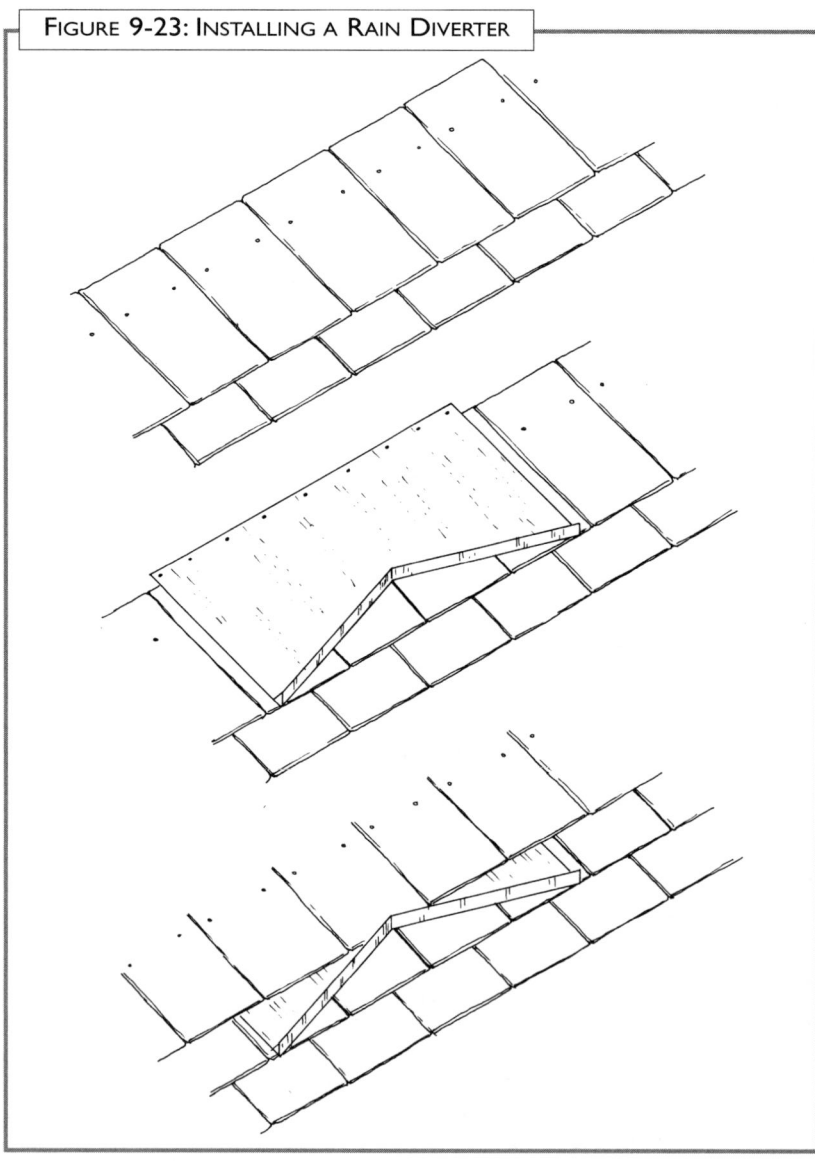

FIGURE 9-23: INSTALLING A RAIN DIVERTER

On a slate roof, the first point of failure is the flashings. Gutters are especially susceptible to failure due to erosion, accumulation of debris, and sliding snow and ice. Gutters installed in snow regions are likely to suffer damage if proper snowguard protection is not provided. The best gutter design allows for repair, maintenance, and replacement of the entire gutter system without disturbing the slate. Whenever possible, use a gutter system that fastens to the facia and is independent of the slate roof (see Figure 9-22).

Rain Diverters

In some cases owners or planners will request a rain diverter. The rain diverter is sometimes used above doorways, for example, in lieu of a gutter. In areas where snow and ice are not a consideration, this detail is relatively straightforward. However, if snow and ice are a consideration, this detail will have to be approached as though it were an above-the-eave gutter.

The problem with a diverter is that it will bear down on the slate below unless the original design provides for proper framing and flashing. In regions with heavy snow and ice, installing a diverter in a slate roof will probably cause unnecessary damage.

If a rain diverter is installed in a snow region, you should consider snow guards to protect the diverter.

CHAPTER 10
Reroofing

Slate roofs last a long time, but they do not last forever. Generally speaking, slate roofs last about 75–100 years. However, there are a number of variables that affect the longevity of these roofs.

For most slate roofs that need to be replaced, it is safe to say that either the slate or the flashings—or both—have failed. There are situations in which natural disasters, such as hurricanes or trees falling, have caused so much damage that reroofing is more sensible than repairing. However, we are approaching a time in history when a tremendous number of buildings that were built around the turn-of-the century now need to be reroofed.

The purpose of this chapter is to help you evaluate the condition of a slate roof in order to decide whether it should be replaced or repaired. When evaluating a slate roof, you will need to consider a number of factors.

Evaluation of Existing Roofs

The first and most important consideration is the condition of the slate itself, including the type of slate and its age. There are tests that can be performed to grade the slate. Slate is graded as an S1, S2, or S3 material. Slate graded S1, the highest rating, has a life expectancy of 75 to 100 years. If you are uncertain what type of slate is on the roof, you can send a piece of the slate to a quarry for identification. In most cases, the quarry will be able to tell you what the slate is and estimate its remaining service life. As an alternative, the slate can be tested by a local lab using ASTM test procedures to determine the current grade.

Based on the results of the slate tests, your inquiries about the age of the building, and the estimated age of the roof, it will generally be easy to determine the best course of action. The choice between repairing a slate roof or reroofing is usually straightforward when considering slate quality.

Flashings

Other than problems associated with broken, damaged, or deteriorated slate, the most common problems are related to the roof flashings. Slate of S1 grade will usually outlast the flashings. Temporary flashing repairs are commonly carried out by tarring over problem flashing areas, but the only way to permanently repair a leak is to replace the flashing.

When evaluating a slate roof that is 70 years old or more, keep in mind that all flashings will need to be replaced. If this means disturbing more than 25% of the slate on the project, it will probably be more cost-effective to replace the entire roof.

Fasteners

A variety of fasteners have been used to install slate. If ferrous fastener were used on a roof that is now more than 70 years old, it may be the fasteners that are failing and not the slate. When the fasteners deteriorate, the slate begins to slide out of the roof. The only reasonable option is to reroof.

Roof Deck

The condition of the roof deck is an important "hidden" factor to consider when evaluating slate roofs. It is important to inspect the deck both from the inside of the structure, if possible, and from the outside by removing the slate in several areas. Look for deck deterioration in areas where current or previous leaks have caused damage. Although these areas may have been repaired properly on the outside, the roof deck may never have been addressed. Simply repairing the slate does not solve the problem. This situation is similar to that of replacing the flashings. If more than 25% of the decking needs to be repaired, replacing the deteriorated roof deck areas and the related slate may be nearly as expensive as totally redecking and reroofing the structure.

It is fairly common, especially on larger public buildings, for the roof deck to have been constructed of more than one material or to have one material covering another. It is a good idea to check several different areas of the roof to ascertain that the same material has been used throughout. It is also fairly common to encounter nailable concrete as a decking material. If this is the case, it is very important to determine whether the concrete is still nailable or whether new wood decking will need to be installed over the existing concrete.

Older buildings rarely had proper ventilation. As a result, moisture condensed on the decking on the inside, causing rot. For this reason, it is possible to have a roof that looks perfect on the outside and has never had a problem, yet should be replaced. Cathedral ceilings are a perfect example of a structure with a roof deck that is often not well vented.

Structure

Keep in mind that the greater the weight on the roof, the greater the load on the supporting building structure. This is especially important to keep in mind if you are considering putting slate on an existing structure that does not currently have a slate roof. The entire structure needs to be evaluated and engineered, if necessary. The damage caused by putting a slate roof on an inadequate structure may not be immediately noticeable at the roof level, but its effects may be felt in the long run.

Previously Repaired Roofs

When evaluating a slate roof, watch for previously repaired areas. Some of these old repairs may not have been done professionally and may need to be reworked.

Budget

Money is always an issue. When a slate roof fails, reroofing is usually the best solution, although repair may seem to be the only affordable option. It is wise to get bids for both repairing and reroofing a structure before making a decision.

There are also a couple of reasonable intermediate options. One is to do the reroofing in stages as budget permits. Second, if the existing slate is of S1 quality, it can often be salvaged and reused. Usually only up to 60% of the materials on the roof can be salvaged, so this alternative will require purchasing additional material. Third, carry out emergency repairs until funding is available. Understand that money put into emergency repairs is generally lost when the roof is later replaced, but this step can prevent costly interior and structural changes.

Bidding

When the evaluation of the roof has been completed, you should have enough information to provide or obtain a fair bid for either replacing or repairing the roof. You will probably need to add unit pricing to the bid to cover the costs of repairing unforeseen damage to the structure of the building. Some of this damage can be identified during the evaluation process, but it is reasonable to assume that not all of it will be detected. By allowing for unit pricing in the bid process, the owner is assured that the roofer will take the time to repair any damage discovered when the existing roof is removed, and the roofer is assured that he will be paid for making repairs that were not seen before the roof was opened up.

CHAPTER 11
Repair, Maintenance, and Troubleshooting

Repair

Every slate roof will at some point need repairs, replacement of shingles, or general maintenance. This chapter provides information about these procedures.

Replacing Slate Shingles

The most common repair situations involve replacing damaged slate shingles. There are some differences of opinion concerning how "broken" a corner has to be before it is replaced. As a rule, a broken corner is one in which the broken area exceeds 1" × 1" on the sides of a triangle. Any slate pieces with missing corners less than 1" × 1" are considered acceptable in a new roof as whole pieces that will not leak or inhibit the overall performance of the roof. However, broken corners, especially on a standard thickness slate roof, are very noticeable. If there appear to be quite a few missing corners, the building owners may want those pieces of slate replaced. Try to avoid using slate with broken lower corners in the field of the roof. These pieces can often be cut and used in the valleys, hips, gable-end half pieces, etc. If the upper corners that will be concealed are within the stated limits, then the pieces can be used without sacrificing the integrity of the roof.

Often, the mill will ship pieces of slate that are missing upper corners as described above. If, at the mill, the slate trimmers can trim pieces of slate with missing upper corners to within acceptable limits, they do so and send them on to the puncher. The puncher is responsible for making sure that the nail holes are punched so that the broken corners will be on the upper or concealed end of the slate.

It is very common for small corners of slate to break off during installation. Develop a habit of marking any broken pieces with an X. Marking these pieces clearly will help you locate these broken pieces later, when it comes time to remove the scaffolding brackets from the roof. If you repair the piece immediately, chances are it will be broken again before that roof surface is completed.

During scaffold removal and subsequent repair process on a new roof, you will obviously need to repair all broken corners that are not within acceptable limits. If the size of the broken corner is close, but questionable, replace the piece. You will save yourself a lot of work if you carry out all repair work while the scaffolding is still in place. This simple gesture also will assure the owner that you take pride in your work. If a small detail like this is overlooked by the installer, the owner will often begin to lose confidence in the installer's ability and may begin to scrutinize the entire project, checking every little detail. Remember that an owner who is willing to spend the money and take the time and to have a slate roof installed is expecting an end product that will last a lifetime. Often it is the noticeable little details, such as broken corners, that an owner will take issue with.

Chapter 11

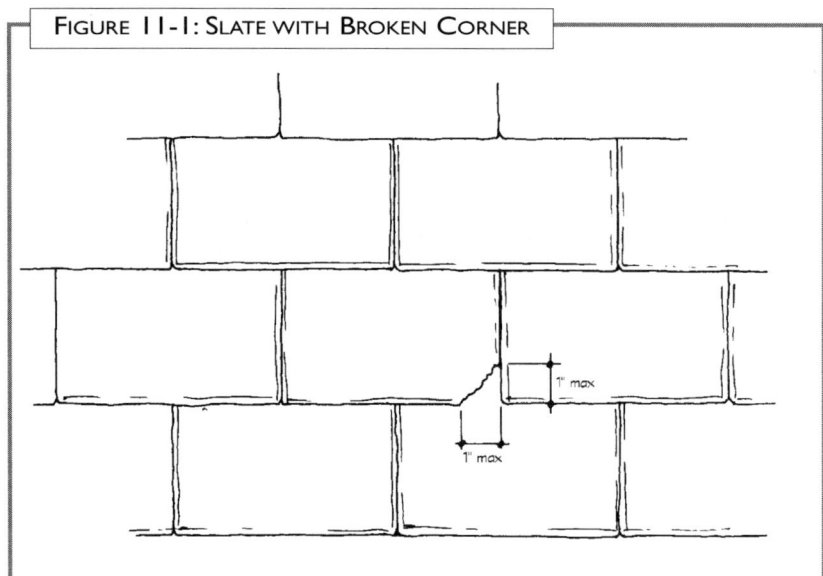

Figure 11-1: Slate with Broken Corner

As a project designer or planner, it is important that you understand this standard for broken corners. If you desire a slate roof without any broken corners, then that requirement will need to be stated in the project specifications. In any event, you may want to specify what percentage of broken corners will be deemed acceptable.

Field Slate

Everyone who is involved with or considering a slate roof needs to be aware that slate does get broken. Because slate is basically a loose-laid material, broken or missing pieces will allow water to infiltrate.

A number of things can cause slate to break. Beyond the breakage that occurs during and as a result of the installation process, some of the more common causes include the following:

1. The roof is walked on by another contractor or repair person, such as a chimney cleaner, painter, gutter installer, window cleaner, etc. Anyone who goes on the roof after it is completed runs the risk of damaging it.
2. Falling trees or large tree limbs can cause damage.
3. Hard balls or rocks thrown onto the roof can break slate.
4. Snow and ice falling from higher roof areas can cause damage to lower roof areas.
5. Homeowners and local tradespeople will sometimes lean ladders against the eave slate and cause damage.
6. Poorly designed snow retention devices as well as some penetrations, such as vent pipes, can cause damage.

Repairing a slate roof is not very complicated, but it should not be considered easy. As with most trades, there are standard techniques that, once learned, make the process seem straightforward. The first thing to do is gain access to the area where the slate is broken.

Scaffolding needs will vary from project to project. Once a safe access is obtained to the roof eaves, the next step is to reach the area where the broken material lies within the roof. If the broken material is within a safe distance to reach from the eave scaffolding, the repair procedure can be carried out. If the slate is too high to reach from the eave access, then a ladder hung over the peak with a ladder hook will usually provide the best access. Naturally, there are other methods of gaining roof access. Each project will vary. Large-scale damage may require regular slate roof scaffolding brackets.

To remove a damaged slate shingle, the mechanic will need to use a slate ripper (see Chapter 7, Slate Tools). The mechanic slides the blade of the slate ripper up under the broken piece and probes to find the fasteners that are holding the slate in place. Once the nails are found, he can manipulate the head of the ripper up and past the shaft of the nail, then carefully pull it back down until the hook part of the ripper head engages the shaft of the nail. By pounding down on the striking surface of the ripper handle, the nail will either be cut off or pulled out.

Ideally, the nail will pull out. This is usually the case when cop-

Repair, Maintenance, and Troubleshooting

per nails have been used. However, it is common for the ripper to cut the shaft of the nail off or for the ripper to bend the nail over and slip off the shaft of the nail. If the shaft is cut or the nail gets bent over, the nail will remain in the void even though the slate comes out. These nails can be released by pushing the ripper blade sharply back up past the nail, then pulling down sharply. By doing this repeatedly, the shaft will either break or the nail will come out. After the nails have been removed, probe around with the ripper to check for any remaining debris. (Quite often, the slate will break above the nails as it is being removed and will remain in place.) If any debris remains in the void, it will make it difficult to install the new piece. Once the bad piece has been removed and the void has been cleared of debris, the repair piece is ready to be installed.

Installing the replacement piece of slate is probably one of the most debated techniques within the slate roofing trade. Every roofer, homeowner, handyman, carpenter, grandfather, and schoolchild seems to have either seen a slate roof repaired or knows through some conventional wisdom the proper way to do it. Despite this apparently common knowledge, there are only two acceptable methods for installing a piece of repair slate.

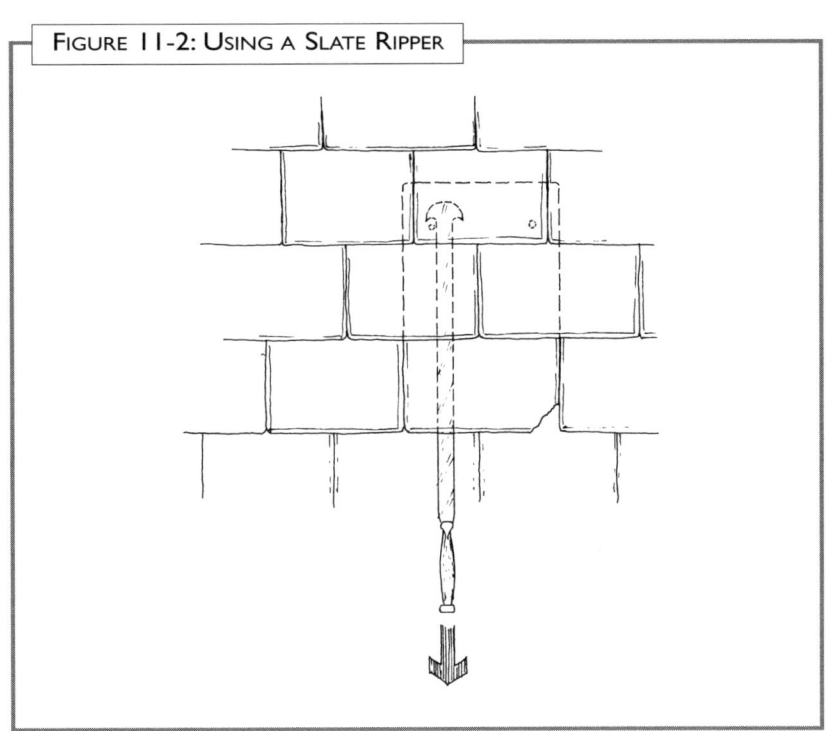

FIGURE 11-2: USING A SLATE RIPPER

Using a Slate Hook

One method of replacing slate is to use a slate hook. The shaft of the slate hook is 3 inches long from the point of the hook to the nail end. (See Figure 11-3.)

This 3-inch dimension relates to the 3-inch headlap necessary to provide proper slate coverage.

To insert the slate hook, determine the location of the hook end so that when the repair piece is slid back into the roof, the butt of the slate will rest against the hook in the same line as the butt of the piece of slate on either side. This repair method works well and will last as long as the hook lasts.

Slate hooks are available in copper, stainless steel, and galvanized steel. The steel pieces are unquestionably the easiest to work with. Copper hooks tend to bend very easily at the nail end as they are being driven into the deck.

An alternative to the standard hook shown in Figure 11-3 is shown in Figure 11-4. It consists of 8-gauge wire formed with an eyelet at one end to receive a nail. The repair slate is slid into place after the wire is nailed in, and the end of the wire is then rolled up around the butt end of the slate to form a hook.

There are two benefits to using this wire method. First, the installer will use a nail of similar material that can be driven into the roof deck with much less likelihood of bending than the shaft of the typical slate hook. This is especially important when working with a hard roof deck. This technique will also allow the use of a screw through the eyelet if the deck is so dense that a nail won't penetrate.

Chapter 11

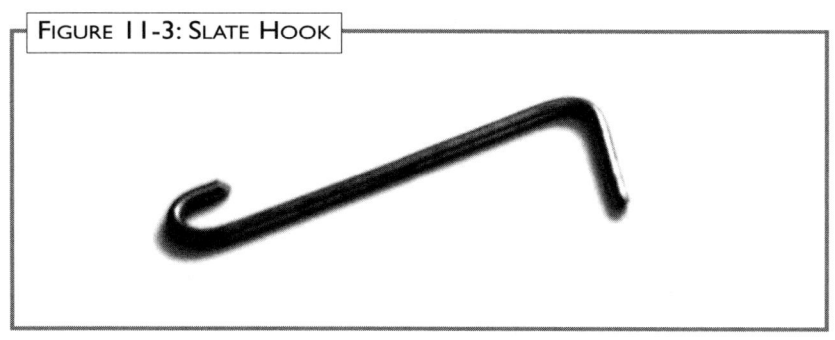

FIGURE 11-3: SLATE HOOK

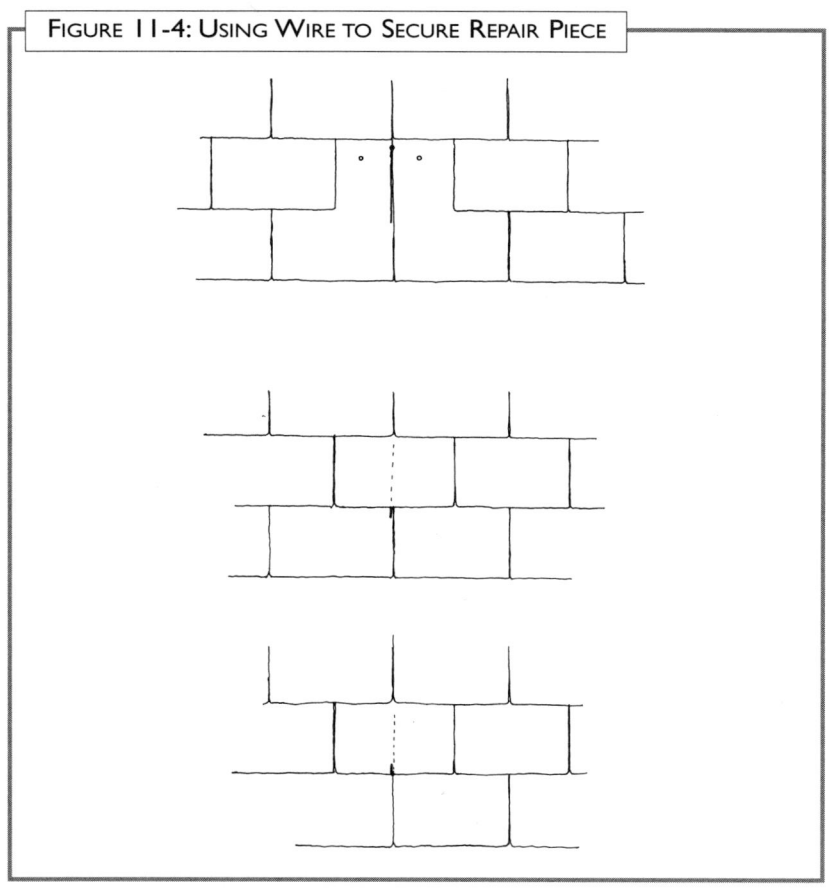

FIGURE 11-4: USING WIRE TO SECURE REPAIR PIECE

The second benefit to this method is that the hook end is not in the way of sliding the repair piece back into place. It may not sound like that big a deal, but the height of the fabricated hook will elevate the exposed end of the repair piece until it is slid up far enough into the void to pass the hook and slide back down. The difficulty arises when the top end of the repair slate is forced down into the roof underlayment material above. The sharp serrated edge of the slate often picks up the underlayment, causing it to crumple and plug the void. As a result, the repair slate will not slide into place properly. By using the wire method, the installer can push down on the butt end while pushing it up the roof. This will keep the head elevated off the deck/underlayment until it is almost all the way back in place.

The only drawback to the wire method is that the wire has to be of sufficient strength that it will not unfold when snow or ice passes over it. When using the wire method in snow regions, the careful application of mastic to the back of the slate prior to repair will help should this problem arise. There are other hook variations available. However, the concept is the same. Be sure to use a hook which will not unfold later and allow the repair slate to slide out of place.

Using the Bib Method

The second acceptable method for replacing slate is the "bib method" (which is sometimes referred to as "babies," "tin," or "slip flashing" in different regions). In this method, the repair piece is slid back into the void after the broken slate is removed. A fastener is then installed through the repair piece, through the vertical joint between the two pieces above. This fastener is usually set approximately 1/2 inch below the butt of the slate in the second course above the repaired piece. (See Figure 11-5.)

The head of the fastener is installed through the vertical gap (often chipping one or both of the adjacent pieces slightly) to a point where the head of the fastener is flush with the surface of the repair slate. A piece of sheet-metal flashing

Repair, Maintenance, and Troubleshooting

of the same material as the repair fastener is slid into place over the top of the repair slate but under the two pieces that form the vertical joint where the fastener has been driven. The flashing then needs to be lifted enough to slide past the fastener head and be pushed up at least 3 inches beyond the butt of the pieces in the second course.

The process of sliding the flashing past the head of the fastener can be very frustrating.

Probably the best way to keep the bib in place is to form a hook on the top of the bib that will ultimately hook over the head of the repair piece. By hooking this hooked edge over the head of your ripper, you can then slide it up into place. The hook will allow the flashing to slide by the nail head more easily, and the ripper can easily be removed once you feel the hooked edge pass over the top of the slate.

It is a good idea to apply some sealant over the head of the fastener prior to installing the bib. The sealant will help adhere the bib to the repair slate and prevent it from later sliding out.

Of the two methods described above, the bib method when properly carried out is the best repair method. The downside of this method is that the repair piece can be broken when the fastener is driven through from above. It is sometimes difficult to see that the piece broke or cracked, because the area where the crack would appear is covered by the slate in the course above. Over a short period of time, wind and vibration will cause this repair piece to slide out. To avoid this, slide the repair piece into place and mark the place where the hole should go with a pencil or nail. Remove the repair piece and punch the nail hole from the back side of the slate. The installer then will know if the repair slate breaks or cracks and can take appropriate action. You can also drill the repair piece with a masonry drill bit. Drilling will allow you to leave the repair piece in place and still create a hole for the fastener without breaking the repair slate.

As a rule, any slate thicker than 3/16 inch – 3/8 inch will have to be

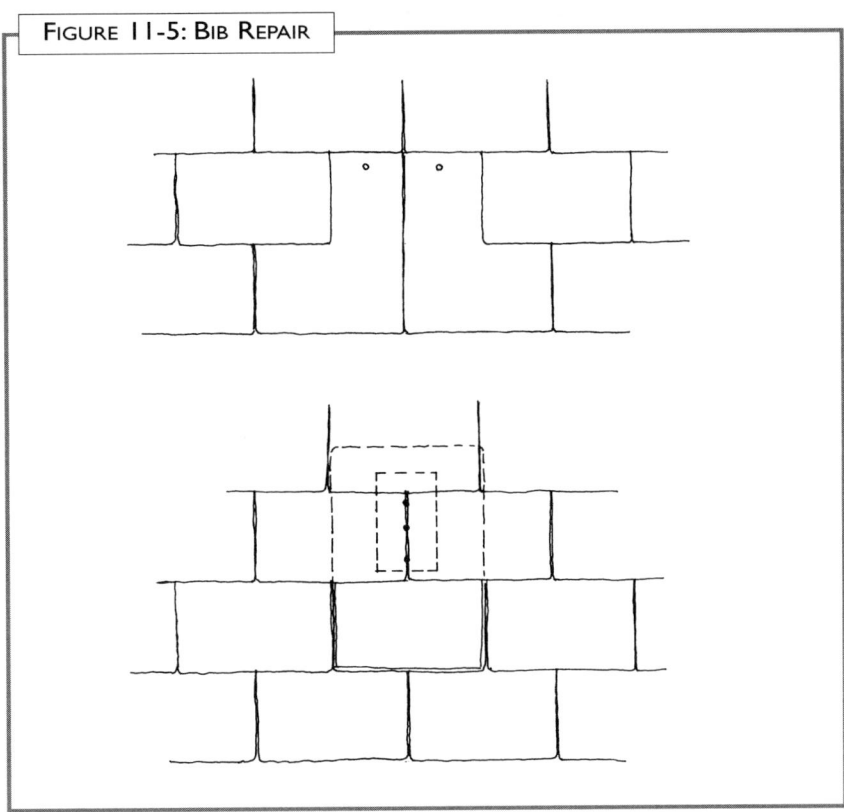

FIGURE 11-5: BIB REPAIR

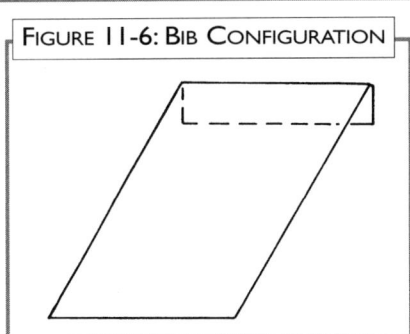

FIGURE 11-6: BIB CONFIGURATION

repaired using the bib technique. The slate hook end is not thick enough to receive the thicker material. In addition, the thicker slate is heavier and may require more than one fastener to hold it in place. The bib method allows for this.

Unacceptable Repair Methods

There are a variety of repair techniques that are common but are just not acceptable. Here are a few.

1) **The bib method without**

157

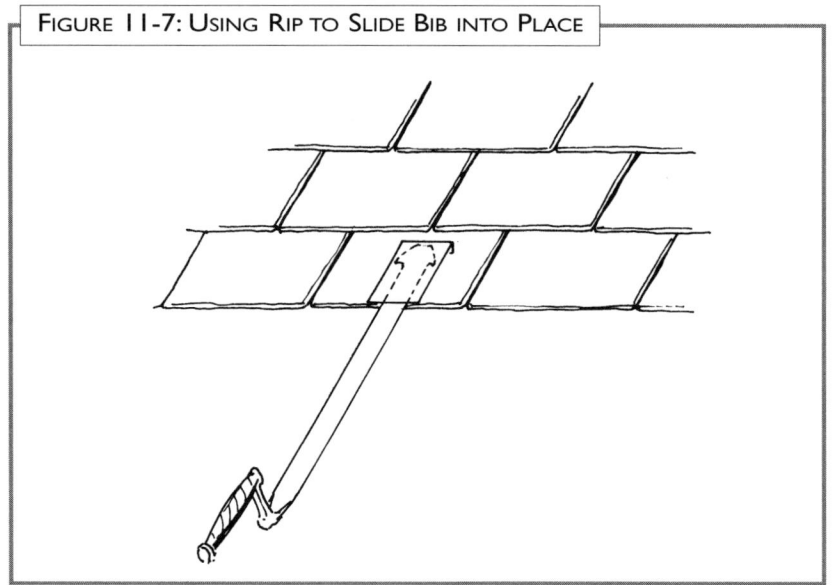

FIGURE 11-7: USING RIP TO SLIDE BIB INTO PLACE

the bib. It is common to see a repair job where the repair is carried out well up to the point that the installer needs to put in the bib flashing. The shortcut is simply to chalk over the fastener head and proceed to the next piece. A repair completed in this way will eventually leak.

2) Face nail through the exposed field area of the slate. In this technique, the slate removal is usually carried out correctly. However, the installer inserts the repair piece back into the void and drives the fastener through the exposed face of the repair slate. The fastener is then sealed with tar or caulking depending upon the quality of the installer. This type of repair will also leak eventually.

3) Construction adhesive method. Most installers will figure out how to get the damaged piece of slate out of the roof. However, if they do not know the proper method of putting a repair piece back in place, they decide to use an adhesive. The biggest problem with this approach is that most adhesives are viscous until they set up. The repair slate will slide out of the void unless it is held in place long enough for the adhesive to set. In addition, many adhesives lose their grip as the slate heats up, and the repair piece will begin to slide. Considering the damage that a piece of slate can create as it comes off a roof, it is not recommended procedure to leave on the roof, any slate that is not held by an accepted fastener.

4) Sheet metal hook method. There are those mechanics who understand how to remove the slate and also the fact that they should avoid nailing through the headlap of the slate exposed in the void. However, if they have never been introduced to the slate hook or bib repair technique, these mechanics will often make a hook out of sheet metal. These hooks are usually about 1 inch wide and 6 inches long. The repair technique they use is the same as the modified wire hook technique described above.

The metal strap is installed, the repair piece is slid into place, and the end of the strap is wrapped up around the bottom of the slate to hold it in place. In theory, this method is okay. However, in snow and ice regions, the sheet metal tab often unfolds as snow and ice slides over it, allowing the repair piece to slide out. Some of the more clever mechanics will notice this unfolding problem and go back to "method 3," adding adhesive. The hook holds the slate in place until the adhesive sets up and the hook is no longer important. This method is unacceptable for the reasons already stated above and also because these sheet metal hooks are quite noticeable and unsightly. Better techniques with longer lasting results are available.

5) Using Cracked Slate. In this method, the installer begins to repair the roof and, for one reason or another, realizes that he will run short of new repair pieces. so, in order to finish the job without incurring additional material cost or lost time, he must somehow reuse some of the existing cracked pieces. At the beginning of the repair project, the installer realizes that some of the pieces with broken corners or substantial cracks on the exposed face are perfectly square

Repair, Maintenance, and Troubleshooting

and whole on the buried end of the slate. Employing any one of the above-mentioned repair methods (usually not one of the two accepted methods), he takes the remainder of the slate removed from the void, turns it upside down, and slides it back into the hole.

Repairs made in this way will obviously leak. The telltale signs of this method are the exposed nail holes on the exposed face of the repair slate. This is a good indication that the installer does not know what he is doing—it is very unlikely that he simply made a mistake and put a good piece in upside down.

Whether you are repairing a new roof or an old roof, be sure that you have all of the proper materials and tools on hand. Make sure you know the size, color, and thickness of the slate. Always check to see what material is used on the roof deck as well as the underlayment. Do not guess at what you cannot see. If you are specifying a repair project, remove a piece of slate on each surface to be repaired to make sure you know what is underneath.

Repairing Valley Slate

In order to replace a valley slate, first remove the full piece of field slate above it. This provides access to the two nails on the outside edge of the valley piece. Remove these nails, slide out the broken valley slate, and replace it

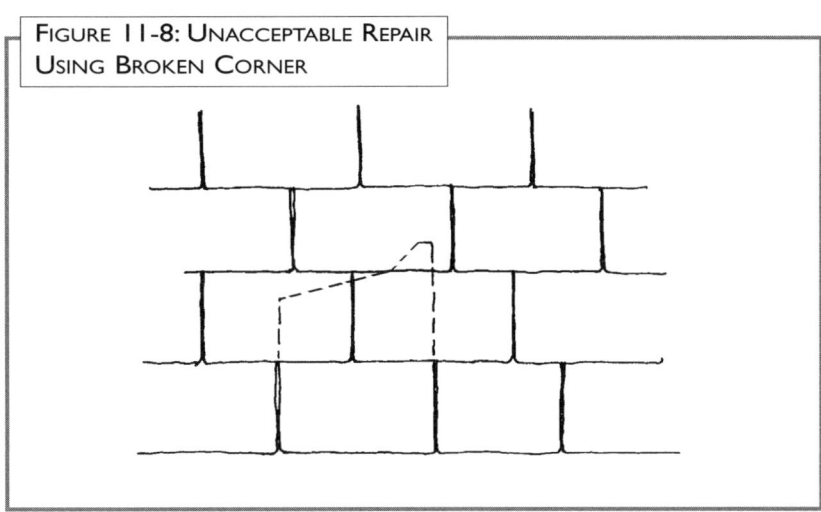

FIGURE 11-8: UNACCEPTABLE REPAIR USING BROKEN CORNER

with a new valley piece. Reattach the new valley slate with two nails, properly situated in relation to the valley flashing. Next, replace the field slate by either hooking or bibbing it.

Never hook the valley slate in place as this will result in the hook's penetrating the valley flashing underneath it. Although a valley slate can be bibbed in place, it is difficult to slide the valley repair piece back in place and the repair fastener most likely will penetrate the valley flashing.

Valley flashing is replaced the same way except that all of the valley slate and the first full piece next to it must be removed on both sides over the entire length. This provides access to the entire valley flashing. Remove the flashing, replace it, and reinstall the slate as described above. Prior to removing the valley slate, mark each piece with a letter or number indicating its location. This will make it much easier to reinstall the valley slate in its original location later.

Repairing Hip Slate

If a saddle hip cap needs to be replaced, first remove the saddle hip cap the same way you would any other broken slate, as described above. Because of the way the hip cap slate is installed there are no vertical joints through which a concealed fastener and bib flashing can be installed. The hip cap slate pieces must be hooked back in place. Use a hook at the butt of the repair hip slate as well as on the side. This creates a pocket in which to cradle the repair piece. In all cases set the hip repair pieces in a bed of mastic, as they will be susceptible to wind uplift otherwise.

If a miter hip slate is broken and needs replacement, the repair can be very difficult. When working on older slate roofs that were installed with no base flashing, simply remove the broken piece, cut a new one to fit properly, and install it using either of the accepted repair techniques.

On a new installation which has been stepped flashed with modified

Chapter 11

bitumen, it can be very difficult to remove the broken piece. You may need to cut the membrane up the centerline of the hip using a utility knife. Although some of the base flashing may still adhere to the concealed portion of the broken hip, it will be much easier to remove the debris if it is not held fast by the base flashing. Be sure to remove all of the debris. Mitered hips are delicate and any obstruction remaining in the void could cause further breakage. Once the repair piece has been properly fastened, caulk the hip joint to prevent water infiltration where the hip base flashing was cut.

Repairing Saddle Ridge Slate

Saddle ridge slate repair is typically carried out in the same manner as saddle hip cap repair. The only difference is that two slate hooks are installed along the bottom edge to help cradle the repair piece.

Repairing Finishing Course Slate

The finishing course slate is typically nailed into a wooden nailer that is usually in close proximity to the ridge cap nailers. This often creates a situation where there is no room to maneuver the head of the slate ripper up and behind the nails holding the piece in place. In this case, slide the leading edge of the ripper up to the nails and drive its cutting edge up and through the nail shaft, cutting off the nail. If the nail is not cut off, but simply bent over and impossible to remove, notch the head of the repair piece so that it will slide around the nail in the void.

Repairing Starter Slate

It is not usually readily apparent that a piece of starter slate is broken unless a corresponding piece of slate in the first full course is also broken. Whenever possible remove the broken field slate first. You will then have access to one of the nails in the starter slate. Simply remove that nail with a hammer. The second nail will not usually be accessible, and you will will need to use a ripper to remove it. Place two to three nail holes in the exposed portion of the starter piece. This slate can then be nailed back in place instead of being hooked or bibbed. Finish the repair by replacing the field slate.

If only the starter slate is broken, it can be repaired and treated the same as any other repair piece. In some cases though the slate hook repair method may not work well for the starter slate and first full course of field slate. It is best to use the bib method on these two bottom courses.

Repairing and Tying in Additions

If a structural addition is being added to a building with an existing slate roof, do your best to salvage as much of the displaced slate as possible. This slate can be used later to make some or all of the repairs when the time comes.

Repairing Gable End Slate

When a piece of slate is broken on the gable end of a building, the repair is carried out as usual. The only difference is that the half piece on the gable cannot be hooked or bibbed back in place. Therefore, you must remove the full piece above the broken half piece, remove the half piece whose fasteners are now exposed, cut and fit the half piece back in place and nail it in the appropriate location, then replace the full piece above using the hook or bib repair method.

Plumbing Vent and Skylight Installation

When installing a plumbing vent, skylight, or any other penetration in an existing slate roof, the biggest difficulty often is locating the area on the outside of the roof that corresponds with the framework on the inside of the building. First, locate the perimeter of the penetration from inside the structure. Then drive a 16d nail out through the roof from the inside. This can be done at all four corners or at the center of the penetration. Indeed this will either break slate or put a hole through it at these locations. The chances are good that these pieces of slate would have been cut or broken anyhow. Remove the slate as needed, relative to the size of the

Repair, Maintenance, and Troubleshooting

penetration. In most cases, removing one foot more of material than the opening requires provides sufficient area to work in and tie back together the related slate.

Snow Guards That Failed

Snow guards have been a necessary slate roofing accessory since the time that slate roofing began. Patents for these products date back to the mid-1800's. Interestingly many turn-of-the-century roofs that still exist today are missing these units. Were they installed in the original roof or not? More than likely they were, but many of these units have failed and been removed. Some classic problems associated with snow guards that have been around for years are:

1) The guard was hooked over the top of the slate shingle but was not nailed to the deck. This results in the shingle's being pulled out under heavy snow load conditions.

2) The guard twists or bends over backwards. When this happens the strap securing the guard to the deck lifts up, causing the slate above it to break. Look for a guard whose snow pad is attached in such a way that it is perpendicular to the roof surface. Any pad which is less than perpendicular to the roof surface will have a tendency to be forced by a snow load into a perpendicular orientation, causing the attachment strap to lift and break the slate above.

3) They are spaced too far apart. This is done in an attempt to save money. However, if the proper quantity is not used, the snow guards will not function properly. They will either break, bend, pull out, or damage the surrounding roof area.

4) The manufacturer does not provide a layout or test data pertaining to the snow guards. It is therefore left up to the roofer to determine the proper quantity and layout pattern for installation. If an incorrect number of snow gaurds are installed, you could potentially run into the problem previously described (see #3).

Pipe-style snow guards work well when installed properly. Look for a design whose base replaces a full piece of slate. This will distribute the load most evenly. Do not space the brackets more than 4 feet apart horizontally, otherwise the pipes will have a tendency to bend, and the load carried by the slate around the base plate will cause these neighboring slate pieces to crush.

Look for a snow guard manufacturer who will assist you in product layout. In many cases one tier of snow guards placed at the eaves is not adequate to support the given snow load. As a result the snow guards fail and often break slate in the process.

Keep in mind that almost any snow guard design available today will function properly if enough snow guards are used. This is the critical part. Costs per snow guard units vary greatly. The least expensive type available today on a per unit basis is probably the wire loop style. However, proper installation may require that one unit be installed between every vertical slate joint on the entire roof. As a result, this guard can be the most costly system available.

If the snow guards in your roof have failed, first try to determine what caused the failure. If they have twisted or bent over backwards, discard them and look for a different style. Do not try to reuse a snow guard that has bent. It will have a tendency to bend again in the same place. In most cases, snow guard failure is the result of insufficient quantities of the product installed on the roof. This is a major issue related to value engineering. If the snow guards fail, not only will they have to be replaced, but they will also most likely create slate roof damage that will have to be repaired.

Gutters

Repairing gutters on a slate roof requires knowledge of both slate and metal repair.

Built-in, Yankee, or Pole Gutter

It is common for maintenance people to use these built-in gutter configurations as scaffolding. They provide easy access for walking, a

Chapter 11

handy place for setting the base of a ladder, or a handy storage area for subcontractors who need a place to set tools or materials. Damage caused by personnel unrelated to the roof is very common.

One of the unfortunate problems that arises with a concealed gutter is the old "out of sight, out of mind" philosophy. Any gutter will accumulate debris. If left unchecked, this debris will eventually build up and clog the downspouts.

Instead of repairing or replacing these gutter systems, many people have chosen to simply slate over the gutter area and install a facia-mounted gutter system.

At-the-Eave Flange Style Gutters

These gutters, if they are severely damaged, can become a big problem. Should sliding snow or ice, someone's ladder during a maintenance operation, or a tree limb, etc., bend rip or otherwise damage this system, the slate in that area will often need to be removed before the damage can be properly repaired. In many cases, these gutters were removed from older buildings by simply cutting off the flange section, leaving it in place under the slate, and replacing the gutter with a new facia-mounted system.

Repair Slate Inventory

After roofing a building with slate, always try to leave one or two percent extra material on site. This will provide enough material of the same color, texture, and variety for any future repairs. It is wise for the owner of a home with an existing slate roof to purchase and have on hand 50-100 pieces of the proper slate for repair purposes. Don't wait until repairs are needed or unforeseen damage is caused before looking into purchasing this material. In the event of an emergency, you can compromise by repairing the roof with different material or accepting the damages that may occur while you wait for the proper replacement pieces to arrive. Ask a slate quarry to help you match your existing slate and have this material available before it is needed.

Repair Tips
Using Ripper Blade as a Guide

In some roof repair situations, it will be difficult to get the repair slate back into place. This may be due to the top of the slate catching on the underlayment or the top of the slate catching on a horizontal batten (if used). Prior to pushing the repair piece back into place, try sliding the ripper blade into the void, and let the repair slate slide up the ripper blade. You can then usually lift the head of the repair slate with the ripper blade. The ripper blade will usually be thin enough to allow the slate to slide up into place.

Cutting Top Backwards

Sometimes it may be next to impossible to get a repair piece to slide back into place. You can try tapping on the piece at the butt end with a hammer, but this will often chip it. Remove the repair piece from the void and examine the top edge. There may be marks on the top edge of the slate that indicate where the piece is getting jammed.

If the underlayment seems to be the problem and the ripper blade sliding technique described above doesn't work, try cutting approximately 1/4 inch of slate off the top edge of the piece. This should be done from the front side of the piece. When the repair piece is cut from the front side, the chamfered edge created is opposite from normal and is more likely to slide past any obstruction. Try not to use this technique unless absolutely necessary. It is best to avoid reducing the length of the slate as this reduction will affect the 3-inch headlap.

Staggered and Pyramid Slate Removal

Flashing deterioration is becoming apparent on many older buildings. As a result, building owners are faced with the question of whether to remove the slate and replace both slate and flashings or to replace the flashings only and leave the slate. If it makes sense to replace just the flashings, try to use one of the following repair procedures.

Repair, Maintenance, and Troubleshooting

Remove slate in a horizontal plane for the purpose of replacing flashings, built-in gutters, or flat roofing materials:

Do not attempt to tuck new flashings or flat roofing material underneath the starter course without first removing slate. You might just as well patch the existing roof with tar. This repair approach will last nearly as long.

You will need to remove the first full course and the starter course. Be aware that when it comes time to replace the slate, the entire first course of slate will need to be set back in place using the bib method of repair. (The hook method, in this case, would cause penetrations through the new base material.)

Start by removing every other piece of slate from the second full course. This will allow you to pull one nail out of every piece of slate in the first full course. The second nail in each piece will need to be removed with the slate ripper. The starter course can then be removed by simply pulling the nails out of each piece with a hammer.

Before removing the slate, number or label each piece as it sits in the roof. This can save a good deal of head scratching later when it comes time to put the roof back together. (Figure 11-9.)

After installing the base flashing material (no higher than the nails in the starter course) reinstall the slate. Once the starters are back in place, install the first full course. Because you have access to one side of each piece, you can double- or triple-nail each piece on that side. Some installers argue that this practice is not a good idea. However, for this type of repair, it is the best and most secure approach. It is also wise to set the pieces in a bed of mastic.

Next, replace the staggered pieces that were removed from the second full course. This replacement will need to be done using one of the two accepted repair methods. The reason for leaving every other piece in the second row will soon become apparent since every other piece is undisturbed and properly nailed, they prevent the repaired pieces from shifting and turning side to side.

Remove slate around a penetration or large repair area:

Whenever multiple pieces of slate need to be removed and repaired/replaced, attempt to do your work in a manner that will allow you to hook or bib in one piece of slate at the top. This can be accomplished in a stepped fashion where one piece is removed, then two pieces below that are removed, then three and so on. Ultimately only the top piece will end up having to be put back in place as a repair piece. (See Figure 11-10.)

Maintenance
Maintenance Related to Other Trades

Try to prevent any tradesperson other than the slate installer

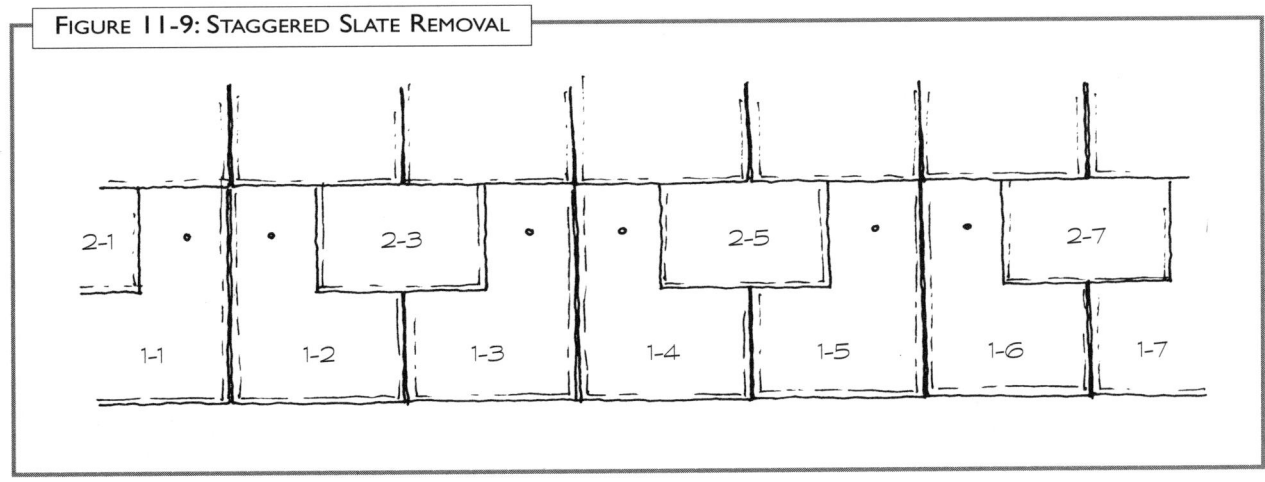

FIGURE 11-9: STAGGERED SLATE REMOVAL

Chapter 11

FIGURE 11-10: PYRAMID SLATE REMOVAL

from going onto a finished slate roof. This is not always possible. Therefore, the building owner needs to understand that allowing people other than the roofing contractor on the roof will very likely cause damage. This should be expected. If, for example, a chimney needs to be cleaned, try to schedule routine roof maintenance on or about the same time or try to find a chimney sweep experienced with slate roofing. Make it clear to any tradespeople who go up on the roof that you expect them to be responsible for any damage they cause. (Suggest that they review this text.)

Accessing the Roof

One of the easiest ways to gain access to an existing slate roof is with a ladder and ladder hook. The hook attaches to the rungs of the ladder and hooks over the peak of the roof. The roofer can then walk up the ladder where it rests on the roof. The fact that the ladder is suspended from the peak and lies flat over the length of the roof, helps to distribute the roofer's weight over the entire length of the ladder.

Always use additional back-up safety lines when working on a steep slope roof, and never tie them to a movable object on the ground.

Annual Maintenance and Inspection

A slate roof can last 100 years if properly maintained. Have the roof inspected at least once a year by a competent mechanic. If you have gutters on your building, make sure that they are cleaned at least once in the spring and again in the fall.

Cleaning Slate

Slate will sometimes become stained. Many things can cause staining. The question is, can slate be cleaned? In some cases you can clean slate, in others you cannot.

1) **Sap Stains from Trees** — Sap stains can be cleaned up reasonably well using a diluted solution of masonry cleaner. Use a sponge or rag and scrub the stained area thoroughly. Be sure to rinse the area well when finished, otherwise a milky residue may appear when the area dries. This residue should wash off, but it may take time. Keep in mind that the tree which originally caused the stain will continue to do so in the future. Avoid cleaning this stain unless the tree, for some reason, is no longer there.

Try to find a non-acid based cleaner. Test the cleaner on a small surface area or on an individual piece before cleaning a large area. Do not use acid where there are vinyl windows or siding. Ask the vinyl manufacturer to recommend a safe cleaner.

2) Moss Growth — Thick moss growing on anything usually indicates a high moisture content present. The moss can be scraped off and the area cleaned as described above, but often the slate itself is in bad condition if this situation has developed.

3) Concrete — Stains resulting from masonry washing, mortar splashes, and stucco spills do not clean up well from slate. Try to avoid doing this work above a finished slate roof. If there is no alternative, cover the slate with plastic or suitable drop cloths to protect it. For these types of stains, the only alternative is to remove the stained slate and replace it.

4) Paint — Paint stains can sometimes be removed by scraping or wiping up spots as soon as they occur. Some paint removers will leave an oily stain that can not be removed, so be sure to protect the finished slate roof properly.

5) Oil will cause a permanent stain that cannot be removed.

Troubleshooting

With any roofing material, problems can and do arise. Resolving these problems will require some troubleshooting.

Roof is one year old, and slate is missing or falling.

Don't panic! A few broken slates should be expected in the first year or two.

There are several factors that contribute to a roof's "shedding." During the installation the slaters rig scaffolding or ladders to keep themselves and their materials on the roof. As they progress up slope so does their scaffolding. After they reach the top and install the ridge, they slowly work back down, removing the scaffolding and repairing any broken, cracked, or missing slates. It is common for the installer to miss some cracks in the slate because the cracks are covered by the slate course above. Wind, building vibration, and sliding snow and ice can dislodge these pieces.

Once the roof is completed, it is common for other contractors to go up on the roof or to lean ladders against the slate. This roof traffic will often break slate. If the slate does not slide out when it is first broken, the factors described above will eventually cause these pieces to slide out of the roof.

Often times, the installer will inadvertently install a thick piece of slate next to a thin one. The piece lying over this rough texture in the next course up will sometimes break as snow and ice slide over it. This will usually happen within the first year or two that the roof has been installed. After the first few years, this "shedding" should end.

It is important that the roofer return to the project and repair any broken slate as soon as possible. Establishing who will be responsible for the repair of broken slate and who will pay for this service should be addressed as early as the initial bidding process.

With minimal yearly maintenance the roof should last a lifetime. If the roof continues to shed for more than two years, there may be a more serious problem that needs to be addressed. For new roof projects greater than 25 squares, no more than one piece of broken slate per square should be acceptable, relative to the installation. Projects less than 25 squares may have more unusual breakage, but they should never exceed 1.25 pieces/square.

The roof is 50 years old, and slate is starting to fall out.

Refer to Chapter 10 Reroofing.

The house is 60 years old and is beginning to develop leaks.

On most slate roofs, the first areas to develop leaks are related to the flashings. The most common flashing used on older buildings in the United States was 16 oz copper. This copper has a service life of approximately 70 years. (See Chapter 10 Reroofing.)

Chapter 11

My roof is leaking, but the slate and gutter seem fine.

Years ago, the drop tubes or downspouts for Yankee, pole, or built-in gutters were connected to the interior plumbing system. Many turn-of-the-century buildings had cast-iron plumbing built into their structures. The copper drop tubes from the exterior gutter system were run in through the roof frame into the interior cast-iron plumbing system. The result was that the copper/cast-iron connection deteriorated due to electrolysis.

Unfortunately this is probably the first point of failure on many older buildings. In the past, when leaks started showing up inside the structure, they were believed to be caused by the slate roof's leaking. The owner would call in a roofer to repair the leak. The roofer could not find any problem on the roof, so the leak continued. It sometimes took several roofers before the real cause of the leak was traced to inside the building's framework, but not before extreme damage had occurred.

Slate is breaking at the eaves.

If an unusually high number of pieces seem to be breaking at the eaves, check to see if there was a cant strip installed. Failure to install a proper cant strip allows the first full course of slate to cantilever over the head of the starter slate. If snow or ice slides across this cantilevered area, these improperly supported pieces will begin to break. Unfortunately this causes a domino effect whereby the progressive courses also begin to do the same thing. The only solution is to install a properly-sized cant strip. Refer to Chapter 5 Roof Construction and Preparation.

Slate has colored or weathered.

Slate can and does weather. Refer to Chapter 2 Design Considerations.

Slate has stained from metal.

Most metals will oxidize. As this oxidation takes place it tends to wash over the roof areas below. In some cases this can cause staining and should be anticipated. Avoid using ferrous metals that will continue to oxidize, creating unsightly stains. Copper and lead-coated copper will both oxidize, but a stain is rarely apparent. Once a protective patina has developed on the surface of the metal, the run-off becomes minimal.

A new house leaks, but only after heavy rains.

Some building owners will observe that their roof doesn't leak everytime it rains; rather it leaks after extended periods of rain. This is most often the result of moisture that is getting in through the masonry. If through-wall flashing has been installed in all masonry, this problem will not exist or will be minimal. However, if surface-mounted counter flashing is used, any moisture that gets in behind the masonry will leak into the structure. Consult with a mason and consider sealing the masonry surface with a chemical designed to prevent such leaks. Understand that any treatment applied to the masonry will need to be reapplied every 2-4 years.

Snow slides off the roof.

To prevent or reduce snow from sliding off a roof, refer to the snow guard section in the installation section of this book. Depending upon the volume of snow that is sliding, it can damage everything in its path. Try to protect doorways, shrubs, walkways, gutters, roof penetrations, and lower roof areas from the effects of sliding and cascading snow and ice. The damage caused by falling snow and ice can far exceed the cost of installing proper snow guards to begin with. Refer to Chapter 8 Installation.

Ice dams occur.

Ice damming can occur on any roof. The formation of ice at the eaves is common. This ice damming can cause water to back up into the structure.

There are several ways to address this problem. First consider the roof ventilation. When large amounts of ice form at the eaves, it is usually due to high amounts of heat loss through the roof. Water runs off the heated roof surface down to the eaves and re-freezes on the lower unheated roof overhang. To solve this problem, try ventilating the roof to get the heat out.

A second solution to this problem was developed several years ago. In this approach, the slate at the eave area is removed, and a standing seam metal snow slide is installed. The theory is that by installing a semi-seamless metal snow belt, you prevent water from backing up. Any water that does back up will not leak into the building. Further, the surface texture of the metal belt is less likely to permit the ice build-up. This method does not completely prevent water back-up. The seams can and do leak. If you choose this method, install modified bitumen under the snow belt before you install the metal. This will prevent any water that may seep through the seams from causing leaks.

A third method is to remove the slate at the eaves up the roof to a point at least 2 feet inside the exterior wall structure. Install modified bitumen over the exposed roof deck and reinstall the slate. In essence you are accepting that ice damming may occur and that the modified bitumen will prevent related leaks. The best solution is to combine the first and third approaches. The ventilation will reduce melting, and the modified bitumen will provide a secure back-up system.

Slate is painted.

On a few occasions, slate roofs have been painted. At one point in history it was believed that slate could be waterproofed by sealing it with a rubberized paint. This does not work. A large sales commission will be earned, but deterioration and leaks will be prevented on a very limited basis.

Metal has been installed over a slate ridge or hip.

Many older buildings have metal hip and ridge caps. Some of these building were designed with this style cap. However, many of them were retrofitted with this cap at a later date, in order to fix leaks that occurred related to the slate hip and ridge details.

At one time, slate hip and ridge details were installed with no base flashings. It was accepted procedure to install the slate cap details and to point the finished joint detail with slater's cement. The slater's or roof cement eventually cracked and fell out. As a result, the detail leaked. The option at that time was to either continually re-point these details as a maintenance item or find a more permanent solution. Most building owners chose to retrofit these details with a metal cap.

On a new building you can match this detail on the ridges. However, the typical metal cap used for this retrofit detail has a very short flange. (See ridge roll in details.) This flange may not be long enough to cover the head of the last full course of slate properly at the ridge in a new installation.

Metal is visible between the slates.

On a properly installed slate roof, there should not be any metal visible between vertical slate joints. However, if repairs have been made using the bib fastening technique, it may be possible to see metal tabs between those related vertical joints. The downside is that it can be unsightly if there are a lot of them. The upside is that if there are a number of these repairs within the roof it is better to have the bibs in place than not there at all.

The slate has paper laced in between courses.

There is no need to lace paper in between the individual courses of slate. If your roof has been installed this way, there is a good chance that it was installed using the "Economy" approach. (See Chapter 2 Design Considerations.) If your roof is less than 25 years old and leaks are developing, it is likely that the paper has worn out. Because the felt paper is the only thing keeping the roof watertight, the only alternative may be to remove the slate and reroof. If you choose to reroof such a building

Chapter 11

with slate, be sure that the structure can handle the weight of a slate roof installed using the proper overlaps.

Bats and bees appear to be living in the slate.

Bees have a tendency to crawl up the vertical joints in slate. They will make their nests anywhere slate is missing or broken. This is not a problem. It should be anticipated. On thick slate roofs such as a 1-inch thick graduated roof, bats too will sometimes hide between the vertical joints. This is not a problem either. However, the roofer should be warned in case he goes poking around with a ripper to do repair work. Unexpectedly encountering bees or bats on a roof "can flat out wreck your day!"

How to isolate a leak.

Do not allow a roofer to tell you that it is impossible to locate the sources of a leak on a slate roof. This can be done using a garden hose. Spray water on the roof, starting at the bottom of the roof in the area that seems to be leaking and slowly work your way up until the leak is re-created within the structure. Do not use a high pressure hose, but rather attempt to simulate the natural water flow.

Once the leak has been isolated and repaired, repeat the test. This will assure the roofer and the owner that the leak was found and has been repaired. This approach will require a person outside running the hose as well as one inside looking for the leak. The cause of the leak will not always be easy to find. It may originate from a gutter backing up or a chimney in the related area leaking. However, using this method combined with patience always works. If this method is done properly and slowly as described, you should never have a problem isolating a leak.

Is slate reusable?

Slate can sometimes be removed from a roof and reused. However, not all slate is worthy of reusing. If you are uncertain about the quality of existing material, consider sending a sample to a quarry for evaluation. You can also have the slate tested locally to determine its grade, relative to the ASTM Specifications. See Chapter 10 Reroofing and Chapter 2 Design Considerations.

Slate has been installed over nailable concrete.

Many slate roofs have been installed over nailable concrete products. When these products are new, they are relatively easy to work with and provide an excellent surface on which to install underlayment and slate. However, if you are involved with a reroofing project that has a concrete roof deck, be sure to evaluate the deck carefully for fastening. Some of these existing concrete products have cured and cannot be nailed into. If this is the case, you may have to install vertical wooden nailer strips over the concrete with appropriate fasteners to provide for the attachment of an alternate sheathing. If the alternate sheathing is solid, the underlayment can be installed over the top of it. If the alternate sheathing is an open batten system, the underlayment should be installed over the concrete and held in place with the vertical nailer strips.

If this approach is used, be sure that the new deck will not be elevated so high that it affects related soffit and facia trim details.

CHAPTER

12

Advanced Situations

Most of this book provides information for standard slate roofing projects. Some slate roofs, however, will be considerably more complicated to install and may require advanced skills and techniques. The purpose of this chapter is to provide information about some of these more advanced applications, such as round valleys, sweeps, cones, eyebrow dormers, and battens and how to handle them.

The primary difference between a standard application and an advanced application is in the layout. Most of the more advanced situations require vertical layout lines to which the slate must be cut and laid.

Round Valley

A round valley is a more detailed form of closed valley. Keep in mind that this valley style does not work on a roof intersection that has two different roof pitches.

In this valley type, the slate follows a radius curve around the valley. This detail usually includes a rounded soffit and facia detail. The basic layout for a roof that contains a round valley is the same as that for any other slate roof. The difference occurs at the valley only.

To lay out the valley:

1. Determine the center of the valley at the eaves and at the peak.
2. Snap a center reference line.
3. Determine the location of the start and finish points of the radius and snap a line from there to the ridge perpendicular to the eaves.
4. Divide the distance from the center line to the line perpendicular, along the peak, into increments equal to the widest slate you have to use. Repeat this for the other side. You should have the same number of marks on each side of the valley center line.
5. Divide the eave radius into the same number of increments as the peak and snap lines from the eaves to the peak joining these marks.
6. Count the number of courses on the main roof surfaces. You should have the same number of courses on each side of the valley. NOTE: In Figure 12-1 you can see that the valley reference lines get longer as you approach the center of the valley. This will be true for most round valleys unless the peak is also radiused and is a large enough radius to keep the roof surface a constant length all the way through the valley.
7. Divide the center reference line and every second or third reference line out from the center into the same number of vertical sections as there are courses on the rest of the roof.
8. Connect each course from one side of the valley to the other by connecting the corre-

Chapter 12

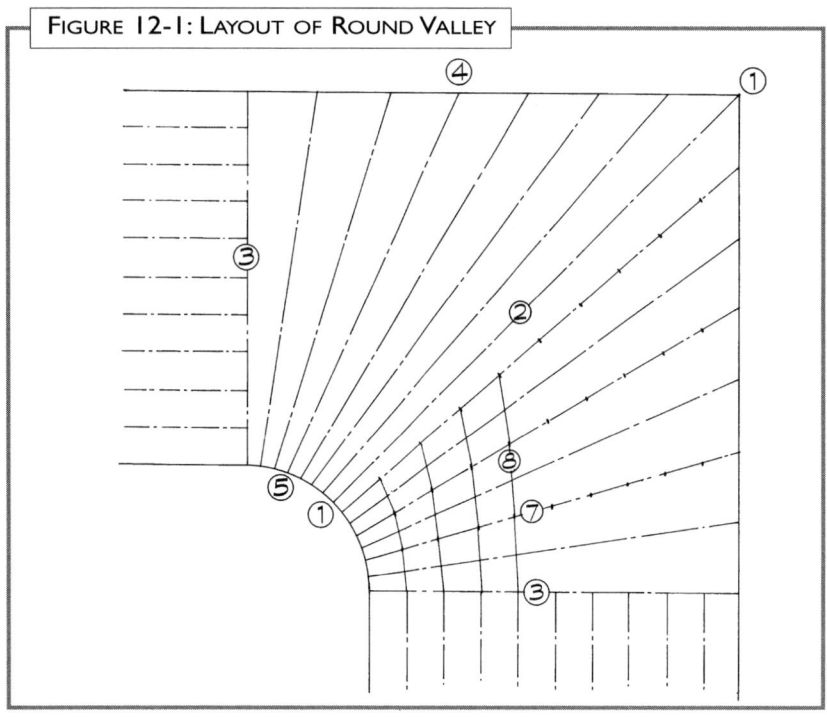

FIGURE 12-1: LAYOUT OF ROUND VALLEY

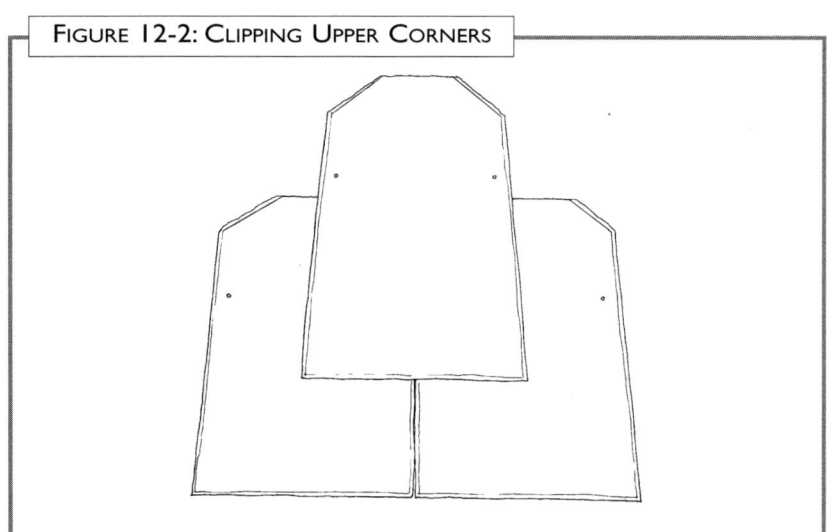

FIGURE 12-2: CLIPPING UPPER CORNERS

sponding mark on the reference lines. As you can see near the center of the valley the course exposure increases. This means that the slate length will also need to increase. Remember to maintain a 3-inch headlap throughout the valley.

Remember you will need to order longer slate for the valley if you choose or encounter a round valley.

As you start to install the slate through this rounded valley, you may need to clip the upper corners of the slate. This allows the pieces to lay flatter through the curve. In any areas where the shingles are not supported or are rocking, try shimming them with a piece of cedar shingle. As you work the courses, from the bottom of the valley to the top, the slate will become progressively wider. The first course of slate will be the same width as the reference line marks at the eaves. However, if the slate is less than 3 inches use every other reference line. When the distance between lines becomes wider than your widest slate, it will become necessary to step down in slate width. For more information refer to page 173 #8 and see Figure 12-7. There will be a slight bump created at the top of the valley that will need to be capped with a metal flashing.

When properly installed with the correct length and width slate, these rounded valleys should not require flashing. However, it is highly recommended that a base layer of modified bitumen be installed up the entire length and width of this type of valley. In addition, each course should be stepped flashed over the entire length of the valley area with modified bitumen or metal. The length of this valley stepped flashing will need to be approximately 1 inch longer than the concealed portion of the slate. This flashing can be copper, but if you prefer, you can use soft copper or lead. Both of these materials will lie well under the slate.

Advanced Situations

Canoe Valley

A canoe valley is the most common type of rounded valley. The only difference between a canoe valley and a round valley is that the valley framing starts at a square point at the eaves associated with conventional framing and ends at a square point at the ridge. Unlike the rounded valley whose framing and rafters follow the contour of the rounded soffit, the canoe valley will require blocking to create support. The blocking will be diamond-shaped with the points at the ridge and the eave. The widest part of the diamond will be at the midpoint of the valley.

1. Install the cricket in the valley. This cricket, at its maximum width, should be as wide as the widest piece of slate being used on the project. The cricket will be widest at the midpoint of the valley and then taper to a point at either end. The framing will have to be well supported with blocking.
2. The entire length and width of this detail should be covered with modified bitumen. Further, each slate course should be laced with modified bitumen as described previously.
3. Start the layout by snapping a vertical valley center reference line.
4. Snap a line perpendicular to the valley centerline at a point midway between the eaves and the

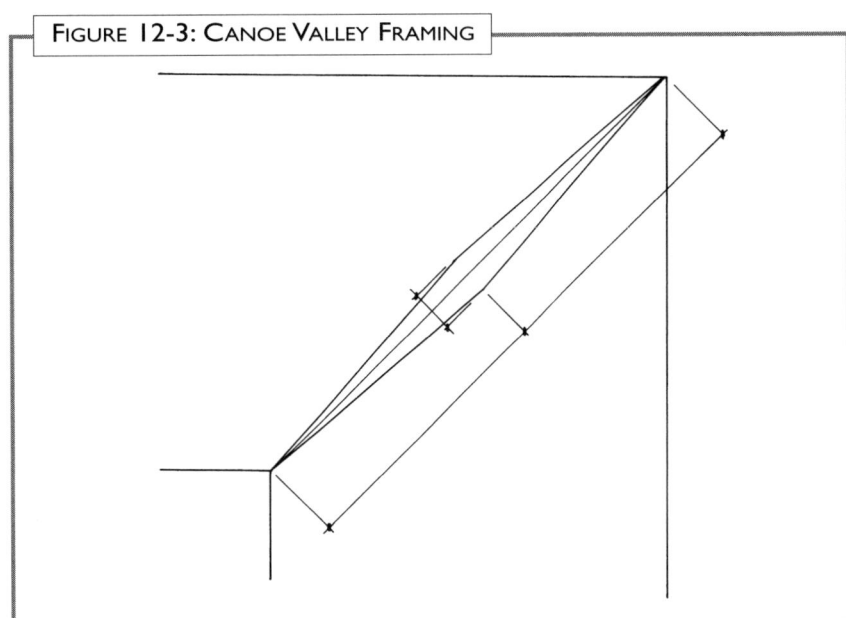

FIGURE 12-3: CANOE VALLEY FRAMING

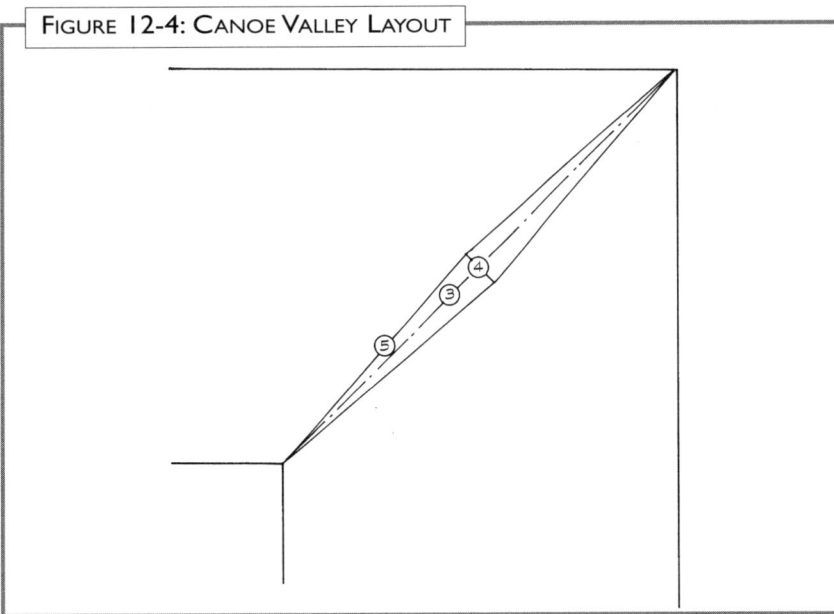

FIGURE 12-4: CANOE VALLEY LAYOUT

peak. The sole purpose for this line is to indicate where the installer should begin tapering the valley up to the point at the ridge.
5. Snap a line along each side of the cricket where the cricket meets the roof surface. These are your cricket reference lines. You are now ready to begin installing the slate.
6. Start by installing the starter slate on either side of the cricket. These starter slates will be cut to the cricket reference lines.
7. Step flash this area by lacing in a strip of modified bitumen and a metal step flashing. Do this for every course of slate.
8. Install the first full piece of slate on either side of the valley, 3 inches out from the cricket reference line.

Chapter 12

9. Cut two valley pieces to finish the first course through the valley. These two symmetrical pieces should be cut to fit between the center valley reference line and the vertical side of the adjoining field slate. Support these pieces with shims as necessary.
10. Start the second full course of slate by installing the center taper-cut valley piece of slate. Center this piece over the two taper-cut pieces in the first full course of slate. This means that the left side of this piece will be resting on the left roof surface, and the right side of the slate will be resting on the right roof surface.
11. Skip the piece on either side of the center taper-cut piece in the valley and install the field slate, leaving a gap no wider than the widest piece of slate available on the job site.
12. Mark, cut, and install the piece of slate to fill the void on either side of the center valley-cut piece. Repeat this procedure all the way to the mid-span.
13. Reverse this procedure as you work from the mid-span back towards the peak.

The canoe shape is formed by alternately installing only two and three tapered pieces of slate per course. The valley becomes wider at the mid-span of the valley, not because more tapered pieces are used, but rather because the pieces used are wider. Therefore it is important to use the narrowest pieces possible at the eave and ridge and the widest pieces possible at the center.

When tying in slate courses between the canoe valley framing and the normal field of the roof, you may need to clip the upper corners of the slate to get it to lay well (see Figure 12-2). A canoe valley requires the use of longer slate in order to maintain proper headlap. This detail will require slate which is 2 inches longer than the field slate, will add additional labor costs, and will add additional flashing, labor, and materials.

Cone Shape

The basic roof layout for a roof which incorporates a cone shape is the same as that for any other slate roof. The only major difference is that vertical reference lines will be snapped from the eaves to the peak around the entire cone. These lines will taper as they progress from eaves to peak. Each piece of slate on a cone shape will have to be cut to follow the tapered reference lines. The cone shape will require additional labor and materials.

1. Start by determining the width of the widest slate on the project. Make sure that there is enough slate of the chosen width to install around the entire perimeter of the base of the cone.
2. Next, snap one line, perpendicular to the eave, on the front side of the cone from eaves to peak.
3. From this first line, measure in each direction the width of the slate chosen to start at the base. Remember that the slate, when installed, will be laid with a 1 1/2-inch overhang at the eave. If a drip edge with a 1 1/2-inch overhang has been installed, then make the marks at the edge of the drip edge. If no drip edge is being used then the marks, which will be at the eave, will need to be about 1/2 inch narrower than the slate used in the first course of slate to account for the 1 1/12-inch projection from the point that the reference marks are made.
4. As you approach the back of the cone, you will have to adjust the width of the layout lines. Ideally, all of the reference marks will be evenly spaced around the cone. If adjustments need to be made in order to produce an even spacing, reduce the vertical layout line marks by no more than 1 inch per mark. For example, if 12-inch wide slate is being used, a few 11-inch wide marks may be needed to

Advanced Situations

produce an even layout around the back.

5. Snap all the vertical lines.
6. To lay out alternating course vertical reference lines, follow the same procedure but use a different colored chalk. This can prove very helpful when you get close to the peak and all of the reference lines are very close together.
7. The slate will be installed to horizontal layout lines as usual. There are several ways to produce horizontal lines on a cone surface. One is to scratch a line into the underlayment using a nail or pencil which is attached to a wire and suspended from the peak on an apparatus which will allow it to swing unobstructed around the entire cone. A small wheel on a shaft works well. It is important that all the horizontal layout lines be parallel with the eave and consistent all the way around the cone.

   Once the cone is laid out, the slate can be installed. On a cone, every piece of slate will need to be cut to follow the taper of the vertical reference lines.
8. The starter slate is cut to a taper and installed following the tapered vertical reference lines.
9. The field slate is cut to a taper and installed using the proper nail and side-lap coverage rules for all slate installation. At the vertical joint alignment, the slate will have a tendency to stick up slightly due to the round shape of the cone. This is to be expected.
10. As the slate is installed up slope it will be cut progressively narrower. Approximately 1/3 of the way up from the eaves, the butt ends of the slate will be slightly less than half the width of the butt ends of the slate at the eaves. At this point, you should install wider slate

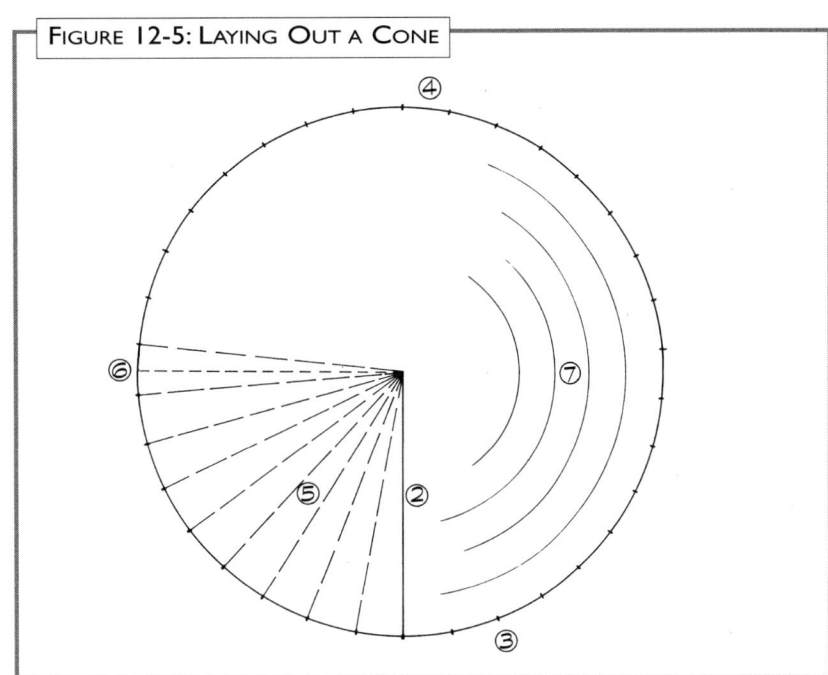

FIGURE 12-5: LAYING OUT A CONE

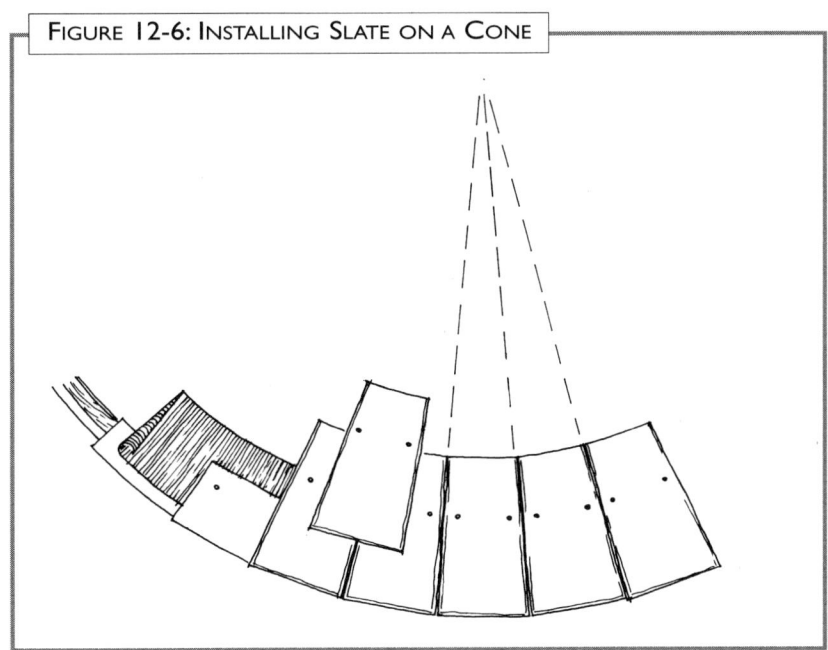

FIGURE 12-6: INSTALLING SLATE ON A CONE

that is cut to align with every other vertical reference line. This reduces the number of pieces in each row by half.

The same thing will happen approximately 2/3 of the way up the cone. If you are uncertain as to when you should make this "jump" in sizes, a good rule to follow is to install slate no less than three inches wide at the butt end.

A cone will require a metal cap at the peak. The size of the cap or finial (usually ornamental) will vary with the overall cone size, roof pitch, and the point at which the slate installation ends. Be sure that this cap covers the last course of slate so as to provide proper head-lap with the slate course below.

Establishing a pattern for each row of slate on the cone may prove helpful, and can be done once the layout has been completed. Count the number of pieces of slate needed for each slate course. On the ground, cut 75% of the slate for each course. Cutting slate on the ground is often faster and reduces unnecessary debris on the roof and scaffolding.

As the slate installation progresses, the precut pieces will not always match the vertical reference lines perfectly. As this happens, borrow a piece from the course before or the course after to adjust the course to line up with the proper reference line. Finish each course by custom fitting the remaining 25%.

Eyebrow Dormers

The eyebrow dormer can be one of the most complicated slate roof details to carry out. Vertical and horizontal line layouts for this detail are critical.

The horizontal line layout for an eyebrow dormer is not as difficult as it might appear. First, lay out all of the horizontal lines on the entire related field area all around the dormer. Obviously these lines will not carry over the top of the eyebrow. In order for this detail to be laid out properly, the horizontal line layout which carries over the top of the eyebrow must match up from side to side. The following method will help you properly join these horizontal lines through and over this complex detail:

1a. All horizontal lines from the field area that intersect the valley area of the eyebrow dormer must be laid out and snapped.

2a. You will need a piece of sheet metal approximately 8-10 feet long and 6-12 inches wide.

3a. Lay this strip of sheet metal onto the first horizontal line which intersects the eyebrow. Approximately 3 feet of this metal strip should be laid so that the bottom edge of the strip lays on and tight to the horizontal layout line in the field of the roof. The remaining length of the strip is then pushed down into the contour of the eyebrow valley and progressively over the top.

4a. Mark a line along the bottom of the strip. This line will be followed as the horizontal line layout.

5a. Repeat this procedure on the opposite side. The lines from each side will not connect over the top of the eyebrow dormer, yet.

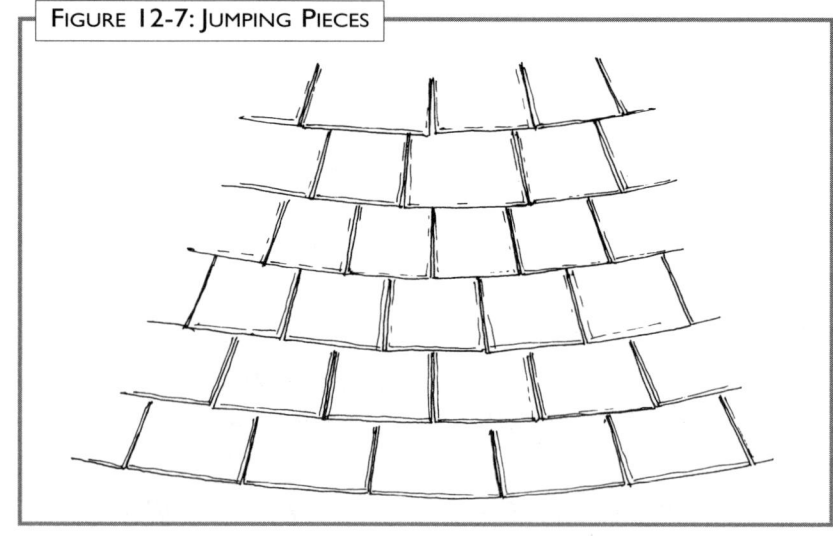

FIGURE 12-7: JUMPING PIECES

Advanced Situations

6a. The first two or three lines scribed into the underlayment this way will actually trail out to the face of the dormer facia detail depending upon the size of the eyebrow. (See Figure 12-8.)

7a. Subsequent courses will begin to intersect over the top of the dormer.

8a. At the top and center of the dormer where the scribed lines intersect, the metal strip will begin to trail off in the wrong direction. Do not follow the metal strip layout beyond the center of the dormer top. In fact, on large eyebrows, this may happen before you actually reach the center. You must watch for this variation and extrapolate between adjoining scribe lines at the apex of the curve.

9a. As you get closer to the top of the dormer, you will see that these scribed lines start to come closer together. Each proceeding course should become straighter, creating a smooth transition into the horizontal layout lines above the eyebrow dormer.

Unlike the horizontal line layout, the vertical line layout is somewhat confusing. (See Figure 12-9.)

1b. Snap a centerline on the top of the dormer from the dormer face to the point at which the eyebrow dormer framing intersects the main roof surface above.

2b. Scribe a reference point into the base of the valley on each side where the dormer framing intersects the main roof.

3b. Snap a valley line from the point described in step #1 above to the point described in step #2.

4b. In the same fashion as for the layout of the base of a cone, lay out the face of the eyebrow from the valley reference point on each side to the centerline reference point. Obviously some adjustments will need to be made at the top of the dormer to create even spacing of the marks.

5b. Snap a line from the top of the center reference point described in step #1 to each of these points along the face.

6b. The layout and application for the field side of the valley will be carried out the same way as that for the canoe valley.

The easy part is over. Installing the slate through an eye-

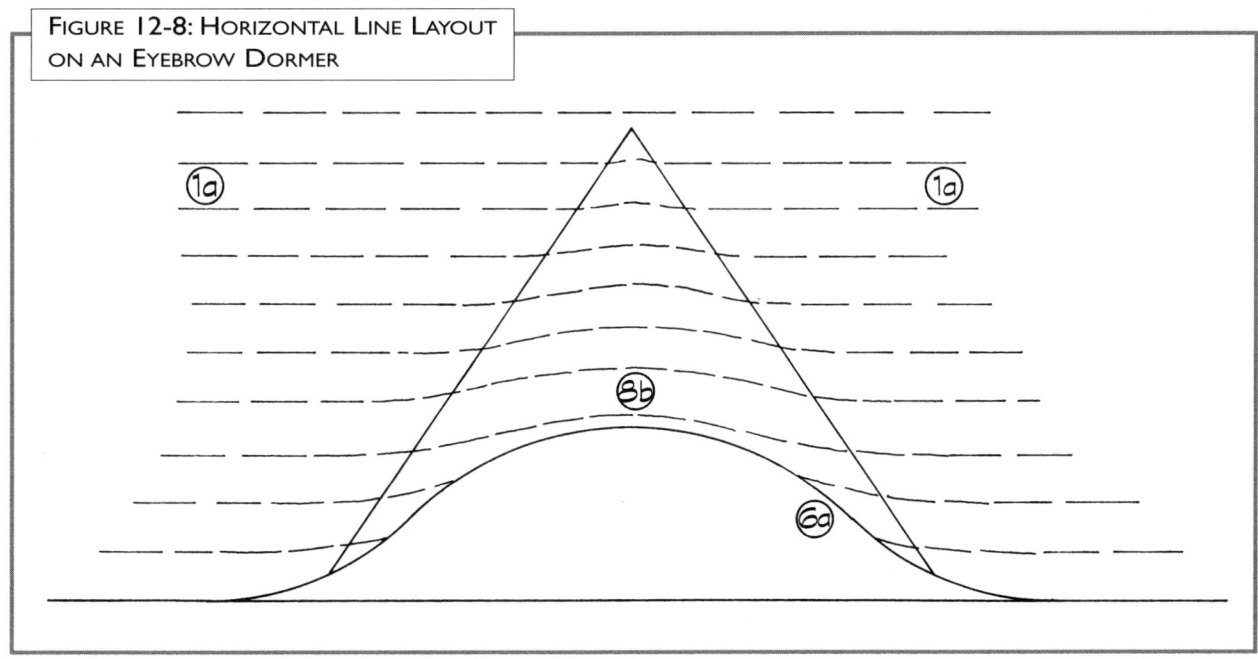

FIGURE 12-8: HORIZONTAL LINE LAYOUT ON AN EYEBROW DORMER

brow dormer requires some shimming, clipping some top corners, aligning reference lines as the slate lays over the complex curve of the valley, lacing in modified bitumen between each course, and working through the sweeping transition at the top of the eyebrow as you go from shallow pitch to the steeper pitch of the adjoining roof surface. As with the canoe valley, you will need longer slate for this entire detail.

This is a very complicated detail to install properly and make watertight. It will take a lot of time and patience. If in doubt about how to proceed, contact a slate quarry and ask them to recommend an experienced installer who might be willing to coach you.

Bell Shapes

A bell shape, when properly installed, is a beautiful slate roofing detail. All of the layout details as described above apply if the base configuration is circular. If the base is multi-sided, vertical lines will not be necessary. The sweeping eave detail will be treated the same as any other sweep.

The complexity arises as you start into the convex curve. The butts of the slate will begin to lift up as you progress into this convex area. To minimize this lifting, you will need to reduce the slate length. However, in order for the detail to appear correct aesthetically, the exposed surface of the slate should not change.

Impossible? Using all the basic practices described thus far, yes it is. You will have to cheat a little to install this detail properly. This is the only situation where an alternative installation method is acceptable.

First, determine the exposed surface dimension that you are using for the bell up to this convex curve area. Then add 2 1/2 inches to that dimension. Cut all of the slate for the convex roof area to this exposed surface plus 2 1/2-inch dimension. Clip the corners off the top edge of each of these pieces. This will allow subsequent courses to lay more evenly. The nail holes for these pieces will have to be punched one inch down from the top edge of the slate. As you install these pieces of slate you will continue with the same exposure as previous courses. However you must lace modified bitumen in between each course so that it covers the nail heads of the previous course and provides the necessary waterproofing. As an added precaution, hook each of these pieces in place with a slate hook. to prevent wind from displacing them. A copper slate hook which will oxidize may work best for this application. Stainless steel hooks may shine and stand out. When the slate at the top becomes narrower than 3 inches, this detail will need to be finished with a metal cap or finial.

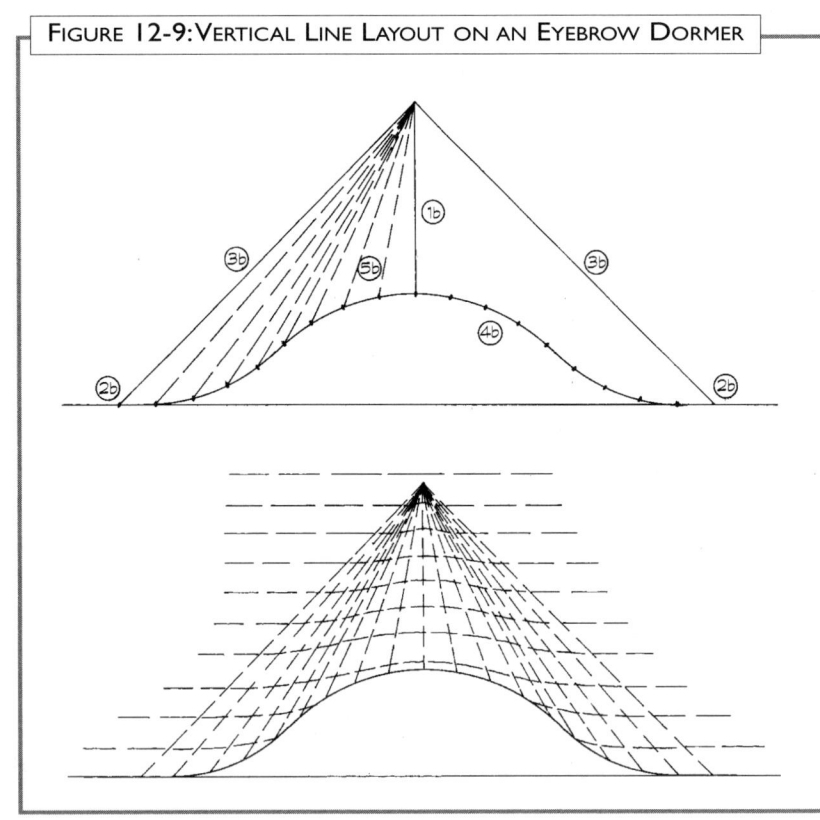

FIGURE 12-9: VERTICAL LINE LAYOUT ON AN EYEBROW DORMER

Advanced Situations

Sweeping Eave Details

Eave configurations that sweep are a very common detail. The complication is that the roof framing changes pitch quickly in a radius fashion. With roofing materials that are flexible, this is not a problem. Slate, however, being rigid will not lay unsupported through a sweep. You will need to shim each course to fully support each piece.

Start by laying out the roof as described in Chapter 6 Roof Layout. For any severe or small radius sweeping detail, the cant strip may need to be eliminated. To determine whether or not a cant strip is required, lay a starter slate and a first full course piece together at the eave. Lay them out to their proper reference line. If the head of the first full course touches the butt of the starter slate and the roof deck at its head without creating a gap at the eave, a cant strip may not be needed. Instead, a shim may be needed between the face of the starter slate and the back of the first full course.

To determine whether or not a shim is required, look back at the starter slate and first full course mock-up at the eaves. The most critical point for shimming is where the butt of the slate from the second full course lays on the first full course. This is the point at which any load exerted from the second full course will bear on the first full course. So, if there is a gap of more than 1/4 inch between the starter slate and the first full course at this point, a shim should be installed.

It is good practice, at least the first time you attempt to do this detail, to install a slate mock-up to the reference lines for all of the slate in the sweep. Lay one piece from each course, one on top of the other with no stagger, just to get a feel for how the sweep will lay. By viewing the mock-up from the side, you will be able to see how much shimming will be required in each course.

As you progress through a small radius sweep, you will need to shim the entire unsupported area under the slate and not just that area where the progressive course butt lies. Notice from your sweep mock-up that the unsupported slate which needs to be shimmed will sometimes reach considerably further up the roof than the framing would indicate. You will need to continue the shimming process until the field slate becomes fully supported as with any standard application.

Shims will vary depending upon the size of the gap that needs to be filled. Wood lath such as that sometimes used for the cant strip will often work. Cedar shim shingles

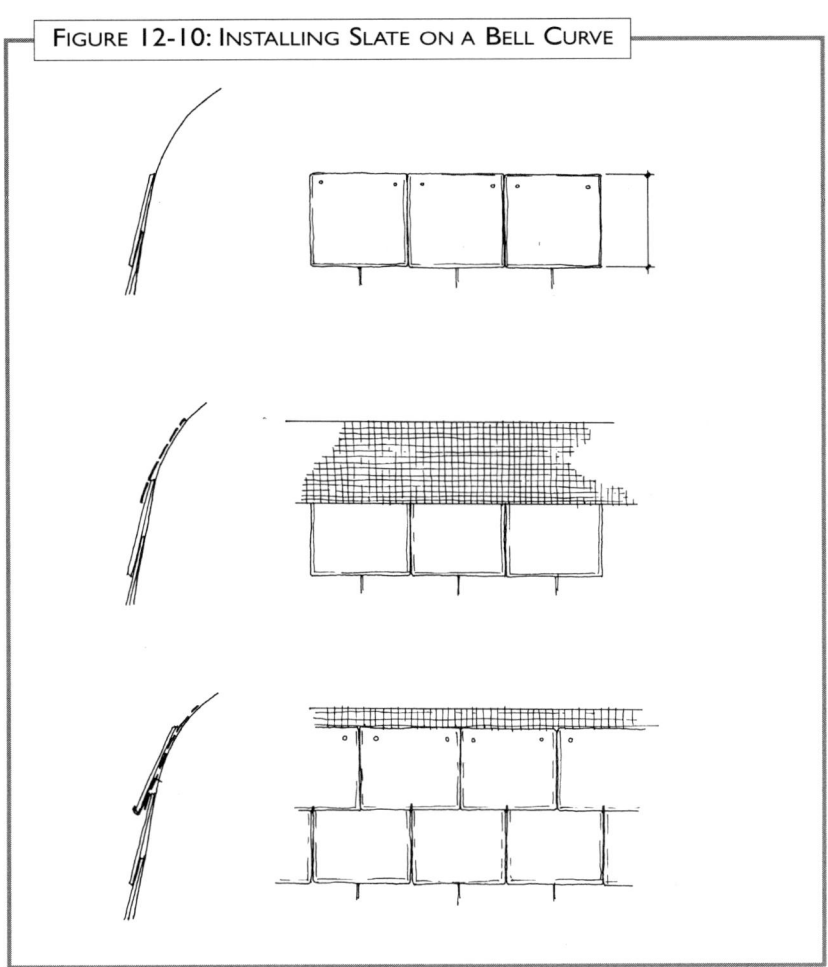

FIGURE 12-10: INSTALLING SLATE ON A BELL CURVE

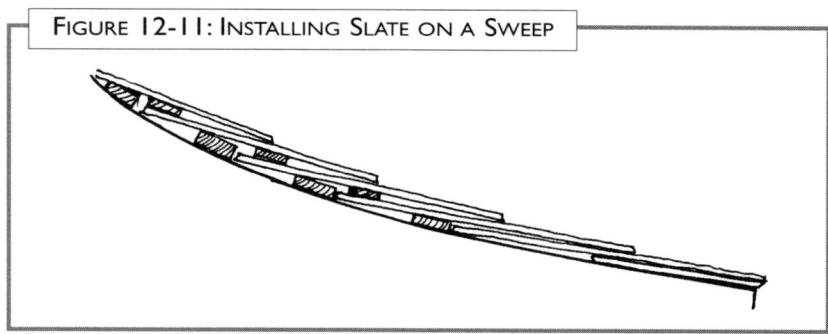

FIGURE 12-11: INSTALLING SLATE ON A SWEEP

will often work well, but they will need to be cut to fit the proper size of the void. One of the best shims for this purpose is a cedar clapboard. Clapboards are available in a variety of sizes and thicknesses that lend themselves well to this application. If the void being shimmed is large, a a useful technique is to insert one clapboard point down, followed by a second clapboard butted to it pointing up.

Although it isn't a requirement, it is good practice to flash each of these shimmed courses with modified bitumen or 30# felt paper. The shims will be exposed through the vertical joints in the slate in some cases. By flashing over the shims they will last longer.

Many installers order very short slate to install through the sweep area. As they progress out of the sweep, they begin to install longer slate. This practice works. The small slate in the sweep requires less shimming and is less susceptible to breakage. However, if you use this technique, be sure to adjust your initial layout accordingly.

As you progress out of the sweep, you must work back into larger slate in one size increment changes at a time. For example, if you install 12-inch long slate in the sweep but are installing 18-inch long slate on the rest of the field of the roof, you must first install a course of 14-inch long slate, then 16-inch long slate, and finally the 18-inch long material. This will assure that a proper headlap is achieved as you make the transition. Failure to do this will create a situation where the butt of the second course of 18-inch long material will not cover the head of the last course of 12-inch long material.

These sweeping details at valleys can be troublesome. If an open valley is specified, do not nail the shim material through the valley flashing. Be sure that the valley pieces are well secured so that snow and ice do not move them. Whenever possible carry out this sweeping valley detail using a closed valley.

At the hips, a saddle cap will work best for this detail. Be sure to build up the hip nailers so that they are at the same elevation as the top surface of the field slate.

In all cases, the sweeping eave detail will require nails that are longer than normal. These nails need to penetrate the shims and related elevated area before penetrating the roof deck. If you need to determine nail length prior to having the slate on site, try building your eave mock-up out of wood. The wood must be the same length and thickness as the slate you will be using.

Try to avoid breaking slate in a sweep detail during installation. Removing debris and installing repair slate around the shims can be difficult. Always use the bib repair technique for repair in this area, as the shaft of a slate hook isn't be long enough to reach the roof deck properly.

Batten Roof Applications

Most slate roofs are installed over a solid deck which is then covered by an acceptable underlayment material such as 30# felt. However, some slate roofs are installed over horizontal nailers or "battens" which are either nailed to the rafters or are nailed to vertical "counter battens" which are installed over the top of the solid decking. If you are working on a project where the battens are nailed directly to the rafters, be sure to install some sort of underlayment. Windblown snow, sand, and dust can and will infiltrate the structure if an underlayment product is not used.

Installing a slate roof over a batten system requires a slightly

Advanced Situations

different layout approach. First you must obtain from the slate producer the exact location of the nail holes within the slate. (This will vary slightly from quarry to quarry.) Provide solid blocking one foot wide along the eaves and valleys. The roof layout will be done the same as usual. However, the battens will need to be placed to accept the nail hole locations of the slate. To accomplish this, use three-inch wide battens. Install them so that the bottom of the batten lays right at the layout line. The bottom of the batten then becomes the reference point for the top of the slate. The three-inch wide dimension should provide for proper nailing without having to punch additional holes in the slate. If your nail holes do not hit a batten, you will have to punch the slate accordingly.

Dealing with stucco in new construction

Installing slate on a project requiring new stucco wall finishes before the stucco goes on can be a huge problem. First, if the stucco gets on the slate the stains are very difficult to remove. Second, access to the areas receiving stucco will be over the finished roof likely resulting in broken slate and long term use of your scaffolding.

To avoid this you can install all of the related flashings using the following method:

1. Install all underlayment and extend the underlayment up the related walls at least the height of the flashings. This will insulate the flashings from any ferrous fastener heads in the wall surface.
2. Do the roof layout for the entire related area.
3. The layout lines are top lines. Measure down from the layout lines the length of the slate to locate the bottoms of the courses.
4. Install the flashings as as described earlier.
5. The flashings will need to be canted and supported with shims the thickness of the slate being used so that they are in the correct plane.

The roof surface is now ready for the stucco contractor to complete his work. Be sure that the stucco installers do not nail through the roof flashings, or remove the shims. It is critical that the shims stay in place because once the stucco is installed you will not be able to move the flashings. If the flashing is not shimmed properly for each course, the slate installer will not be able to lay the slate close to the wall, the way it is intended to be laid. Also, make sure that the stucco installers insulate their ferrous wire lath from any non-ferrous roof flashings. If this installation is carried out properly, the stucco becomes the counter flashing and no further counter flashing is required.

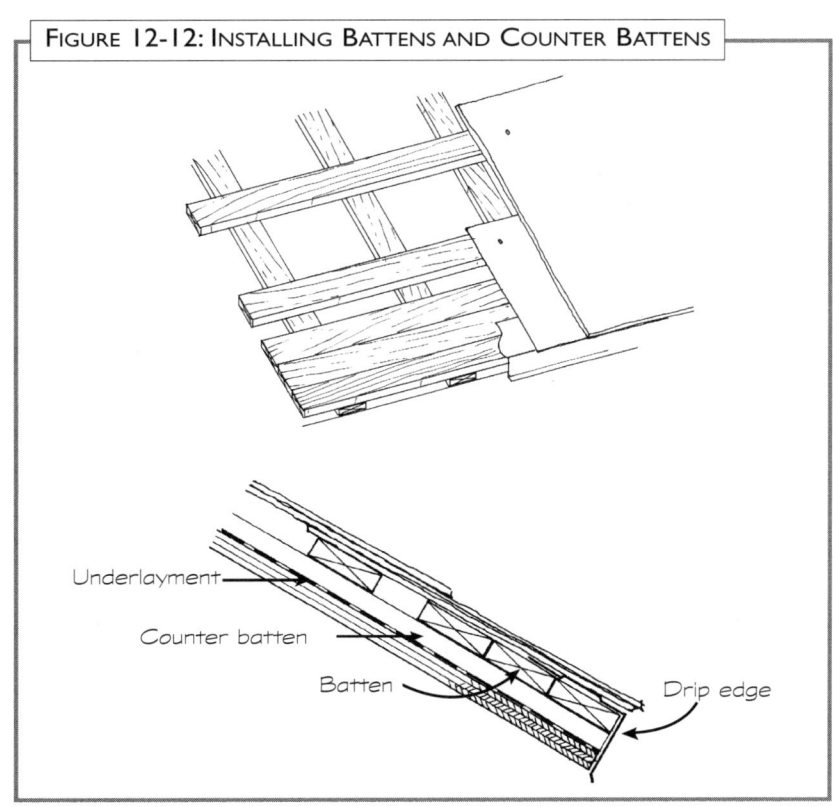

FIGURE 12-12: INSTALLING BATTENS AND COUNTER BATTENS

AFTERWORD

Our purpose in writing this manual was to create a benchmark reference source that all parties involved in a slate roofing project could use to better communicate with each other. There is a tremendous amount of information set forth herein that has never before been assembled in one roofing text. Although we have presented this information to the best of our ability, there are bound to be questions and debate arising from some portions of this manual. We welcome the opportunity to share your thoughts, questions, and suggestions. Through an open forum of debate and discussion, we hope to continue to revise and rework this manual in a way that will continue to best serve the industry as a whole.

ACKNOWLEDGMENTS

The following people, businesses, and organizations were very helpful in the production of this manual.

Michael Priestley, who helped us take the first steps toward organizing our initial concepts into manual form.

Heather Palmer of Creative Pages, who had the patience to work with three former roofers to organize a comprehensive manual.

Bob Parker, who did all of the artwork and drafting.

Joe Hale of HDH, who provided one of the specifications shown in Appendix C and offered moral and professional support along the way.

RCI (Roof Consultants Institute), who encouraged us to follow through and complete this work.

Dan Moriarty and Sandra Gray, who provided the second sample specification in Appendix C.

The former National Slate Roofing Association, which organized and published the original "Slate Book." Without the information contained in that original manual, we would have had difficulty getting started in the slate roofing business to begin with.

We would also like to thank our family and friends who have supported us along the way.

PART 3
APPENDICES

APPENDIX A   **CHEMICAL COMPOSITION OF SLATE: PAGE 182**

    The optical and physical constants of the mineral constituents of slate.

APPENDIX B   **CHECKLISTS: PAGE 183**

    Checklist 1: Slate Roof Design Considerations
    Checklist 2: Estimating
    Checklist 3: Slate Quotation Form

APPENDIX C   **SAMPLE SLATE SPECIFICATIONS: PAGE 187**

    Sample A: Slate Roofing (Section 07310)
        Part 1: General
        Part 2: Products
        Part 3: Execution
    Sample B: Slate Shingles (Section 07310)
        Part 1: General
        Part 2: Products
        Part 3: Execution
    Sample C: Slate Roofing (Section 07310)
        Part 1: General
        Part 2: Products
        Part 3: Execution

APPENDIX D   **SNOW GUARD SPECIFICATION: PAGE 199**

    Gussetted Snow Guard Specifications
    Wire Loop Snow Guard Specifications

APPENDIX E   **LIST OF TABLES AND FIGURES: PAGE 201**

    List of tables and figures used in this book

APPENDIX F   **ASTM SPECIFICATIONS: PAGE 204**

    The American Society for Testing and Materials (ASTM) Specifications

APPENDIX G   **ROOF PARTS NOMENCLATURE: PAGE 205**

    Figures and labeled roof parts

BIBLIOGRAPHY: PAGE 206

APPENDIX A

The Optical & Physical Constants of the Mineral Constituents of the Slates

NAME	CHEMICAL COMPOSITION	OPAQUE COLOR BY INCIDENT LIGHT	ISOTROPIC	ANISOTROPIC COLORLESS UNI	ANISOTROPIC COLORLESS BI-AXIAL	COLORED NON-PLEOCHROIC UNI	COLORED NON-PLEOCHROIC BI-AXIAL	COLORED PLEOCHROIC UNI	COLORED PLEOCHROIC BI-AXIAL	FORM OF OCCURRENCE	REFRACTIVE INDEX AVERAGE	MAXIMUM BIREFRINGENCE	PLEOCHROISM	ORIENTATION	OPTIC AXIAL ANGLE	HARDNESS MHOS SCALE	SPECIFIC GRAVITY	MICRO-CHEMICAL REACTIONS
PYRITE	FeS_2	BRASS YELLOW	—	—	—	—	—	—	—	Crystals & Irregular Grains	—	—	—	—	—	6–6.5	4.9–5.2	Non-magnetic; insoluble in HCL. Readily soluble in HNO_3; when tested, gives off SO_2 fumes.
PYRRHOTITE	FeS	BRONZE YELLOW	—	—	—	—	—	—	—	Granules	—	—	—	—	—	3.5–4.5	4.4–4.6	Magnetic
MAGNETITE	Fe_3O_4	BLACK	—	—	—	—	—	—	—	Octahedrons, Cubes, Irregular Grains	—	—	—	—	—	5.5–6.5	4.9–5.2	Strong magnetic; not readily soluble in HCl.
GRAPHITE	C	BLACK	—	—	—	—	—	—	—	Flakes-Needles	—	—	—	—	—	3.5–4.5	4.4–4.6	Insoluble in acids; often contains Fe_2O_3 & Al_2O_3
HEMATITE	Fe_2O_3	RED - Rarely Crystallized						Rarely Hexagonal Neg.		Hexagonal Crystals or as Rim Around Magnetite	If Crystallized 2.50	0.250	O = Brownish Red, E = Yellowish Red	—	—	5–6.5	4.2–5.3	Not readily soluble in acids; limonite is yellow; magnetite black; hematite red.
APATITE	$3(Ca_3P_2O_8) \cdot Ca(ClF)_2$	COLORLESS			Hexagonal Negative					Small Prisms Large Crystals	1.63	0.004	Sometimes Pleochroic	—	—	5	3.2	Readily soluble in H_2SO_4. This solution gives yellow precipitate with ammonium molybdate.
QUARTZ	SiO_2	COLORLESS			Hexagonal Positive					Granules Crystals	1.54–1.58	0.009	Sometimes Pleochroic	—	Anomalous 2E = Up to 25°	7	2.5–2.8	Insoluble in common acids. Soluble only in HF; SiO_2 as Na_2SiF_6
ZIRCON	$ZrO_2 \cdot SiO_2$	COLORLESS			Tetragonal Positive					Small Stout Crystals	1.83–1.99	0.050 to 0.062	Very Weak but Brilliant Interference Colors	—	—	7.5	4.4–4.7	Not soluble in acids; only HF.
KAOLIN	$H_4Al_2Si_2O_9$	COLORLESS AMORPH WHITISH			If Cryst. Neg. Monoclinic					Flour-like White Opaque Flakes, Seldom in Scales	1.56–1.57	0.005		If Crystallized Extinction Angle = 13°	—	1–1.5	2.2–2.6	Not readily soluble in acids. Highly refractory; gives off water above 330°C.
FELDSPAR	ANORTHITE-PLAGIOCLASE $CaO \cdot Al_2O_3 \cdot 2SiO_2$ / ORTHOCLASE $K_2O \cdot Al_2O_3 \cdot 6SiO_2$	COLORLESS			Triclinic Neg. / Monoclinic Neg.					Crystals Granules	1.53–1.54	0.005		$a:\mu = 5°$	$2V = 43°–53°$	6–6.5	2.7–2.8	Soluble in HCl; Ca as $CaSO_4 + 2H_2O$
MUSCOVITE SERICITE	$K_2O \cdot 3Al_2O_3 \cdot 6SiO_2 \cdot 2H_2O$	COLORLESS			Monoclinic Neg.					Prisms Granules	1.53–1.54	0.005		$c:\mu$	$2V = 69°–43'$ $2E = 121°–6'$	6–6.5	2.5–2.6	Gelatinizes in HCl. To differentiate from quartz: etch with dilute HF for 10 seconds — quartz remains clear; feldspar becomes cloudy.
					Monoclinic Neg.					Bird's-Eye Maple Appearance, Flakes, Leaves & Shreds	1.56–1.60	0.037		$c:\mu = 0°$ to $2°$	$2V = 30°$ to $45°$ $2E = 60°$ to $90°$	1–2	2.8–3.1	Not soluble in HCl & H_2SO_4, but readily in HF; K as K_2PtCl_6
RUTILE	TiO_2	YELLOWISH BROWN				Tetragonal Pos.		Sometimes Neg. Rhombic		Grains Acicular Crystals	2.50–2.90	0.260	Seldom Noticeable in Thin Sections	$c:\alpha$	—	6–6.5	4.2–4.3	Not soluble in acids; fused with Na_2O_2, T_1 as $Rb_2T_1F_6 + H_2O$
CHLORITE = CLINOCHLORE	$H_8Mg_5Al_2Si_3O_{18}$	GREENISH					Monoclinic Pos. Sometimes Neg.			Flakes Leaves	1.59–1.60	0.011	$b:\mu$ = Green I = Yellowish	$c:\mu = -2°$ to $-9°$	$2E = 32°–90°$	2–2.5	2.6–2.8	Soluble in concentrated H_2SO_4
TOURMALINE	Na Al B Mg Cr Fe Plus SILICATES	REDDISH BROWN TO BLUISH VIOLET					Hexagonal Neg.			Prisms and Acicular Crystals	1.65–1.68	0.032	Strongest at Right Angles to Vibration of Lower Nichol	—	—	7–7.5	3–3.2	Not readily soluble in acids. Fuses with Na_2CO_3.
BIOTITE	$(KH)_2 (MgFe)_2 (AlFe)_2 Si_3O_{12}$	GREENISH BROWN TO RED-BROWN					Monoclinic Neg.			Bird's-Eye Maple Appearance, Flakes, Leaves & Shreds	1.56–1.60	0.040	Strong	$C:\mu$ $0°$ to $-7°$	$2E = 0°–50°$	2.5–3	2.8–3	Soluble in concentrated H_2SO_4; Mg as $NH_4MgPO_4 + 6H_2O$
CALCITE	$CaCO_3$	COLORLESS GRAY					Hexagonal Neg.			Granules & Cavity Fillings	1.49–1.66	0.130	Weak	—	—	3	2.6–2.8	Readily soluble in HCl & $HC_2H_3O_2$
ANDALUSITE	$Al_2O_3 \cdot SiO_2$	COLORLESS PINK					Rhombic Neg.			Prisms, Irregular Granules	1.63–1.64	0.011	Weak	$c:\mu$ $a:\kappa$	$2V = 85°$	7–7.5	3.1–3.2	Soluble only in HF.
RHODOCHROSITE	$MnCO_3$	DARK RED TO BLACK					Hexagonal Neg.			Flakes & Granules	1.83	0.010	Weak	—	—	3.5–4.5	3.3–3.6	Soluble in $HCl(CO_2)$; Mn as $MnC_2O_4 + 3H_2O$
BARITE	$BaSO_4$	GRAYISH WHITE					Rhombic Pos.				1.64–1.65	0.010	Weak	—	$2E = 63°$	3–3.5	4.3–4.7	Insoluble in HCl & HNO_3, fused with Na_2CO_3, Ba as BaS_1F_6

182

APPENDIX B

Checklist 1: Slate Roof Design Considerations

Slate Roof Types (choose one)
    Commercial Standard Slate   p.18 _____
    Standard Thickness Slate   p.18 _____
    Textural Slate   p.18 _____
    Intermingled Thickness   p.18 _____ (Specify Thickness)
    Heavy   p.20 _____ (Specify Thickness)
    Graduated   p.20 _____ (Specify Thickness)
    Special Order   p.20 _____
    Used or Salvaged   p.20 _____

Colors* p.21

	Weathering	Unfading	%
Black	____	____	____
Gray	____	____	____
Purple	____	____	____
Green	____	____	____
Blue-black	____	____	____
Blue-gray	____	____	____
Mottled Purple & Green	____	____	____
Red	____	____	____

    Total = 100%

Width (choose one)
    Single- or One-Width   p.22 _____
    Random   p.22 _____

Length (choose one)
    Single Length   p.23 _____
    Staggered Butts   p.23 _____
    Graduated Length   p.23 _____
    Cut Butts   p.23 _____
    Patterns   p.23 _____

Valley Styles (choose one)
    Open Valley   p.26 _____
    Closed Valley   p.26 _____
    Round Valley   p.26 _____
    Canoe Valley   p.27 _____

Hip Styles (choose one)
    Saddle Hip   p.27 _____
    Boston Hip   p.27 _____
    Mitered Hip   p.27 _____
    Fantail Hip   p.28 _____
    Metal Hip   p.28 _____
    Tile Hip   p.28 _____

Ridge Styles (choose one)
    Saddle Ridge   p.29 _____
    Comb Ridge   p.29 _____
    Strip Ridge   p.29 _____
    Metal Ridge   p.30 _____
    Tile Ridge   p.30 _____
    Vented Ridge   p.55 _____

Fasteners (choose one)   p.50
    Copper _____
    Stainless Steel _____

Flashing Material (choose one) p.41
    Copper _____
    Lead-coated Copper _____
    Other _____ _____

Accessories (all may apply)
    Snow Guards   p.31 _____
    Lightning Protection   p.31 _____
    Gutters   p.32 _____
    Finials   p.32 _____
    Vents   p.32 _____
    Rain Diverters   p.33 _____
    Snow Slides   p.33 _____
    Plumbing Vents   p.55 _____

* Manufacturers' trade names will vary from those colors listed above.

Appendix B

Checklist 2: Estimating

Slate

Field Slate	p.37	_____ sq ft ÷ 100 =		_____ squares
Cutting loss	p.37	_____ squares of field slate × _____ % =		_____ squares
Starter Slate	p.41	_____ ft of eaves =	_____ ln ft	
Finish Slate	p.41	_____ ft of ridge =	_____ ln ft	
Hip Slate				
Hip approach	p.40	_____ ln ft × 1 sq ft ÷ 100 =		_____ squares
Hip cap	p.40	_____ ln ft × (see formula page 39) =		_____ squares
Ridge Slate				
Strip	p.40	_____ ln ft × 1 sq ft ÷ 100 =		_____ squares
Comb	p.40	_____ ln ft × 1 sq ft ÷ 100 =		_____ squares
Saddle	p.41	_____ ln ft × (see formula page 39) =		_____ squares
Valley slate	p.39	_____ ln ft × 1 sq ft ÷ 100 =		_____ squares
Chimney	p.41	_____ sq ft (only deduct if larger than 50 square feet)		_____ squares
Skylight	p.41	_____ sq ft (only deduct if larger than 50 square feet)		_____ squares
Dormer	p.41	_____ sq ft (only deduct if larger than 50 square feet)		_____ squares

Total Lineal Feet of Starter and Finishing Slate Estimated = _____
Price of Slate per Lineal Foot = × _____

A. Total Lineal Foot Slate Cost = ▭

Total squares of slate estimated = _____
Add 1% of total squares of slate estimated for future repair and maintenance = _____
Total squares of slate to order = _____
Price of slate per square delivered on pallets (see slate quotation form on page 186) − × _____

B. Total Square Slate Cost = ▭

A + B = TOTAL SLATE MATERIAL COST $ ▭

Slate Labor (page 36)

D. Total lineal feet of starter & finishing slate estimated = _____ × _____ labor/lineal foot to install = ▭

E. Total squares of slate estimated = _____ × _____ labor per square to install = ▭

D + E = TOTAL SLATE LABOR COST $ ▭

Flashings

Drip Edge	p.43	_____ ln ft of eave and gable × width in feet =		_____ sq ft
Valley Flashing				
Open	p.43	_____ ln ft (see Table 3-12 p. 44) × width =		_____ sq ft
Closed				
Conventional	p.44	_____ ln ft (see Table 3-13 p. 44) × width =		_____ sq ft
Point-to-Point	p.44	_____ ln ft (see Table 3-14 p. 45) × width =		_____ sq ft
Hip Flashing				
Mitered	p.44	_____ ln ft of hip × (see Table 3-15 page 45) =		_____ sq ft
Saddle	p.44	_____ ln ft of hip × (see Table 3-15 page 45) =		_____ sq ft
Metal	p.45	_____ ln ft of hip × width of hip cap =		_____ sq ft
Ridge Flashing				
Saddle	p.45	_____ ln ft of ridge × (see Table 3-15 page 45) =		_____ sq ft
Metal	p.45	_____ ln ft of ridge × width of ridge cap		_____ sq ft
Stepped Flashing	p.46	_____ ln ft of wall × (see Table 3-16 page 46) =		_____ sq ft
Apron Flashing	p.46	_____ ln ft × width =		_____ sq ft
Counter Flashing	p.47	_____ ln ft × width =		_____ sq ft
Transition Flashing	p.48	_____ ln ft × width =		_____ sq ft
Cricket	p.49	_____ sq ft (will vary) =		_____ sq ft
Cap Flashing	p.48	_____ ln ft × width =		_____ sq ft
Gutters	p.49	_____ ln ft × width =		_____ sq ft

Total square feet of flashing estimated = ▭
Add 5% of total square feet of flashing estimated for miscellaneous flashing details = ▭
Total square feet of flashing needed = ▭
Price of flashing material per square foot = × ▭

TOTAL FLASHING MATERIAL COST $ ▭

Appendix B

Checklist 2: Estimating, *continued*

Flashing Labor (Chapter 9) Use one of the two options below for estimating flashing labor costs.
1. Total hours estimated to install flashing = _____ × $ _____ flashing labor cost/hour = $ [_____]
OR
2. Total sq. feet of flashing estimated = _____ × _____ flashing labor per sq. foot to install = $ [_____]
   TOTAL FLASHING LABOR COST = $ [_____]

Fasteners

Underlayment*	p.50	_____ pounds × $_____/lb =	$ _____
Field Slate	p.50	_____ pounds of _____ length ×$_____/lb =	$ _____
Hip Slate	p.51	_____ pounds of _____ length ×$_____/lb =	$ _____
Ridge Slate	p.51	_____ pounds of _____ length × $_____/lb =	$ _____
Scaffolding Brackets	p.57	_____ pounds of _____ length × $_____/lb =	$ _____

*consult local building codes)   **TOTAL FASTENER COST = $** [_____]

Underlayment

Felt Paper	p.71	_____ sq ft of roof area ÷ _____ sq ft / roll = _____ rolls × $_____ /roll = $_____
Modified Bitumen	p.73	_____ sq ft of roof area ÷ _____ sq ft / roll = _____ rolls × $_____ /roll = $_____
Hip & Ridge	p.45	_____ sq ft of roof area ÷ _____ sq ft / roll = _____ rolls × $_____ /roll = $_____

TOTAL MATERIAL COST OF UNDERLAYMENT = $ [_____]

Cant Strip p.52   _____ Lineal ft × $ _____ per Lineal ft.   Total Cost = $ [_____]
Nailers   p.79   _____ Lineal ft × $ _____ per Lineal ft.   Total Cost = $ [_____]

Accessories

Roof Vents	p.54	_____ units × $ _____ per unit = Total Cost = $ _____
Snow Guards	p.53	_____ units × $ _____ per unit = Total Cost = $ _____
Ridge Vents	p.55	
Finials	p.32	_____ units × $ _____ per unit = Total Cost = $ _____
Lightning Protection	p.31	_____ units × $ _____ per unit = Total Cost = $ _____

TOTAL MATERIAL COST OF ACCESSORIES = $ [_____]

Other - The following are items that vary in need and cost from job to job. This list may not be complete. Each job will have specific needs and requirements that should not be overlooked. It is the responsibility of the estimator to see that these items are properly accounted for.

Ground-to-Eave Scaffolding	p.71	Cost = $ [_____]
Material-Handling Equipment	p.62	Cost = $ [_____]
Insurance		Cost = $ [_____]
Overhead and Profit		Cost = $ [_____]
Permits and Fees		Cost = $ [_____]
Tools	p.95	Cost = $ [_____]
Travel		Cost = $ [_____]
Lodging		Cost = $ [_____]
Cost of Mock-up	p.24	Cost = $ [_____]
Clean-up and Removal of Debris		Cost = $ [_____]
Hiring Design Professionals	p.35	Cost = $ [_____]
Evaluating Structural Design	p.35	Cost = $ [_____]

TOTAL PROJECT ESTIMATE = $ [_____]

Appendix B

Checklist 3: Slate Quotation Form

Job Name _____ Slate Manufacturer's Name _____
Date _____ Contact Person _____
Approximate Job Size _____ Phone Number _____
Quote (circle one)   1st Choice   2nd Choice   3rd Choice   Job Zip Code or Region (for freight cost) _____

Slate Roof Types (choose one)		Thickness	Availability	Unit Cost/Sq*
Commercial Standard Slate	p.18 ____	3/16"	____	____
Standard Thickness Slate	p.18 ____	1/4" ± 1/16"	____	____
Textural Slate	p.18 ____	3/16" to 3/8"	____	____
Intermingled Thickness	p.18 ____	____" at ____%	____	____
		____" at ____%		
		____" at ____%		
Heavy	p.20 ____	____"	____	____
Graduated	p.20 ____	from ____" at eave to ____" at ridge.	____	____
Special Order	p.20 ____	____"	____	____

* Delivered to site on slate pallets

Colors p. 21	Weathering	Unfading	%	Manufacturer's Trade Name
Black	____	____	____	____
Gray	____	____	____	____
Purple	____	____	____	____
Green	____	____	____	____
Blue-black	____	____	____	____
Blue-gray	____	____	____	____
Mottled Purple & Green	____	____	____	____
Red	____	____	____	____
			Total = 100%	

Width (choose one)   Desired
   Single or One Width   p.22   _____
   Random   p.22   _____
Length (choose one)
   Single Length   p.23   _____
   Staggered Butts   p.23   _____
   Graduated Length   p.23   _____ from _____" at eave to _____" at ridge.
   Cut Butts   p.23   _____
   Patterns   p.23   _____
Starter Slate
   Size   p.41   _____" × _____"
   Lineal Feet of Eaves   _____ ft
Finishing Slate
   Size   p.41   _____" × _____"
   Lineal Feet of Ridge   _____ ft
Hip Styles (for Saddle Hip ONLY)
   Size   p.40   _____" × _____"
   Lineal Feet of Hip   _____ ft
Ridge Styles (for Saddle Ridge ONLY)
   Size   p.41   _____" × _____"
   Lineal Feet of Ridge   _____ ft

In ft _____
In ft _____

TOTAL $ _____

Submit separate forms for 1st Choice and alternate quote requests.

APPENDIX C

Sample Slate Specifications

In this section there are three sample specifications provided. The first shows a somewhat generic specification while the next two are actual specifications related to projects.

Every project will have variables related specifically to that project. These sample specifications are provided as guidelines only. The specifier will have to tailor the specifications for each given project. Referencing sections and/or page numbers from this book will help to clarify specific details later.

SAMPLE A: Section 07310 Slate Roofing

PART 1 GENERAL

1.01 SECTION INCLUDES
A. Slate And Underlayment (chapter 2)
B. Slate Application (chapter 8)
C. Incorporation of Sheet Metal Flashing and Roofing Accessories into the Roof System (chapter 9)

1.02 RELATED SECTIONS
A. All Documents Listed in Table of Contents are a Condition of this Section

1.03 STANDARDS
A. Standards of the following associates and current publications shall apply to materials furnished under this section:
   1. American Society for Testing and Materials (ASTM) Philadelphia, PA (215) 299-5585
   2. *The Slate Book: How to Design, Specify, Install, and Repair a Slate Roof* – Etals Publishing, Moscow, VT 05662. (888) 766-4273.
   3. Sheet Metal and Air Conditioning Contractors National Association (SMACNA), 5811 Amaya Drive #100, La Mesa, CA 91942 (619) 460-5362.
   4. *Copper & Common Sense*. Revere Copper Products, Inc., P.O. Box 300, Rome, NY 13440 (315) 338-2022.
   5. *Copper in Architecture*. Copper Development Association, Inc. (CDA), 260 Madison Ave, New York, NY 10016. (212) 251-7200 phone; (212) 251-7234 fax.

1.04 DESCRIPTION OF WORK
A. The extent of the work is shown on the Drawings, and as specified herein. The work includes all material, labor, supervision, etc. associated with the proper execution of the slate roofing work, including but not limited to installation of new 30# felt underlayment, installation of ice and water shield, installation of slate roofing, and installation of roof related flashings.

1.05 SUBMITTALS
A. Comply with pertinent provisions of Section 01300.
B. Submit the following items within ten (10) days after Notice to Proceed:
   1. Three (3) samples of each slate color. (page 21)
   2. Three (3) four (4") inch × six (6") inch samples of felt. (page 71)

Appendix C

3. Three (3) four (4") inch × six (6") inch samples of ice and water shield. (page 73)
4. Three (3) samples of each length and gauge nail to be used for each detail. (pages 50-52)
5. Three (3) samples of proposed snow guard. (page 53)
6. Descriptive list of the materials proposed for use.
7. Manufacturer's product data sheets for all materials.
8. Shop drawings including but not limited to:
    a. Roof assembly including underlayment (chapter 8)
    b. Slate installation at dormer wall intersections with sloped roof (chapter 9)
    c. Slate installation at valleys, eaves, and ridges
    d. Slate installation at hips (chapter 8)
    e. Slate installation related to gutters (chapter 9)

1.06 QUALITY ASSURANCE

A. The Contractor shall have a minimum of five (5) years' experience in successfully installing slate roofing. Submit list of completed projects; include names, addresses, and names of architects and owners. Also include names of key craftspeople who will work on this specific job, their experience, and jobs worked on.

B. Use skilled workmen who are trained especially in properly laying and nailing slate and who are completely familiar with the specified requirements and methods needed for proper performance of the work of this section

C. Conform to regulations of public agencies, including any specific requirements of the city and/or state of jurisdiction.

D. Written guarantee shall be furnished by the contractor, that materials are in accordance with these specifications and that all repairs required on the roof due to defective or workmanship furnished under this contract shall be made, without cost to the owner, for a period of one year.

PART 2 PRODUCTS

2.01 SLATE SUPPLIERS

A. Slate producers must be actual quarries that have been in existence for at least twenty (20) years. Provide slate by one of the following companies:
    1.
    2.
    3.

2.02 MATERIALS

A. Roofing Felt: 30# ASTM D226 asphalt-saturated organic felt (page 71)
B. Cant Strip: Wood or metal shall match the thickness of the field slate. (page 52)
C. Ice & Water Shield: W.R. Grace & Co. or approved equal (page 73)
    1. Thickness: 40 mils
    2. Tensile strength - psi: 250 min. ASTM D412
    3. Elongation: 250 minimum ASTM D412
D. Slate
    1. Slate shall be ASTM C406-89 Grade S1, smooth-textured, hard, dense, sound rock, machine-punched for two (2) nails each except where hand punching of holes in fitting hips etc. is necessary. (chapter 2)

Appendix C

2. For Standard Thickness Roofing Slate (page 18)
    a. Slate shall be of standard thickness smooth texture, or standard thickness rough texture, or all ___" thickness or ___" and ___" intermingled thicknesses.
    b. State whether one length or graduated lengths, and one width or random widths.
    c. State whether one color, or the percentage of each color required.
    d. State whether unfading or semi-weathering material and what percentage of each is required.
    e. A certificate shall be furnished to the roofing contractor by the quarrier certifying that the roofing slate furnished is in accordance with these specifications and/or approved layout.
3. For Graduated Slate Roof (page 20)
    a. Slate shall be machine-punched for two holes, and varying in thickness from ___" at eave to ___" at ridge; the percentage of each thickness to be respectively ___. The thicknesses shall be intermingled in the various courses, modulating from the heavier and thicker slates in the lower courses of the roof to the thinner slates at the ridge, in such a way and manner as will develop the best architectural effect. A detailed roof layout for the application of the slate shall be furnished by the supplier and approved by the architect before material is fabricated.
    b. All slate shall be in random widths and graduated in length from ___" at eave to ___" at ridge, applied with standard 3" headlap and standard exposures.
    c. See (c) and (d) as listed under "For Standard Thickness Roofing Slate" specifications.
4. Hips (pages 27-28)
    a. All hips shall be laid to form "saddle" hips. (Specify "Boston," "Mitered," or "Metal" hips if preferred.)
    b. Specify hip slate length, width, and thickness.
    c. Specify hip nailer width and thickness (where applicable).
5. Ridges (pages 29-30, 55)
    a. All ridges shall be laid to form "saddle" ridges. (Specify "Comb," "Strip," or "Metal" ridge if preferred.
    b. Specify ridge slate length, width, and thickness.
    c. Specify ridge nailer width and thickness (where applicable).
6. Starter - Specify starter slate length and thickness. (page 41)
7. Finishing Slate - Specify slate length (page 41)
    a. Specify finishing slate length and thickness.
    b. Specify finishing nailer width and thickness (where applicable).
8. Valleys - All valleys shall be laid to form "open" valleys. (Specify "closed" valleys if preferred.) (pages 26-27)

E. **Sealant**
  1. For slate: Red Slater's Cement ASTM D-4586 Type I asbestos-free or approved equal.

F. **Nails**
  1. For slate: All slate shall be fastened with two large-head 10 gauge solid copper slating nails. Use 1 1/2" length for standard 3/16"–1/4" thickness field slate and 2" nails for hip and ridge slate. (Specify longer nail length for slate thicker than 1/4 inch.) (page 50)

Appendix C

PART 3 EXECUTION

3.01 PREPARATION (Chapter 5)

A. Contractor shall inspect all surfaces prepared for slating by other contractors, point out all defects to the proper authority, and shall not proceed with the laying of felt, flashings, or slate until the necessary corrections have been made.

B. The roofing contractor shall furnish his own scaffold or rigging, or arrange with the general contractor for the use of scaffolds furnished by others.

C. Thoroughly sweep off the sheathing to remove all loose nails and debris.

3.02 INSTALLATION

A. Underlayment (Chapter 5)
   1. Ice and Water Shield
      a. Install ice and water shield around the entire perimeter of the building, extending from the eaves up to a point two feet inside of the exterior wall of the building.
      b. Install ice and water shield in all valleys. At valleys, center full-width roll and press in place, working from the center of the valley outward in each direction. Apply membrane starting at the low point and working upwards.
      c. Side laps shall be minimum of 3 1/2" and end laps 6 inches.
   2. Install 30# felt
      a. Felt shall be laid in horizontal layers with joints lapped toward eaves at least 2 in. and at ends at least 6 in. and well secured along laps and at ends as necessary to properly hold the felt in place and protect the structure until covered by the slate. All felt shall be preserved unbroken, tight and whole.
      b. The felt shall lap over all hips and ridges at least 12 inches.
      c. Felt shall be lapped 2" over the metal of any valleys or built-in gutters.

B. Slating (Chapter 8)
   1. The entire surface of roof, where indicated on the drawings, shall be covered with slate in a proper and watertight manner.
   2. Slate shall be fastened with large-head slaters' copper nails to adequately penetrate roof boarding. Care shall be taken to avoid exposing the nails on cornice, soffits, overhanging eaves, etc.
   3. Nails shall not be driven in so far as to produce a strain on the slate.
   4. No through joints should occur from the roof surface to the felt. The joints in each slate course should be well separated from those below; otherwise, water may migrate through the joints and cause felts to disintegrate and leaks to develop in the roof.
   5. Where slates of random width are used, the overlapping slate should be laid jointed as near the center of the underlying slate as possible and not less than 3 inches from any underlying joint.
   6. The heads of slating nails should just touch the slate and should not be driven "home" or draw the slate, but left with the heads just clearing the slate so that the slate hangs on the nail.
   7. All nails shall penetrate the sheathing and not the joints between boards.
   8. The slate shall project 1 1/2-inch at the eaves and 1 1/2-inch at all gable ends, and shall be laid in horizontal courses with the standard 3" head lap, and each course shall break joints with the preceding one. Slates at the eaves or cornice line shall be installed with a proper starter slate and cant strip.

9. Cover all exposed nail heads with approved sealant. Hip slates and ridge slates shall be laid in sealant spread evenly over unexposed surface of under courses of slate and nailed securely in place.
10. Exposed nails shall be permissible only in top courses where unavoidable and shall be covered with approved sealant.
11. Neatly fit slate around any pipes, ventilators, etc.
12. Build-in and place all flashing pieces required for proper performance of the roof. Slates overlapping sheet metal work shall have nails placed so as to avoid puncturing the sheet metal.
13. Any nail holes that must be field punched are to be punched from the back of the slate so as to provide the proper countersink of the fastener head. The only exception to this rule is holes in starter slate.
14. Cutting slate in the field-
    a. All field slate shall be cut from the back side so as to produce the proper orientation of the beveled edge.
    b. All slate for open valleys shall be cut from the back side so as to produce the proper orientation of the beveled edge.
    c. All slate to be cut for a Closed Valley shall be cut from the front of the slate so as to create a mitered joint.
    d. All slate to be cut for a Mitered, Boston and Fantail Hip shall be cut from the front of the slate so as to create a mitered joint.
    e. All slate to be cut for a Saddle Hip shall be cut from the back side so as to produce the proper orientation of the beveled edge.
    f. All slate to be cut around penetrations shall be cut from the back side so as to produce the proper orientation of the beveled edge.

C. Ridges (Chapter 8)
   1. Install field slates extended to the proper layout line near the ridge.
   2. Install the finishing course slates set in sealant spread evenly over unexposed surface of under courses of slate and nailed securely in place.
   3. Install "Saddle" Ridge (Comb, Strip, Tile, Metal, or Vented ridge can be substituted).

D. Hips (Chapter 8)
   1. Install hips with slates cut accurately to form tight joints.
   2. Nail holes of each slate shall fall under the succeeding hip slate or hip cap.

E. Repair (Chapter 11)
   1. On completion all slate must be sound, whole, and clean, and the roof shall be left in every respect tight and a neat example of workmanship.
   2. Slate repair shall be performed using the "Bib" method ("Slate Hook" method may be substituted).
   3. No slate with a broken corner larger than 2 inches by 2 inches shall remain in the finished roof (specify exact tolerances).

3.04 FINAL INSPECTION
A. The Architect shall perform a final inspection of work and supply Contractor with punch list if necessary. All punch list items shall be completed prior to approval by Architect of final payment.

Appendix C

SAMPLE B: Division 7 - Section 07310 - Slate Shingles

PART 1 GENERAL

1.1 RELATED DOCUMENTS

Drawings and General Provisions of Contract, including General Conditions, Supplementary General Conditions and Special Conditions and Division 1 - Specifications Section, apply to work of this Section.

1.2 DESCRIPTION

A. Work includes installation of 30# felt and installation of slate shingles.

1.3 QUALITY ASSURANCE

A. Acceptable manufacturers: Buckingham or an approved equal to match existing.

B. Install slate shingles to meet requirements of published manufacturer's instructions.

1.4 SUBMITTALS

A. Samples:
   1. Slate Shingles: Two (2) each.

B. Manufacturer's literature: material description

1.5 PRODUCT DELIVERY, STORAGE, AND HANDLING

A. Deliver materials with manufacturer's labels intact and legible.

B. Store materials on raised platforms and protect with coverings at outdoor locations.

C. Do not stack bundles of slate shingles more than the height recommended by the manufacturer.

D. Store rolled goods on end.

1.6 ENVIRONMENTAL CONDITIONS

A. Do not install underlayment or slate shingles on wet surfaces.

1.7 WARRANTY

A. Workmanship Applicator warranty against defects for two (2) years and recommended installation procedure.

1.8 JOB CONDITIONS

Examine substrate and the conditions under which the slate will be installed and notify the Architect of any unsatisfactory conditions. Do not proceed with installation until unsatisfactory conditions have been corrected in an acceptable manner.

Appendix C

PART 2 PRODUCTS

2.1 SLATE

A. New slate shall match existing slate. Grade S-1 as described in ASTM C406-58.

B. Slate shall be genuine unfading. Slate of sizes and thicknesses to match existing. (Contractor shall be responsible for verifying sizes and thicknesses of existing slate prior to bidding the work.)

C. All slate shall be hard, dense, sound, rock punched for two nails. No cracked slate shall be used. All exposed corners shall be partially full. No broken corners on covered ends, which sacrifice nailing strength or the laying of a watertight roof will be allowed.

D. Average properties of slate:

PROPERTY	TEST METHOD	VALUE	
Classification	Macroscope	Blue-black micaceous, fine grain slate	
Absorption of Water	ASTM C121	0.104%	
Bulk Density	ASTM C97	174.4 lb/ft3	
Compressive strength	ASTM C170	25,800 lb/in2	
Modules of Rupture (flexible strength of structural and electrical slate)	ASTM C120	13,000 lb/in2	
Modules of Elasticity	ASTM C120, IITRI	7,014,000 lb/in2	
Toughness	IITRI	56 in-lb/in2	
Flat Slab Strength	RP477	4840 lb load	
Coefficient of Thermal Expansion	ASTM E228		
Coefficient of Friction	IITRI	STATIC	KINETIC
	Natural Cleft:		
		Dry	
	0.73		0.69
		Wet	
	0.73		0.69
	Sand Rubbed:		
		Dry	
	0.73		0.69
		Wet	
	0.73		0.69
Adhesive Resistance	ASTM C241	13.2 Ha 5-6 mon's	
Fire Resistance	IITRI	1800° F bloat non-combustible	
Acid Resistance	ASTM C217	0.0018 in	
Weather Resistance (atmospheric durability)	IITRI	29% modules loss	
	ASTM CC215	-10% Strength loss	
Durability (wet-dry cycling)	RP477	0.004% weight loss - 10% strength loss	
Life Expectancy	observation of existing structure	no visible deterioration of fading apparent	

2.2 MISCELLANEOUS MATERIALS

A. Roofing felt shall be 30 pound unperforated, asphalt-saturated felt.

B. Nails shall be large head diamond point, smooth shaft 3d hard copper slating nails 1 1/2 inches long or sufficient length to adequately penetrate the roof sheathing.

Appendix C

PART 3 EXECUTION

3.1 INSPECTION

A. Assure that surfaces to which state shingles are to be applied are uniform, smooth, clean, dry, and free of irregularities.

B. Verify that installation of copper flashings has been completed.

C. Verify that work of other trades that penetrates roof deck has been completed. deterioration of fading apparent

D. Do not start work until unsatisfactory conditions are corrected.

3.2 SLATE INSTALLATION

A. All broken, slopped, or missing slate shall be replaced.

B. All areas shall be covered with slate as herein specified, in a proper and watertight manner.

C. Slate shall be flush with copper eave strips, and shall be laid in horizontal courses with 3" headlap, and each course shall break joints with the pre-existing one by at least 3 inches. Slates at the eaves shall be doubled using same thickness slate for undereaves at first exposed course. Undereave slate to be approximately 3" longer than exposure of first course.

D. Slates overlapping metal work shall have the nails so placed as to avoid puncturing the copper metal where possible.

E. Neatly fit slate around all pipes, ventilators, and other vertical surfaces.

F. When the slate shingles are nailed. the heads of the slating nails should just touch the slate and should not be driven "home" or draw the slate, but should be left with the heads just clearing the slate so that the slate hangs on the nail.

G. All slating nails should penetrate the sheathing and not the joints between boards.

H. Upon completion, all slate must be sound, whole, clean, and the roof shall be left watertight and neat in every respect and subject to the Architect's approval.

3.4 ADJUST AND CLEAN

A. Replace any state shingles that are damaged during the project.

B. All debris or excess slate determined by the Owner as salvageable will be carefully retained and stored for the Owner in an area designated for storage.

C. Remove any debris from project area.

SAMPLE C: Section 07310 Slate Roofing

PART 1 GENERAL

1.01 SECTION INCLUDES
A. Slate And Underlayment
B. Slate Application
C. Incorporation of Sheet Metal Flashing and Roofing Accessories into the Roof System

1.02 RELATED SECTIONS
A. All Documents Listed in Table of Contents are a Condition of This Section

1.03 STANDARDS
A. Standards of the following associates and current publications shall apply to materials furnished under this section:
   1. American Society for Testing and Materials (ASTM) Philadelphia, PA (215) 299-5585
   2. National Roofing Contractors Association (NRCA) - Current Roofing Manual Rosemont, IL (708) 299-9070

1.04 DESCRIPTION OF WORK
A. The extent of the work is shown on the Drawings, and as herein specified. The work includes all material, labor, supervision, etc. associated with the proper execution of the slate Roofing work, including but not limited to installation of new double layer No. 30 felt underlayment, installation of ice and water shield, installation of slate roofing and slating dormer walls.

1.05 SUBMITTALS
A. Comply with pertinent provisions of Section 01300.
B. Submit the following items within ten (10) days after Notice to Proceed:
   1. Three (3) samples of each slate color.
   2. Three (3) four-inch (4") × six-inch (6") samples of felt.
   3. Three (3) four-inch (4") × six-inch (6") samples of ice and water shield.
   4. Descriptive list of the materials proposed for use.
   5. Manufacturer's product data sheets for all materials.
   6. Shop drawings including but not limited to:
      a. Roof assembly including underlayment
      b. Slate installation at scupper
      c. Slate installation at dormer wall intersections with sloped roof
      d. Slate installation at valleys, eaves, and ridges
      e. Slate installation at hips
   7. Shop drawing of roof from slate producer showing slate layout, including slate pattern and color.

Appendix C

1.06 QUALITY ASSURANCE

A. Bidder shall have at least five years' experience working with historic structures, including those on the National Register. Submit list of completed projects; include names, addresses, and names of architects and owners. Also include names of key craftspeople who will work on this specific job, their experience, and jobs worked on.

B. The Contractor shall have a minimum of ten (10) years' experience in successfully installing slate roofing.

C. Use skilled workmen who are trained especially in properly laying and nailing slate and who are completely familiar with the specified requirements and methods needed for proper performance of the work of this section.

D. Conform to regulations of public agencies, including any specific requirements of the city and/or state of jurisdiction.

PART 2 PRODUCTS

2.01 SLATE SUPPLIERS

A. Slate producers must be an actual quarry that has been in existence for at least twenty (20) years. Provide slate by one of the following companies:
   1.
   2.
   3.
   4.

2.02 MATERIALS

A. Roofing Felt: No. 30 ASTM D226 asphalt-saturated organic felt

B. Ice & Water Shield: W.R. Grace & Co. or approved equal
   1. Thickness: 40 mils
   2. Tensile strength - psi: 250 min. ASTM D412
   3. Elongation: 250 minimum ASTM D412

C. Slate
   1. Slate shall be Vermont Domestic ASTM C406-89 Grade S1 Commercial Standard, smooth-textured, hard, dense, sound rock, machine-punched for two (2) nails each except where hand punching of holes in fitting hips etc. is necessary.
   2. Size of slate:
      a. Field slates: Sixteen (16) inches long × eight (8) inches wide at 3/16 inch thick.
      b. Decorative courses: Sixteen (16) inches long × eight (8) inches wide with thirty-degree (30°) clipped corners, at 3/16 inch thick.
   3. Color of slate:
      a. Field slates: Unfading Vermont Black
      b. Decorative courses: Unfading Green, Commercial Standard min. 1/4" smooth

D. Plastic Cement
   1. For slate: Red Slater's Cement ASTM D-4586 Type I asbestos free

E. Nails
   1. For slate: Large flat-head slaters solid copper wire nails. 4d (1 1/2" long) and 6d (2" long) for slates on hips and ridges

Appendix C

PART 3 EXECUTION

3.01 PREPARATION

A. Check substrate to ensure that all boards are securely nailed in place and all asphalt shingles and underlayments have been removed. Remove or drive "home" any projecting nails.

B. Replace any deteriorated wood lumber at a square foot unit price.

C. Thoroughly sweep off the sheathing to remove all loose nails.

3.02 INSTALLATION

A. Underlayment
   1. Install two (2) layers of No. 30 felt over all boards to be covered with slate.
   2. Felts shall be laid shingle fashion with joints lapped toward eaves and at ends minimum three (3) inches. Felts shall be well secured along laps and at ends as necessary to properly hold the felts in place and protect the structure until covered by the slate.
   3. Lap felt over all hips and ridges and two (2) inches over the valley metal and built-in gutter.
   4. Install ice and water shield over entire gutter ledge, in valleys at roof eaves min. 3'-0" up roof 12 inches around scupper, at hips, and at dormer roof and wall intersections with sloped roof.
      a. At valleys, center full-width roll and press in place working from the center of the valley outward in each direction. Apply membrane starting at the low point and working upwards.
      b. Side laps shall be minimum of 3 1/2 inches and end laps 6 inches.

B. Slating
   1. The entire surface of roof and sides of dormer walls where indicated on the drawings, shall be covered with slate in a proper and watertight manner.
   2. Slate shall be installed in combination of colors and pattern indicated on the drawings and approved layout drawing furnished by slate producer.
   3. Slate shall be fastened with large-head slaters' copper nails to adequately penetrate roof boarding. Care shall be taken to avoid exposing the nails on cornice, soffits, overhanging eaves, etc.
   4. The heads of slating nails should just touch the slate and should not be driven "home" or draw the slate, but should be left with the heads just clearing the slate so that the slate hangs on the nail.
   6. All nails shall penetrate the sheathing and not the joints between boards.
   7. Slate shall be laid with a three-inch (3") head lap
   8. Each course shall break joints with the preceding one. The overlapping slate should be jointed as near the center of the underlying slate as possible and not less than three (3) inches from any underlying joint. Slates at the eaves or cornice line shall be doubled.
   9. The slate shall project two (2) inches at the eaves and one (1) inch at all gable ends.
   10. Nails shall not be driven in so far as to produce a strain on the slate.
   11. Cover all exposed nail heads with elastic cement. Hip slates and ridge slates shall be laid in elastic cement spread thickly over unexposed surface of under courses of slate, nailed securely in place, and pointed with elastic cement.
   12. Build-in and place all flashing pieces required for proper performance of the roof. Slates overlapping sheet metal work shall have nails placed as to avoid puncturing the sheet metal. Neatly fit slate around scupper.

Appendix C

   C. Ridges
      1. Install regular roofing slates extended to the ridge. On top of the last roofing slate, install a 9 1/2" course of slate. Fill joints on top of ridge formed by the butted edges of slate with slater's plastic cement.
      2. Metal ridge rolls shall be installed as detailed.
   D. Hips
      1. Install mitered hips with slates cut accurately to form tight joints.
      2. Fill joints in with plastic cement.
      3. Nail holes of each slate shall fall under the succeeding hip slate.

3.04 FINAL INSPECTION
   A. The Architect shall perform a final inspection of work and supply Contractor with punch list if necessary. All punch list items shall be completed before Architect gives approval for final payment.

APPENDIX D

Gusseted Snow Guard Specification

PART 1 - GENERAL

1.1 SUMMARY
A. WORK INCLUDES
   1. Gusseted snow guard that attaches directly to the roof deck.
   2. Coordinate with the installation of the roof to assure proper placement of the snow guards.
   3. Provide appropriate snow guard and fasteners for the roof system.

B. RELATED SECTIONS
   1. Section 07600:   Flashing and Sheet Metal.
   2. Section 07310:   Shingles.
   3. Section 07320:   Roofing Tiles.
   4. Division 7:   Thermal and Moisture Protection.

1.2 SYSTEM DESCRIPTION
A. COMPONENTS:
   1. Snow guard system consists of individual metal snow guards.
   2. Fasteners: All snow guards should be fastened using fasteners compatible with the snow guards and roof structure.

B. DESIGN REQUIREMENTS:
   1. Snow guard to be of a gussetted support design.
   2. Horizontal spacing not to exceed 24 inches.
   3. Vertical spacing not to exceed 24 inches.
   4. Minimum 2 fasteners per snow guard.
   5. It is important to design new structures or assess existing structures to make sure that they can withstand retained snow loads.

1.3 SUBMITTAL
Submit manufacturer's specifications and recommended layout, standard detail drawings and installation instructions.

1.4 QUALITY ASSURANCE
Installer to be experienced in the installation of specified roofing material and snow guards for not less than 5 years.

1.5 DELIVERY / STORAGE / HANDLING
Inspect material upon delivery and order replacements for any missing or defective items. Keep material dry, covered and off the ground until installed.

PART 2 PRODUCTS

2.1 MANUFACTURER - Snow guards by Vermont Slate & Copper Services, Inc., P.O. Box 430, Stowe, VT   Toll Free (888) 766-4273.

2.2 MATERIALS
A. Copper strap, hood and gusset are 16 oz. Cold Rolled copper.
B. Lead-coated copper strap, hood and gusset are 16 oz. Cold Rolled copper sheet stock lead-coated and cut to size.

2.3 FINISH - Mill finish.

PART 3 - EXECUTION

3.1 EXAMINATION
A. Substrate: Inspect roof system to be properly attached and installed to withstand additional loading incurred. Notify General Contractor of any deficiencies before installing Vermont Slate & Copper Services Inc. snow guards.

3.2 INSTALLATION
A. Comply with architectural drawings for location and with Manufacturer's instructions for installation.

Appendix D

Wire Loop Snow Guard Specification

PART 1 GENERAL

1.1 SUMMARY
A. WORK INCLUDES
 1. Wire Loop snow guard that attaches directly to the roof deck.
 2. Coordinate with the installation of the roof to assure proper placement of the snow guards.
 3. Provide appropriate snowguard and fasteners for the roof system.
B. RELATED SECTIONS
 1. Section 07600: Flashing and Sheet Metal.
 2. Section 07310: Shingles.
 3. Section 07320: Roofing Tiles.
 4. Division 7: Thermal and Moisture Protection.

1.2 SYSTEM DESCRIPTION
A. COMPONENTS:
 1. Snow guard system consists of individual metal snow guards.
B. DESIGN REQUIREMENTS:
 1. Snow guard to be of a wire loop design.
 2. Horizontal spacing not to exceed 24 inches.
 3. Vertical spacing not to exceed 24 inches.
 4. It is important to design new structures or assess existing structures to make sure that they can withstand retained snow loads.

1.3 SUBMITTAL
Submit manufacturer's specifications, manufacturer's recommended layout, standard detail drawings and installation instructions.

1.4 QUALITY ASSURANCE
Installer to be experienced in the installation of specified roofing material and snow guards for not less than 5 years in the area of the project.

1.5 DELIVERY / STORAGE / HANDLING
Inspect material upon delivery and order replacements for any missing or defective items. Keep material dry, covered and off the ground until installed.

PART 2 PRODUCTS

2.1 MANUFACTURER
2.2 MATERIALS - Copper wire of .145-inch diameter or greater.
2.3 FINISH - Mill finish.

PART 3 EXECUTION

3.1 EXAMINATION
A. Substrate: Inspect roof system to be properly attached and installed to withstand additional loading incurred. Notify General Contractor of any deficiencies before installing snow guards.

3.2 INSTALLATION
A. Comply with architectural drawings for location and with Manufacturer's instructions for installation.

APPENDIX E

List of Tables and Figures

Foreword
U.S. Slate Quarries 10

Chapter 1
Figure 1-1: Splitting Slate to a Manageable Size 11
Figure 1-2: Hand-Splitting for Desired Thickness 12
Figure 1-3: Nail Holes 13
Figure 1-4: Chamfering or Countersinking Nail Holes 14
Figure 1-5: Slate Parts 15
Figure 1-6: Slate Parts in Section 15
Figure 1-7: Slate Parts in Plan 16
Figure 1-8: Slate Pallet Ready for Shipping 16

Chapter 2
Table 2-1: Slate Roof Types and Variables 19
Figure 2-1: Commercial Standard Thickness 19
Figure 2-2: Textural 19
Figure 2-3: Intermingled Thickness 19
Figure 2-4: Standard Thickness 19
Figure 2-5: Heavy 19
Figure 2-6: Graduated Length and Thickness 19
Figure 2-7: French or 45-degree Angle Method 24
Figure 2-8: Economy Installation 24
Figure 2-9: Open Valley - Finished 26
Figure 2-10: Closed Valley - Finished 26
Figure 2-11: Round Valley - Finished 26
Figure 2-12: Canoe Valley - Finished 27
Figure 2-13: Saddle Hip/Boston Hip - Finished 27
Figure 2-14: Mitered Hip - Finished 27
Figure 2-15: Fantail Hip - Finished 28
Figure 2-16: Metal Hip - Finished 28
Figure 2-17: Tile Hip - Finished 28
Figure 2-18: Saddle Ridge - Finished 29
Figure 2-19: Strip Ridge - Finished 29
Figure 2-20: Comb Ridge - Finished 29
Figure 2-21: Metal Ridge - Finished 30
Figure 2-22: Tile Ridge - Finished 30
Figure 2-23: Dormer Side 30
Figure 2-24: Snow Guards 31

Figure 2-25: Lightning Protection 31
Figure 2-26: Gutters 32
Figure 2-27: Finials 32
Figure 2-28: Vents 32
Figure 2-29: Diverters 33
Figure 2-30: Snow Slides 33

Chapter 3
Table 3-1: Weight per Square 36
Table 3-2: Standard Slate Sizes 37
Table 3-3: Slate Exposure 38
Table 3-4: Number of Slates per Square (3-in. headlap) ...38
Figure 3-1: Slate Exposure 38
Table 3-5: Converting Quantity for Headlap 39
Table 3-6: Length of Valley and Hip 40
Table 3-7: Slate Size for a Saddle Hip or Ridge 40
Table 3-8: Starter Length 41
Table 3-9: Finishing Slate Length 41
Table 3-10: Nobility of Common Metals 42
Table 3-11: Weight and Thickness of Flashing Materials ...42
Figure 3-2: Drip Edge Configurations 43
Table 3-12: Open Valley Flashing Width 44
Table 3-13: Conventional Stepped Flashing for a
   Closed Valley 44
Table 3-14: Point-to-Point Flashing for a Closed Valley
   or Mitered Hip 45
Table 3-15: Estimating Hip Flashing 45
Table 3-16: Estimating Stepped Flashing 46
Figure 3-3: Stepped Flashing 46
Figure 3-4: Apron Flashing 46
Figure 3-5: Proper Through-wall Counter Flashing 47
Figure 3-6: Replacing Through-wall Flashing 47
Figure 3-7: Continuous Flashing (least desirable) 47
Figure 3-8: Roof Transitions 48
Figure 3-9: Cricket 49
Table 3-17: Nail Length Formula 50
Table 3-18: Nail Dimensions and Weight 51
Figure 3-10: Wood Supported Cant Strip 52
Figure 3-11: Metal Cant Strip 53

Appendix E

Table 3-19: Roof Pitch vs. Angle of Incline 54
Figure 3-12: Vented Ridge in Section 55
Figure 3-13: High Hat Detail - Plumbing Vents 55
Figure 3-14: Roof Scaffolding Bracket 57

Chapter 4
Figure 4-1: Stacking Slate 66

Chapter 5
Figure 5-1: Trim and Sheathing 68
Figure 5-2: Dormer Eave to Roof Trim 70
Figure 5-3: Triangular Gap at Slate Overlap 72
Figure 5-4: Installing Modified Bitumen
   (Various Applications) 74
Figure 5-5: Installing Modified Bitumen around Penetrations . . 74
Figure 5-6: Installing Modified Bitumen with Drip Edge . . . 75
Figure 5-7: Basic Underlayment Installation 75
Figure 5-8: Wood Lath Cant Strip Without Drip Edge 76
Figure 5-9: Elevated Facia May Cause Water Back-up 76
Figure 5-10: Consequences of Not Using a Cant Strip . . . 76
Figure 5-11: Metal Cant Strip as Part of Apron 77
Figure 5-12: Metal Cant Strip Soldered to Apron 77
Figure 5-13: Clapboard as Cant Strip 77
Figure 5-14: Hip Nailers 78
Figure 5-15: Finishing Nailer Needed 79
Figure 5-16: Finishing Nailer Not Needed 79
Figure 5-17: Wood Cant Strip Under Drip Edge 79
Figure 5-18: Steps Shown in Table 5-1 for Standard
   Thickness Slate 80
Table 5-1: Roof Preparation Sequence 80

Chapter 6
Figure 6-1: Actual Ridge 85
Table 6-1: Finish Slate Length 85
Figure 6-2: Reference Lines 87
Figure 6-3: Tying in Around Gable Additions 88
Figure 6-4: Creating Perpendicular Line 89
Figure 6-5: 3-4-5 Method 90
Figure 6-6: Graduated Slate Roof Layout 91
Figure 6-7: Roof Pitch Transitions 92
Figure 6-8: Vented Ridge in Section 92
Figure 6-9: Tying in Around Dormers 93
Figure 6-10: Intersecting Roof Layout 94

Chapter 7
Figure 7-1: Slate Hammer 96
Figure 7-2: Punching Machine 96
Figure 7-3: Trimming Machine 97
Figure 7-4: Asbestos Board Shear Cutter 98
Figure 7-5: Concave Handle Slate Cutter 98
Figure 7-6: Sickle Handle Cutter 98
Figure 7-7: Hand Slate Nibbler 98
Figure 7-8: Slate Ripper 99

Chapter 8
Figure 8-1: Roof Scaffold Bracket 102
Figure 8-2: Proper Nailing 104
Figure 8-3: Wiring Slate Technique 104
Figure 8-4: Installing Slate with Hooks 105
Figure 8-5: Using Slate Hooks on the Apex of a Bell . . . 105
Figure 8-6: Notching Slate 107
Figure 8-7: Relief Cutting Hips and Valleys 107
Figure 8-8: Vertical Joint Alignment 109
Figure 8-9: Improper Vertical Joint Alignment 109
Figure 8-10: Varying Horizontal Dimension of
   Valley Butt 111
Figure 8-11: Proper Intermingled Thickness 112
Figure 8-12: Graduating Widths Technique 1 113
Figure 8-13: Graduating Widths Technique 2 113
Figure 8-14: Notching Last Full Course of Field Slate . . . 114
Figure 8-15: Valley Slate Pieces 115
Figure 8-16: Nail Holes in a Valley Slate 115
Figure 8-17: Marking Open Valley Slate 116
Figure 8-18: Curving Valley Made of Straight Cuts 117
Figure 8-19: Installing Closed Valley Continuous Flashing . . . 117
Figure 8-20: Installing Closed Valley with Stepped Flashing . . . 118
Figure 8-21: Two Piece Valley Cuts 118
Figure 8-22: Curving Valley Made of Curving Cuts 119
Figure 8-23: Eliminating Concealed Slate from the Valley . . . 119
Figure 8-24: Notching Starter Slate at Hip 121
Figure 8-25: Field Slate at a Hip 121
Figure 8-26: Preparing a Saddle Hip in Section 122

Figure 8-27: Installing Saddle Hip 122
Figure 8-28: Steps for Installing Saddle Hip Cap 123
Figure 8-29: Hip Cap Coverage over Field Slate 123
Figure 8-30: Saddle Hip to Ridge Detail 124
Figure 8-31: Installing a Mitered Hip 125
Figure 8-32: Mitered Hip Using Approach Slate 126
Figure 8-33: Nail Holes in Hip Slate 126
Figure 8-34: Lowest Possible Nail Hole on Hip Slate 127
Figure 8-35: Notching Approach Slate Below Mitered Hip ..127
Figure 8-36: Point-to-Point Bent Hip Flashing 128
Figure 8-37: Installing a Fantail Hip 129
Figure 8-38: Installing a Boston Hip 129
Figure 8-39: Strip Ridge in Section 130
Figure 8-40: Ridge at Closed Valley 131
Figure 8-41: Ridge at an Open Valley 132
Figure 8-42: Finishing a Saddle Ridge at a Gable 132
Figure 8-43: Starting a Ridge at a Hip 133
Figure 8-44: Triangular Hip Finishing Slate 133
Figure 8-45: Fastening Triangular Hip Slate 133
Figure 8-46: Last Ridge Slate at a Hip 134
Figure 8-47: Vented Ridge in Section 135
Figure 8-48: Plumbing Vent in Plan - Steps 135

Chapter 9

Figure 9-1: Stepped Flashing at Wall Intersection 140
Figure 9-2: Stepped Flashing 140
Figure 9-3: Nailing Stepped Flashing 140
Figure 9-4: Steps of Apron Flashing 141
Figure 9-5: Apron Hold-down Clips 141
Figure 9-6: Open Valley "V" Installation 142
Figure 9-7: Open Valley "W" Installation 142
Figure 9-8: Cross Section Open Valleys 142
Figure 9-9: Installing First Piece of Open Valley Flashing ..143
Figure 9-10a: Nailing, Clipping, and Overlapping Valley ...144
Figure 9-10b: Valley Cleat 144
Figure 9-11: Installing Continuous Flashing in a Closed Valley 145
Figure 9-12: Installing Stepped Valley Flashing 145
Figure 9-13: Stepped Valley Flashing at Two Different Pitches 146

Figure 9-14: Stepped Valley Flashing Bent Point-to-Point ..146
Figure 9-15: Chimney Cricket 147
Figure 9-16: Complex Valley Cricket 148
Figure 9-17: Installing Built-in or Box Gutter 148
Figure 9-18: Installing a Yankee or Pole Gutter 148
Figure 9-19: Installing Built-in Gutter at the Eave 149
Figure 9-20: Installing Hanging Gutter at the Eave 149
Figure 9-21: Installing Facia-mounted Gutter with Flange on the Roof 149
Figure 9-22: Installing Facia-mounted Gutter 150
Figure 9-23: Installing a Rain Diverter 150

Chapter 11

Figure 11-1: Slate with Broken Corner 152
Figure 11-2: Using a Slate Ripper 153
Figure 11-3: Slate Hook 154
Figure 11-4: Using Wire to Secure Repair Piece 154
Figure 11-5: Bib Repair 155
Figure 11-6: Bib Configuration 155
Figure 11-7: Using Rip to Slide Bib into Place 156
Figure 11-8: Unacceptable Repair Using Broken Corner .157
Figure 11-9: Staggered Slate Removal 161
Figure 11-10: Pyramid Slate Removal 162

Chapter 12

Figure 12-1: Layout of Round Valley 170
Figure 12-2: Clipping Upper Corners 170
Figure 12-3: Canoe Valley Framing 171
Figure 12-4: Canoe Valley Layout 171
Figure 12-5: Laying Out a Cone 173
Figure 12-6: Installing Slate on a Cone 173
Figure 12-7: Jumping Pieces 174
Figure 12-8: Horizontal Line Layout on an Eyebrow Dormer 175
Figure 12-9: Vertical Line Layout on an Eyebrow Dormer ...176
Figure 12-10: Installing Slate on a Bell Curve 177
Figure 12-11: Installing Slate on a Sweep 178
Figure 12-12: Installing Battens and Counter Battens 179

APPENDIX F

ASTM

The American Society For Testing And Materials (ASTM) has established several tests related to roofing slate:

ASTM Designation: C 406 - 89   Standard Specification for Slate Roofing

This standard addresses the material characteristics, physical requirements, and sampling appropriate to the selection of slate for use as roof shingles. Based on ASTM Designations, C120, C121, and C217 listed below, slate is classified as Grade S1, S2, or S3.

Slate graded as S1 is the only material that should be used as roofing slate.

ASTM Designation: C 120 - 90   Standard Test Method of Flexure Testing of Slate

This standard addresses determination of the modulus of rupture and modulus of elasticity of slate by means of flexure testing.

ASTM Designation: C 121 - 90   Standard Test Method for Water Absorption of Slate

This standard addresses determination of the water absorption of slate.

ASTM Designation: C 217 - 85   Standard Test Method for Weather Resistance of Slate

This standard addresses two procedures for weather resistance of slate in all outdoor installations by determining the depth of softening by an abrasive or by hand scraping.

To obtain a copy of any or all of these specifications you can:

Write to ASTM at   American Society for Testing and Materials
1916 Race Street
Philadelphia, PA 19103

Phone (610) 832-9500
Fax   (610) 832-9555
E-Mail Service@ASTM.org
WebSite http://www.ASTM.org

NOTE: These specifications may be available through your local library.

APPENDIX G

Roof Parts Nomenclature

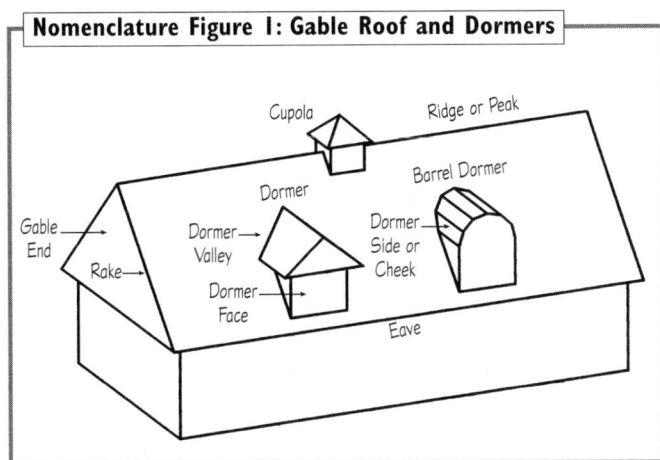

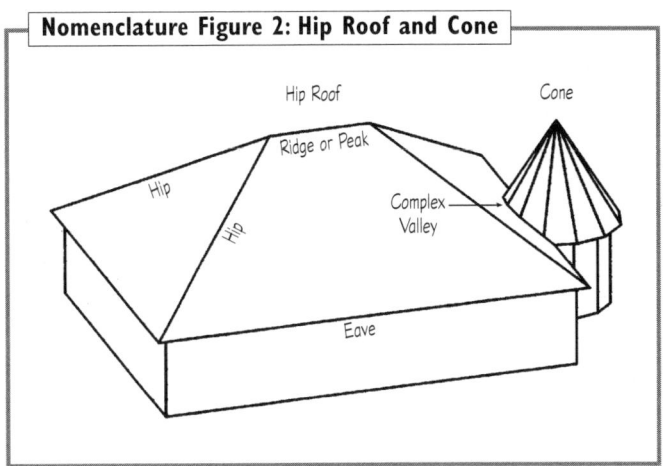

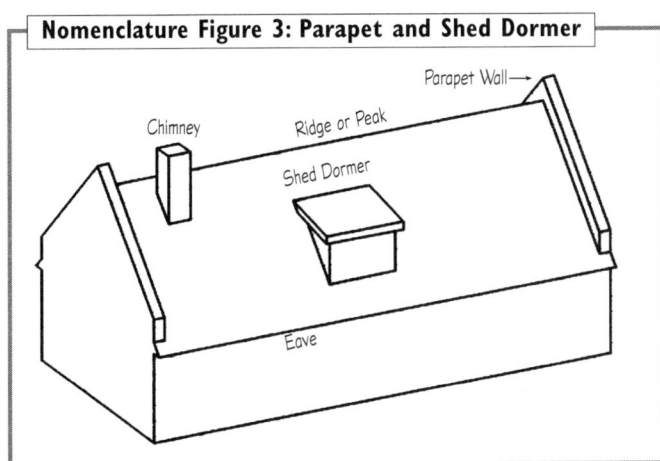

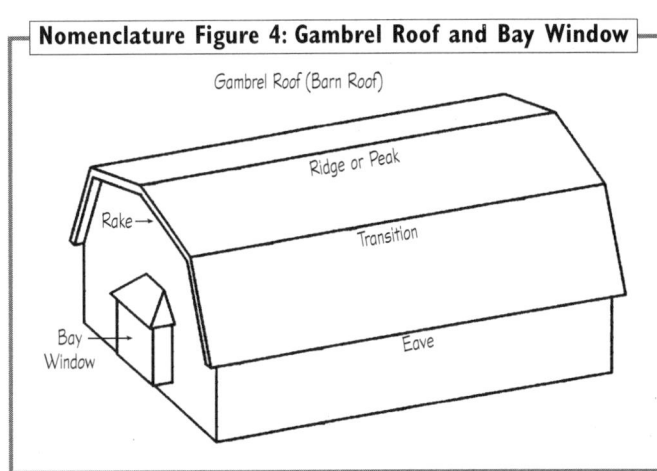

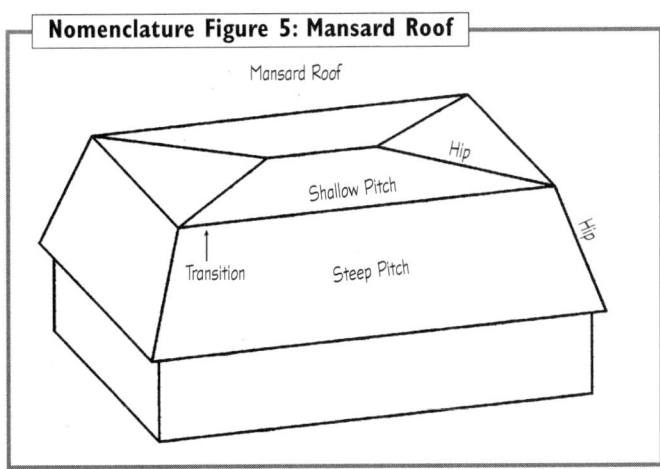

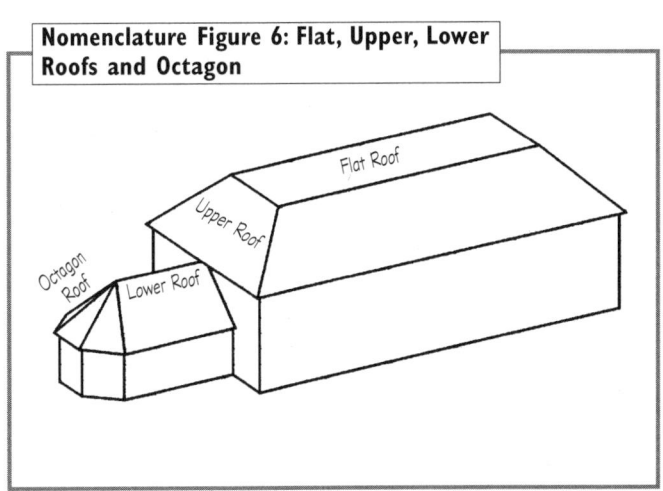

BIBLIOGRAPHY

The following are a few of the more important and interesting sources of information related to slate roofing from 1866 to the present. We have compiled a list of over 400 related articles. Please contact us if you are interested in receiving a complete list of slate-related publications.

Aston, Michael. *Stonesfield Slate.* England Oxfordshire County Council, 1974.

Auld, David, Jr. *The Slate Roofer.* A book comprising roofing slate tables, rules for measuring slate work, designs for ornamental work, nails required for a square, rules and information for beginners, and other valuable information. Cleveland, OH: Auld & Conger, 1891.

Baker, Earl Robert. "The Strength and Weathering Qualities of Vermont Roofing Slates." Montpelier, VT: 1912.

Behre, C.H., Jr. "Slate in Pennsylvania." Pennsylvania Geological Survey, 11th Series, Bulletin No. 16, 1933.

Bleekman, George. "The Story of Slate and the Slate Industry." Harrisburg, PA: The Telegraph Printing Company, 1911.

Bureau of Mines. "Minerals in the Economy of Vermont." Slate Mineral Profiles, 1979.

Copper & Common Sense. Revere Copper Products, Inc., P.O. Box 300, Rome, NY 13440. 315-338-2022.

Copper in Architecture. Copper Development Association, Inc. (CDA), 260 Madison Ave, New York, NY 10016. (212) 251-7200 phone; (212) 251-7234 fax.

Cullen, Mary K. "Slate Roofing in Canada." Ottawa National Historic Parks and Sites, Parks Service, Environment Canada, 1990.

Dale, Thomas Nelson. "Slate Deposits and Slate Industry of the United States." Washington: Government Printing Office, 1906.

Day, W.C. "Slate: U.S. Geological Survey Mineral Resources." 1894–1899 (published as parts of Sixteenth to Twenty-first Ann. Repts.).

Dickson, T. "Slate—Rebuilding the Markets." *Industrial Minerals.* October 1979: 45–53.

Eckel, E.C. "On the Chemical Composition of American Shales and Roofing Slates." *Jour. Geology,* Vol. 12, 25–9, 1904.

Greenfeld, S.H. "Hail Resistance of Roofing Products." National Bureau of Standards, Washington, DC, Building Research Division, August 1969: 15.

Hower, Franklin M. "Hower's Lightning Slate Reckoner on 33 Practical Sizes Roofing Slate"...Cherryville, PA; Allentown, PA: S. A. Woolever, Printer, 1884.

Levine, Jeffery S. "The Repair, Replacement, and Maintenance of Historic Slate Roofs." Preservation Briefs 29, U.S. Department of the Interior, Washington, DC, 1993.

Kessler, Daniel William. "Physical Properties and Weathering Characteristics of Slate." Washington, DC: U. S. Government Printing Office, 1932.

Merill, G.P. *The Collection of Building and Ornamental Stones in the United States National Museum, a Handbook and Catalogue.* Smithsonian Institute Annual Report, Pt. 2, June 30, 1886. Early History of Slate Industry in United States: 291.

Merriman, Mansfield. "The Strength and Weathering Qualities of Roofing Slates; with Discussion: Am. Soc. Civil Eng. Trans., Vol. 27, No. 3, September, 1892: 331–49; No. 6, December, 1892: 685; also Vol. 32, 1894: 529–39.

National Slate Association. *Slate Roofs.* Vermont Structural Slate Company, Inc., Fair Haven, VT, Report No. AIA File Number 12-D, March 1977.

Parrish, -----. "History of the Slate Trade in America." *American Journal Mining,* Vol. 2, 1866–1867. In Johns Hopkins University Library.

Reverdin, F., and De La Harpe, C. "The Examination of Roofing Slates." *Chem. Zeitung,* Vol. 14, 64–5, 94–5, 126–7, Abstract in English in *Journal Soc. Chem. Ind.,* Vol. 9, London: 30 April, 1890: 394.

Shaler, N.S. "Slates, Description of Quarries and Quarry Regions." Tenth Census U.S., Vol. 10, Pt. 2, 1880: 168-74.

Sheet Metal and Air Conditioning Contractors National Association (SMACNA). 5811 Amaya Drive #100, La Mesa, CA 91942. 619-460-5362.

Stafford, Horatio N. "Slate Tables Designed for the Use of Slaters, Quarrymen, Architects and Dealers." New York: J. Galt & Sons, 1896.

The Quarryman's Slate Tables; comp. and arranged especially for the use of slaters and quarry clerks for computing the production and shipments of roofing slate. Bangor, PA: W.J. Williams, 1901.

United States National Recovery Administration. Code of Fair Competition for the Slate Industry as Approved on January 22, 1934. Washington, DC: U. S. Government Printing Office, 1934.

Watkins, Cyril Mercer. "The Durability of Slates for Roofing." London: H. M. Stationery Office. Printed by Harrison and Sons, Ltd., 1932.

2-inch headlap, 39
3-inch headlap, 38, 39
4-inch headlap, 39
45-degree angle installation, 24

A
Actual ridge, 85
Aluminum, 41, 42
American Society for Testing and Materials (ASTM), 10, 204
Approach slate, 40, 126, 127
Apron flashing, 46, 139–141
At eave gutters, 149–150, 162

B
Base flashing, 79
Battens, 179
Bell shape roof, 105, 176
Bevel. See chamfer.
Bib repair, 156–158
Bidding, 58, 59
Boston hip
　description of, 27
　flashing, 129
　installing, 129
Bottom lines, 87, 88
Box gutter, 148
Budget, 34, 36, 152
Built in gutter, 148, 161–162
Butt, 15, 88

C
Canoe valley
　description of, 27
　flashing, 171
　installing, 171
　layout of, 171
Cant strip
　description of, 15, 52
　estimating, 52, 53
　installing, 75–78
Cap flashing, 48, 184
Cap slate, 27, 29, 44, 45, 121–135
Chalk line, 95
Chamfer, 12
Change of pitch, 92
Clapboard as cant strip, 77, 178
Cleaning slate, 164–165
Clean-up, 58
Cleavage, 9
Clips, 144
Closed valley
　description of, 26
　estimating of, 44, 45
　flashing of, 144–146
　installing, 117-120
　repair of, 159
Color. See Slate color.
Comb ridge
　description of, 29
　estimating of, 40
　flashing, 45
　installing, 130
Commercial standard, 18
Complex valley
　flashing, 147
　installing, 119
Cone shape
　description of, 172
　flashing, 172
　installing, 172
　layout of, 173
Copper, 31
Copper Development Association (CDA), 139, 187, 206
Counter battens, 69, 179
Counter flashing, 47, 70,
Counter sink, 14, 103–104
Cricket, 49, 147–148
Cut butts, 23
Cutting slate, 97, 107

D
Diverter. See Rain diverter.
Dormer
　description of, 30
　layout of, 93–94
Dormer sides, 30, 136
Drilling slate, 106
Drip edge
　description of, 43
　estimating of, 43
　installing, 79

E
Economy installation, 24
Exposure, 38
Eyebrow dormer
　description of, 169, 174–176
　flashing, 169, 174–176
　installing, 169, 174–176
　layout of, 169, 174–176

F
Facia-mounted gutter, 162
Fantail hip
　description of, 28
　flashing, 129
　installing, 129
　layout of, 129
Fasteners
　estimating, 50, 52
　size of, 50–51
Felt paper
　estimating, 49
　installation, 71–73
Field slate
　estimating, 37 -39
　installing, 107–114
　layout of, 84–86
Finials, 32
Finding a contractor, 58
Finishing course
　description of, 41
　estimating, 41
　installing, 114
　layout, 85
Finishing course nailer, 78, 79
Forklift, 57–58, 63-64, 66
French Method installation, 24

G
Gable end slate
　installation, 109
　repair, 160
Galvanic reaction, 42
Graduated slate roofs
　description of, 20
　installing, 112–113
　layout of, 90–92
Grain, 10
Gutters, 32, 148–150, 161-162

H
Handling slate, 65–66
Headlap
　description of, 15, 38
　calculating, 38
Heavy slate 20
Hip approach slate. See Approach slate.
Hip cap slate. See Cap slate.
Hip cut slate, 120–121
Hip flashing
　estimating, 40, 44, 45
　installing, 121, 128, 129
Hip nailer. See Nailers.
Hold down clips, 141

I
Ice damming, 166–167
Intermingled thickness
　description of, 18–19
　installing, 112
Intersecting roofs
　layout, 87–88, 94

L
Labor, 36
Ladder hook, 154
Lath. See Cant strip.
Lead-coated copper, 31
Lightning protection, 31

M
Marking valley slate, 116
Masonry, 46–47, 70
Material handling equipment, 56–57, 62–63, 66
Metal cant strip. See Cant strip.
Metal hip
　description of, 28
　estimating, 45
　installing, 146–147
Metal ridge
　description of, 30
　estimating, 45
　installing, 146–147
Mineral composition, 182
Mitered hip
　description of, 27
　estimating, 44
　installing, 125–129
　layout 85–86
Mock-ups, 24–25
Modified bitumen
　description of, 83

estimating field, 49
estimating hip and ridge, 44–45
installing, 73–75

N

Nail holes, 13, 103–106
Nail holes in valley slate, 115
Nail set, 95, 97
Nailers
    hip, 78
    ridge, 78
    finishing course, 78
Nailable concrete, 168
Nailing slate, 103–104
Nails. See Fasteners.
National Slate Association, 5
Non-weathering, 21
Notching slate, 107, 114

O

open valley
    description of, 26
    estimating, 44
    installing, 115–117
    layout, 85–86
    size, 44
Ordering slate, 61-62
OSHA, 56, 71, 101

P

Patterns, 19, 23
Perpendicular lines, 89–90
Planks. See Scaffold plank.
Plumbing vent, 55, 135–136, 160
Point-to-point bent flashing
    description of, 44
    estimating, 45
    installing hip, 128
    installing valley, 146
    size, 45
Pole gutter, 32, 148–149, 161–162
Punching holes in slate, 12–14, 105–106
Pyramid slate removal, 163–164

Q

Quarrying slate, 11

R

Rain diverter, 33, 150
Random width slate
    description of, 22
    estimating, 37
    installing, 110–112
    layout, 86
Reference lines, 86–87
Repairing slate, 153–157, 159–160
Ribbons, 10
Ridge flashing
    description of, 40, 44–45
    estimating, 40, 44–45
    installing, 130–135
    size, 40, 44–45
Ridge nailer. See Nailers.
Ridge vent
    description of, 54
    estimating, 54–55

installing, 134–135
layout, 92–93
Ripper, 95, 99
Roof consultant, 17, 35
Roof pitch, 54
Roof preparation, 38, 71–80
Roof scaffold bracket, 56, 57, 101–102
Roof sheathing, 68–69
Roof vent, 32, 54
Rosin paper, 74–75, 141
Round valley
    description of, 26
    installing, 169–171
    layout, 169–171

S

S1, 151, 204
S2, 151, 204
S3, 151, 204
Saddle hip
    description of, 27
    estimating, 40
    flashing, 45
    installing, 121–125
    size, 40
Saddle ridge
    description of, 29
    estimating, 40
    flashing, 45
    installing, 130–134
    size, 40
Salvaged slate, 20
Samples, 23
Scaffold plank, 57, 71
Scaffolding. See Roof scaffold bracket.
Scheduling, 58
Schist, 9
Sheetmetal and Air Conditioning
    Contractors Association (SMACNA), 139
Shim, 170, 177–178
Semi-weathering, 21
Side wall flashing. See Stepped flashing.
Sidelap, 109
Single width slate
    description, 22
    installation, 110
Skirt flashing. See apron flashing.
Skylights, 41, 160
Slate color, 21. See also color insert.
Slate cutter, 95, 97–98
Slate hammer, 95, 95–96
Slate hook, 155–160
Slate parts, 15
Slate per square, 38
Slate repair. See Repairing slate.
Slate sizes, 37
Slate tool list, 95
Slater's "T", 98
Snow guard, 31, 53, 54, 137, 166
Snow slides, 33
Special order, 20
Splitting, 11, 12
Stacking slate, 66
Staggered slate removal, 163

Stainless steel, 42
Standard thickness slate, 18
Starter slate
    description of, 41
    estimating, 41
    installing, 108
    layout, 84
    size of, 41
Stepped flashing
    description of, 139
    estimating, 46
    installing, 139, 140
    size, 46
Storing slate, 64–65
Stucco, 179
Strip ridge
    description of, 29
    estimating, 40
    installing, 130
    size of, 29
Structural design, 35
Sweeping eave
    description of, 177–179
    installing, 177–179

T

Textural slate roof, 18, 112
Through wall flashing, 47
Tile hip
    description of, 28
    installing, 137
Tile ridge
    description of, 30
    installing, 137
Tin snips, 95
Top lines, 87
Transition flashing
    description of, 48
    estimating, 48
    layout, 92
Trim, 69–70
Trimming slate to size, 12

U

Underlayment
    estimating, 49
    installing, 71–75
Unfading slate, 21
Used slate. See Salvaged slate.

V

Vented ridge. See Ridge vent.
Vents. See Roof vents, Plumbing vents or
    Ridge vent.
Vertical joint alignment, 109

W

Weathering slate, 21
Weight per square, 35
Wet saw, 99
Wiring slate, 104

Y

Yankee gutter, 32, 148–149, 161–162

Z

Zinc, 31

Register Today!

Please take the time to register your book. By doing so, you will be qualified for the following:

- 1 hour of free phone consultation with the authors
- Advanced notice of new publications
- Discounts on future publications
- 1-year free subscription to the *Slate Roof Quarterly*

✂

Registration #: _____	Your comments and suggestions would
Name: _____	be greatly appreciated:
Company: _____	_____
Address: Home Work (circle one)	_____
_____	_____
_____	_____
City: _____	_____
State: _____ Country: _____	_____
Zip: _____	_____
Phone: _____	_____
Fax: _____	_____
Email: _____	_____

How would you prefer to be contacted with future updates? mail   phone   fax   email   (circle one)

Please return by mail or fax to: Vermont Slate & Copper Services, Inc.
P.O. Box 430
Stowe, VT 05672
Phone 1-888-766-4273
Fax 1-802-253-4369

Printed in Canada